What Remains to

A Touchstone Book
Published by Simon & Schuster

Be Discovered

MAPPING THE SECRETS OF THE UNIVERSE,

THE ORIGINS OF LIFE,

AND THE FUTURE OF THE HUMAN RACE

John Maddox

TOUCHSTONE
Rockefeller Center
1230 Avenue of the Americas
New York, NY 10020

First Touchstone Edition 1999

TOUCHSTONE and colophon are registered trademarks
of Simon & Schuster, Inc.

Designed by Kim Llewellyn

Manufactured in the United States of America
10 9 8 7 6 5 4 3 2 1

The Library of Congress has cataloged the Free Press
edition as follows:

Maddox, John Royden, date.
    What remains to be discovered : mapping the secrets
of the universe, the origins of life and the future of the
human race / John Maddox.
        p. cm.
    Includes bibliographical references and index.
    1. Discoveries in science.   2. Science—History.
3. Research—Miscellanea.   I. Title.
Q180.55.D57M33   1998
500—dc21   98-29137   CIP

ISBN 0-684-82292-X
ISBN 0-684-86300-6 (Pbk)

To my ex-colleagues at *Nature* for their constant stimulation,
enthusiasm and comradeship.

Contents

PART ONE: MATTER

*. . . in which the origins of the universe and of matter are explored,
as well as the prospects for a theory of everything*

PART TWO: LIFE

*. . . in which the origin of life is considered as well as biological machinery,
the riddle of the selfish gene, and the next human genome projects*

PART THREE: OUR WORLD

*. . . in which the nature of our brain is explained, as well as our greatest
invention, mathematics, and how we will avoid the catastrophes of the future*

Preface

This book springs from a question first asked by my son Bruno: "If you're editor of *Nature*, why can't you say what will be discovered next?" In 1995, when I knew I would be leaving that thoroughly international science journal after nearly 23 years as editor-in-chief, it seemed that it would be useful to set down in simple language an account of what scientists are hoping to achieve. One of the joys of being such an editor is listening to researchers eagerly enthuse about the significance and possible outcomes of their work, while knowing at the same time that they would never be so enthusiastic about their research in print. Why not distill that unrestrained chatter into an account of where science is heading, of what remains to be discovered?

I was much helped when the Brookhaven National Laboratory invited me, in June 1996, to deliver its annual Pegram Lectures. It was an opportunity to see whether my ideas about the future of science hung together, and, having completed the lecture series, I was encouraged and emboldened. As work on the manuscript of this book proceeded, my son Bruno came to my aid. He raised serious and thoughtful objections to an early version, and I am deeply grateful for his perceptive criticism ever since.

What remains to be discovered is not, of course, the same as what will be discovered. It is possible to tell what loose ends are now dangling before us, but not how they will eventually be pulled together. People who knew that would take themselves off to a laboratory confident that a Nobel Prize would soon be on the way.

Science at present is a curious patchwork. Fundamental physics is perhaps the oddest: the research community is divided into those who believe that there will be a "theory of everything" very shortly and those who suspect (or hope) that the years ahead will throw up some kind of "new physics" instead. History is on the side of the second camp, to which I belong. By contrast, exuberant molecular genetics seems in a state in which any problem that can be accurately defined can be solved by a few weeks' work in a laboratory. There, it is more difficult to tell what problems will emerge—as they certainly will.

I am aware that many important fields of science are not touched by this survey of outstanding problems. The most obvious is that of the solar system. The closing third of this century has seen a quite remarkable transformation of our views of how the Earth is built. The doctrine of plate tectonics (continental drift) has been firmly established. Superficially, the matter appears to have been tidied up. But a little reflection shows that to be an illusion. The mechanism that drives the tectonic plates over Earth's surface is far from clear. More to the point, it remains to be seen how the same ideas can be applied to the understanding of other solid objects in the solar system—both planets such as Venus and satellites of Jupiter such as the strange object Io. And exactly how were the planets formed from the solar nebula, anyway? These are all absorbing questions but no new principles are involved.

. . . . . .

Two close friends of mine have read the penultimate version of this text. Professor Maxime Frank-Kamanetsky, a molecular biologist whom I first met in Moscow in 1986 and who is now a member of the faculty at Boston University, and Dr. Henry Gee, *Nature*'s resident paleontologist who has a catholic interest in all of science, have both made valuable and constructive suggestions. I owe them a great debt, although the errors and the omissions that remain are my own responsibility.

I am also grateful to my publishers, who have put up with my vacillation, and particularly to Stephen Morrow of The Free Press, who has helped enormously to shape this text by providing a stream of pointed and detailed comment on its successive versions, always with intelligence and good humor.

. . . . . .

And the message? Despite assertions to the contrary, the lode of discovery is far from worked out. This book provides an agenda for several decades, even centuries, of constructive discovery that will undoubtedly change our view of our place in the world as radically as it has been changed since the time of Copernicus. Indeed, the transformation in prospect is likely to touch the imagination of all of us dramatically. How shall we feel when we know the true history of the evolution of *Homo sapiens* from the great apes? And when there are found to be, or even to have been, living things elsewhere in the galaxy?

But that is merely the tip of the iceberg of future discovery. The record shows that generations of scientists have been repeatedly surprised by discoveries that were not anticipated and could not have been guessed at by much earlier versions of a book like this. Who, at the end of the nineteenth century, could have smelled out the way in which physics would be turned on its head by relativity and quantum mechanics, and how the structure of DNA would have made life intelligible? And who, now, dares say that the days of surprise are over?

London, January 1998

Introduction:
The River of Discovery

This century has been so rich in discovery and so packed with technical innovation that it is tempting to believe that there can never be another like it. That conceit betrays the poverty of our collective imagination. One purpose of this book is to take up some of the questions now crying out for attention, but which cannot yet be answered. The record of previous centuries suggests that the excitement in the years ahead will spring from the answers to the questions we do not yet know enough to ask.

It is an abiding difficulty in science that perspective is distorted because the structure of scientific knowledge makes its own history seem irrelevant. We forget that modern science in the European tradition is already 500 years old, dating from the time of the Polish astronomer Copernicus. The Copernican revolution was a century-long and, ultimately, successful struggle to establish that the Sun, not Earth, is at the center of the solar system.

The contributions of the ancient Chinese, the extinct civilizations of the Indus Valley, the Babylonians and Greeks, naïve though they may now seem, cannot be scorned. Copernicus and those who followed profited from them. And it remains a source of wonder that Chinese records of past astronomical events have been indispensable in the interpretation, within the past 30 years, of the exploding stars known as supernovae; that only now is a systematic search being made of the same records for plants that may be sources of still-useful therapeutic drugs; that the Athenian Greek Erastothenes made a good estimate of the circumference of Earth more than 2,000 years ago; and that the concept of zero in arithmetic, of which even Euclid was innocent, was formulated on the Indian subcontinent soon after the beginning of the modern era. Even the alchemists who distracted the Middle Ages with their search for the philosopher's stone that would turn anything it touched into gold should not be despised; they understood that chemical reactions can turn one substance into another that looks very different.

There is nevertheless a clear distinction between modern science and its precursors: the interplay between observation and explanation was formerly less important than it is now. A theory qualifies as an explanation only if it can be and has been tested by observation or experiment, employing when necessary measurements more sensitive than the human senses can yield. A further novelty of the modern idiom is that each phenomenon—the existence of the Universe, the fact of life on Earth, and the working of the brain—demands a physical explanation.

Copernicus is explicitly acknowledged in what came to be called the Copernican principle, the rule that in trying to understand the world, a person should not assume that he or she occupies a privileged position. The Copernican principle also applies to the history of discovery: How can we suppose that science has reached its apogee in the twentieth century? We are right to marvel at what has been accomplished in the past 100 years—but we forget that there would have been the same sense of achievement at the end of each of the three preceding centuries.

Looking back at the seventeenth century from 1700, an historian could fairly have boasted that there had never previously been a century of scientific prosperity like it, and certainly not since the resurgence of Greek science at Alexandria in the second century. In Britain, the century had begun with a forceful argument in favor of experiment by Francis Bacon. His compatriot, William Harvey, had fitted action to word by dissecting animals and people for almost the first time since Galen 1,200 years earlier, discovering the functions of the heart, the arteries, and the blood. In France, already recognizable as the cradle of modern mathematics, prolific René Descartes had embarked on his astonishing philosophical account of the nature of the world. A closet Copernican, he argued that the solar system and the fixed stars beyond are but a kind of machine driven by God. His legacy to succeeding centuries was the system of geometry in the language of algebra, still called Cartesian geometry.

Galileo was the first to use the modern idiom of science and, in that sense, was the first scientist. Whether or not he used the Leaning Tower of Pisa as a laboratory in the 1580s, he established that acceleration creates force, as when one's body is pushed back into one's forward-facing seat in an aircraft taking off. From that principle, Galileo saw, the mass of an object measured by its weight on Earth must be identical with the mass inferred from what happens when it collides with other objects anywhere in the universe or is accelerated by some force. This conclusion is called the *equivalence principle*. It is a discovery of the first importance about the nature of gravitational attraction, and it is one of the foundations of modern theories of relativity. And famously, Galileo discovered the satellites of Jupiter with a marvelous new invention known as the telescope.

Isaac Newton, at the University of Cambridge, showed, in the 1680s, that the orbits of the planets are the result of an attractive force between the Sun and each planet. Thus gravity was discovered. Newton's law of gravitation is universal. It accounts not only for the orbits of the planets and those of artificial satellites about Earth, but also for the attraction between Earth and, say, a falling apple, as well as for the roughly spherical shape of Earth, the Moon, the planets and the Sun and stars. Newton also spelled out the rules that specify how objects move under the influence of mechanical forces and, for good measure, he devised a novel mathematical technique, known as differential calculus, for calculating the orbits of planets and other trajectories. His synthesis of two centuries of intelligent speculation about the nature of the world was first published in 1687; it is now known simply as the *Principia*. The second edition, in Latin like the first, appeared in 1713 and circulated widely on the European mainland.

Newton had to invent the mathematics as he went along; the full importance of his accomplishment was evident to others only after the differential calculus had been made into a usable tool by French and German mathematicians. The consequences of these developments were profound. Newton's mechanics, as originally defined, amounted to a series of statements about the behavior of "bodies" that were essentially single points endowed with mass and subjected to external forces (such as gravitational attraction). The mathematicians generalized Newton's system in ways Newton could not have foreseen.

Newton's genius set the agenda of science for the two centuries to follow. Could there ever again be a century of such marvelous accomplishment?

There was. While a small army of mathematicians in France and Germany were busy turning Newton's clumsy version of the differential calculus into a usable tool, science was beginning to come to terms with electricity and magnetism. The discoveries soon included the following: that electricity can have either a positive or a negative charge, that these charges can neutralize each other if put together, that a sufficient amount of electric charge can cause a spark to travel through the air (lightning), that like charges repel and unlike charges attract, that steady currents of electrical charge can be made to flow through metal wires, that currents of electricity can affect the direction in which a nearby magnet points and that, as Luigi Galvani found in 1794, a current or a shock of electricity will make the muscle of a dead frog twitch.

Although electricity and magnetism preoccupied the eighteenth century, by the end of it Carl von Linné in Sweden (better known as Linnaeus) had devised the system by which animals and plants are still classified today. Antoine Lavoisier had laid the foundations of modern chemistry by the time of his execution in 1794 during the French Revolution. The astronomers, notably William Herschel in Britain, were building ever more powerful telescopes. And salesmen were trudging through the industrial enterprises of western Europe, offering newly designed steam engines as replacements for traditional water wheels (not to mention human drudgery) as a source of motive power. Was that not a century to boast about?

But then came the age of certainty. In variety and subtlety, the discoveries of nineteenth-century science outdid those of all previous centuries. The idea that matter is made of atoms, indivisible particles [whose properties on the tiniest scale mirror those of the same matter on the scale of everyday reality] was firmly established by John Dalton, a teacher in the north of England, within the first two decades. He proposed that the only strictly indivisible atoms are those of substances such as carbon and copper, which he called *elements*. All other substances, carbon dioxide and copper oxide, for example, are not elements but combinations of atoms. Dalton also concluded that each different kind of atom has a different weight: hydrogen atoms are the lightest, and carbon atoms are each roughly twelve times heavier.

The nineteenth century also put the concept of energy on a firm foundation. Since Galileo's time, it had been understood that one form of energy can sometimes be converted into another: Lift an object a certain height above the ground, against the downward pull of gravity, and then let it fall. It will reach the ground again with a speed that increases with the height to which it is lifted.[1] The energy of the object's motion is its *kinetic energy*. Galileo was the first to show that (in the absence of friction and other disturbances) this energy increases in proportion with the vertical distance the object falls. He inferred that, at its greatest height, the object must somehow be endowed with an equal quantity of potential energy in virtue of its displacement against the gravitational attraction.[2] But heat, light, electricity and magnetism are also phenomena that embody energy, all of which can be converted into other forms. Heat, for example, will power a steam engine that can be used to generate mechanical energy, and electricity will make a lightbulb glow. What rules govern these conversions?

That question was answered by James Prescott Joule, also from the north of England, whose careful measurements demonstrated by 1851 that, if disturbing influences are avoided, no energy is lost in the conversion of one form of energy into another. The doctrine is that energy is conserved. Support for this idea remains unanimous.

In 1865 the German Rudolf Clausius introduced the idea of *entropy,* which is a measure of the degree to which the internal energy of an object is not accessible for practical purposes, and which is in the present day also equated with the degree of internal disorder on an atomic scale. That loose set of concepts eventually became known as the *second law of thermodynamics:* other things being equal, there will be a tendency for the entropy or the disorder of an isolated system to increase. The principle that energy is conserved became known as the *first law of thermodynamics.* Linked with that is the simple truth, amply borne out by common observation, that heat does not flow spontaneously from a lower to a higher temperature. This set of nineteenth-century concepts has a deep significance. Uniquely among the laws of physics, the second law specifies the directions in which systems change with the passage of time; it defines what has been called the "arrow of time," the capacity of most physical systems to evolve in one direction only.

Elsewhere surprises came thick and fast. Charles Darwin's theory of the evolution of species by natural selection, published in 1858, was one of them. The geologists and fossil hunters had learned enough of the fossil record to know that many once-successful forms of life had disappeared from the surface of Earth, to be replaced by others. The notion that there may have been a patterned evolution of life-forms was not new, but had been suggested by the apparent gradation with time of the fossils found in successive layers of sedimentary rocks. Darwin's proposal was that the evolution of living things is driven by the interaction between the environment of a population of animals or plants and the variations of form or fitness that arise naturally, but unpredictably, within the population.

The theory created a sensation, not only by its assertion of a link between human beings and the great apes or even because it is godless theory. It changed the terms of the debate about humans' place in the world by emphasizing that people are part of nature, certainly in origin and perhaps even for the rest of time. Here is the Copernican principle at work again.

Many of the developments of the age of certainty marked an important trend in the practice of fundamental science—that of bringing together phenomena of different kinds under a single umbrella of explanation. In the 1860s, James Clerk Maxwell, a Scot then teaching in London, put forward a mathematical scheme for describing in one set of equations both electricity and magnetism. His prize was not just a coherent account of unified electromagnetism, but an explanation of the phenomenon of light. A ray of light is indeed a wave phenomenon, and the speed of light in empty space is simply related to the electrical and magnetic properties of empty space. Maxwell's wave theory is an explanation of all kinds of electromagnetic radiation, most of which have since been discovered.

At the end of the nineteenth century, Maxwell's triumph raised a conceptual difficulty. A ray of light, or some other form of radiation, may be a pattern of oscillating electromagnetic fields, but it can have an existence independent of its source; the flashes of light from an exploding star, for example, keep on traveling outward long after their source has vanished. What could sustain the vibrations of such a disembodied flash? Maxwell took the view that there must be something, which he called the *lumeniferous æther,* filling all of empty space. How else could one part of an electromagnetic field influence its neighboring elements? Only after a quarter of a century of fruitless and fanciful searching did people appreciate that they were looking for a will-o'-the-wisp: the lumeniferous æther was no different from empty space, or the vacuum as it is called. It has taken almost a whole century since to reach some (imperfect) understanding of the subtlety of the vacuum.

The discoveries of the last two decades of the nineteenth century came in a sensational flurry. In France, Louis Pasteur showed that the fermentation that turns milk into cheese is accomplished by bacteria, and went on to demonstrate the germ theory of infectious disease. In the 1880s, Heinrich Hertz, then at Karlsruhe in Germany, generated invisible Maxwell waves in the radio-frequency range; two decades later, Guglielmo Marconi spanned the Atlantic with these waves, founding the global communications industry. In 1895, W. K. Röntgen at Würtzburg in Bavaria discovered that an electrical discharge in a tube from which as much as possible of the atmospheric gas had been extracted would lead to the emission of a novel kind of radiation, capable of penetrating people's flesh but not their bones—X rays. The following year, Antoine Bequerel in Paris found that similar radiation is given off by such sub-stances as uranium; they were the first elements to be called *radioactive*. Just a year later, at Cambridge (England), J. J. Thompson proved the existence of the atom of negatively charged electricity, now called the *electron*.

Thompson thus brought the nineteenth century full circle. It began with the proof of the reality of atoms, a different kind of atom for each kind of chemical element, and it ended with the demonstration that even atoms are not indivisible. Maxwell's definition, that an "atom is a body which cannot be cut in two,"[3] was found to be false. For the atoms of electricity were evidently but components of atoms from which they had been separated. The fragility of the atom became a question for the twentieth century.

Maxwell's search for the æther, misguided though it proved to be, was a search for a mechanism to account for electromagnetism; 200 years earlier, Newton had been content to describe the gravitational attraction between objects without asking why nature relies on his and not on some other law. Similarly, in biology, what distinguished Darwin's work from earlier speculations was not the fact of evolution, but that he offered a mechanism for it. In the nineteenth century, "Why?" rather than "How?" became the overriding question.

The nineteenth was also the century in which mathematics became the handmaiden of scientific inquiry. By the end of the century, any problem in which physical objects are influenced by external and internal forces, and respond as Newton prescribed, could be restated as a problem in mathematics. Exuberantly people set out to tackle the deformation of solid objects by external forces, the motion of fluids such as water when driven by mechanical pumps and the propagation of seismic signals from distant earthquakes through the solid Earth. Thus were numerous problems of importance in engineering solved and the twentieth century equipped with many of the mathematical tools that would be required in those changed circumstances; fluid dynamics spawned aerodynamics once aircraft had been invented, for example. So as the century ended, there emerged a tendency to believe that once a problem had been stated mathematically, it had been solved; the idea that the underlying physics might be in error was hardly entertained.

The century thus ended on a triumphant note. Not only had fundamental physics been reduced to a series of problems in mathematics that would in due course be solved, but the closing decades of the century were made prosperous by technology resting on science that was itself the product of the same century. The dyestuffs industry and the chemical industry more generally were the products of the atomic theory and what followed from it. The electrical industry (harbinger of the communications industry) had already begun to change the world. For science and technology, the nineteenth century was certainly the best there had yet been. Only now do we know that it was merely a beginning.

. . . . . .

At the great mathematical congress held in Paris in 1900, David Hilbert, the outstanding mathematician of his time, produced a list of the problems still to be dealt with in mathematics. One was to find a proof of Fermat's last theorem (accomplished only in 1995);[4] another was nothing less than to find a systematic procedure for demonstrating the truth or falsity of all propositions in mathematics. (As will be seen, the second project had an unexpected sequel in the 1930s.) In 1900 we had many achievements in all fields of science behind us and many apparent contradictions before us. With the benefit of hindsight, it is not too difficult to reconstruct the outline of a book with the title *What Remains to Be Discovered (in 1900)*. What scientific puzzles then in the minds of courageous people would the author have identified? There were several.

What is space made of? What does energy have to do with matter? In the 1880s two U.S. scientists, Albert A. Michelson and Edward W. Morley, set out to find direct proof of the reality of James Clerk Maxwell's lumeniferous æther, which was supposed not merely to be fixed in space, but to be in some sense space itself. Then Earth in its orbit about the Sun must be rushing through the æther at great speed. Michelson and Morley erected in Ohio a piece of equipment that, with the help of mirrors, sent the same beam of light traveling along two paths at right angles to each other and then reflected the two beams back to their common starting point. Because one of these paths would usually be more in line with Earth's supposed motion through the æther than the other, the speed of light along the two paths should be different. The measurements showed it to be the same.

At the time, many scientists were disconcerted. By showing that there is no lumeniferous æther, the experiment had thrown doubt on the mechanism of Maxwell's electromagnetic waves. George Fitzgerald at Trinity College, Dublin, had immediately taken the Michelson-Morley experiment as a sign that the dimensions of quickly moving objects appear to contract (in the direction of the motion),[5] a suggestion quickly taken up by Hendrik Lorentz at Leiden in the Netherlands. Meanwhile, Henri Poincaré of Paris was openly drawing attention to the need for a redefinition of space and time. (In 1904, Poincaré would carry the message across the Atlantic on a lecture tour of U.S. universities.) The author of our apocryphal work could not have known, of course, that Einstein would put the issue to rest just five years after the turn of the century with the publication of his special theory of relativity.

Albert Einstein, first at Zurich, then Berlin and finally at Princeton in the United States, was to make an incomparable contribution, both as an innovator and a critic, to the deepening of understanding that marks the twentieth century. His theory of relativity is, operationally, a correction of Newtonian mechanics for objects traveling at a significant fraction of the speed of light. In the familiar low-speed world, a fixed force applied for some fixed but short length of time to an object that is free to move will produce a fixed increase of the velocity of the object (or of its speed in the direction of the force). Not so in special relativity: the greater the speed of the object, the smaller will be the increase of the velocity. What then happens to the remainder of the energy expended by the force? It ends up as an increase of the mass of the moving object. The energy turns into mass, or substance.

Several counterintuitive conclusions follow. First, nothing can travel faster than light, whose velocity therefore has a special status. Second, energy can be converted into mass[6] and mass into energy (whence the energy of exploding nuclear weapons). Third, the Newtonian notions of absolute position and time are devoid of meaning; only relative distances and times make sense (in which respect, Einstein dutifully followed the precepts of Ernst Mach, one of the founders of the positivist school of philosophy, that theories should refer only to quantities that are measurable). Fourth, there can be no lumeniferous æther, for that would allow the determination of absolute speeds or distances. Fifth, we live in a four-dimensional world, because the three dimensions of ordinary space are conjoined with the extra dimension of time.

Beginning in 1905, it has been commonplace to describe the counterintuitive character of special relativity as "paradoxical." Turning mass into energy sounds a little like turning water into wine, and was indeed often represented as a near-miracle in the 1950s, the early days of the nuclear power industry. But this now-common happening is not a miracle, merely a fact of life. Mass is energy and energy is mass. Their equivalence is both a discovery about the nature of mass and an extension of Galileo's equivalence principle. There is no paradox in special relativity, all of which has been amply confirmed by experiment. The theory runs against our intuition because our senses lack experience of objects moving almost at the speed of light.

The importance of this key development in our understanding of space-time became clear only in succeeding decades. And even now the character of empty space is still not fully resolved. What the Michelson and Morley experiment showed was that the æther could not serve as a means by which space and time can be given absolute meaning. It is an open question whether the abandonment of action at a distance compels the notion that space and time have an internal structure of their own. The notion of the ether was perhaps still tenable until 1905, but with a modicum of foresight on the part of our hypothetical author, speculation about the structure of space and time would surely have featured in *What Remains to Be Discovered (in 1900)*.

Our author would also have had to ask: what is the nature of heat? Difficulties that had arisen during the nineteenth century in the treatment of radiation mostly centered on the question why the quality of the radiation emitted by a hot object changes in a characteristic way as the temperature is increased. It would have been a matter of common observation that radiation from objects such as the human body is invisible but still sensible by the hands, that a smoldering wood fire is a more prolific source of essentially similar radiation, that pieces of iron can be heated in a blacksmith's fire until they glow red and that the radiation from the Sun consists of "white light." By the end of the nineteenth century, the facts had been clearly established: all objects emit radiation that spans a spectrum of frequencies, and that spectrum is shifted to higher frequencies as the temperature is increased. In the 1880s, the German physicist Ludwig Boltzmann had even been able to show that there must be a universal mathematical description of how the intensity of the radiation in such a spectrum varies with the frequency, but all attempts to discover what form that description takes ended in failure.

The question was eventually decided by Max Planck at Berlin in 1900 —it would have been great news in our hypothetical book. Planck's objective was to explain why the amount of energy radiated at a particular frequency increases with the temperature, but that, at any temperature, there is a frequency at which the intensity of radiation is a maximum. Earlier, in 1893, the German physicist Wilhelm Wien had shown that this maximum frequency increases with the absolute temperature—the temperature measured in relation to the absolute zero.[7] Planck's radical proposal was that radiation of a particular frequency exists only as *quanta*, as Planck called them. The greater the frequency, the greater the energy of the quanta concerned. Planck acknowledged that "the physical meaning [of quanta] is not easily appreciated," a confession his contemporaries were only too ready to accept. What Planck had done was to discover that even energy consists of atoms. Then Einstein (in 1905) proved the point by explaining why a certain minimum frequency (and thus energy) of light is required to extract electrons from metals or to make semiconductors carry an electrical current—now the basis of photographers' light meters.

The discovery of the electron and of radioactivity would certainly have been prominent features of *What Remains to Be Discovered (in 1900)*. J. J. Thompson, who had shown that electrons have the same properties whatever atoms they are torn from, acknowledged the subversiveness of his discovery by saying in a public lecture in 1897[8] that, "The assumption of a state of matter more finely subdivided than the atom of an element is a somewhat startling one . . ."—before going on to defend it. Bequerel's discovery of radioactivity (in 1896) also led people to the conclusion that atoms are not indivisible, although the point was not proved until the new century had begun. The two discoveries together led directly to the series of experiments in which Ernest Rutherford, a New Zealander itinerant between Cambridge, Montreal (McGill University) and Manchester, established (in 1910) that all atoms are constructed from a number of electrons and a nucleus carrying positive electrical charge and embodying most of the mass of the atom. Typically, the dimensions of the nucleus are one ten-thousandth of those of the atom as a whole.

Our hypothetical book could not have anticipated these developments, although it would no doubt have had much to say concerning the continuing uncertainty about the dimensions of atoms as a whole (resolved at about the time of Rutherford's model of the atom). Nor could its author have guessed that, in 1913, Niels Bohr (settled briefly at Manchester with Rutherford) would set out to reconcile the properties of the simplest atom, that of hydrogen, with Planck's discovery that energy is transferred between atoms only in quanta.

The connection between atomic structure and the quantal character of energy is that particular atoms emit radiation with a precisely defined frequency—the so-called spectral lines. An author of a book about discoveries yet to be made could not have failed to note that by 1860, characteristic spectral lines had been recognized as a means of analyzing the chemical composition of stars, the Sun included.[9] By the 1890s, a detailed investigation of the spectral lines of hydrogen (many of which are in the ultraviolet) had revealed mathematical regularities in the frequencies at which these appear. In 1900, the Swedish physicist Johannes Rydberg first proposed that the frequency of the spectral lines of hydrogen is most simply expressed as the difference between two numbers for which a simple mathematical formula can be given.[10] In 1913, that and later developments provided Bohr with an essential clue to the structure of the hydrogen atom: the electrons travel around the nucleus much as if they are planets revolving about a star, except that only a restricted set of orbits is allowed, each with a well-defined energy. Then the regularities in the frequencies of the spectral lines is understood; each spectral line arises when electrons change from one allowed orbit to another, and its frequency corresponds to the difference between the two energies.

Bohr's discovery that only some orbits are stable is not a restriction of what is physically possible imposed by the then-new quantum theory, but, on the contrary, is a kind of exemption license from classical expectations. In Maxwell's theory, electrons or any other electrically charged particles should not be able to travel in tight orbits with the dimensions of atoms without losing all their energy as radiation—so much had been clear since the time of Lorentz. Far from discovering a novel restriction, Bohr had hit upon conditions in which an electron in a tiny hydrogen atom can remain indefinitely in its orbit without losing energy. He called these orbits *stationary states*. Disappointingly, for Bohr and his contemporaries, atoms other than that of hydrogen remained inexplicable. A decade (and the First World War) passed before that deficiency was made good. The recognition that radiation, say a beam of light or the heat from a domestic radiator, can be understood by supposing that it consists of indivisible quanta characterized by frequencies that may span a considerable range is superficially at odds with the idea that radiation is a wavelike phenomenon, first advocated by Huyghens in the eighteenth century and substantiated by Maxwell in the 1860s; it amounts to a return to Newton's notion that a beam of light consists of "corpuscles" of different colors. By 1913, when Einstein published a second crucial paper on quantum mechanics (which, among other things, included the principle on which lasers function), physics had no choice but to accept that radiation is *both* wavelike and corpuscular; the corpuscles were christened *photons*, the indivisible atoms of radiation.

In 1924, the French scientist Louis de Broglie turned the argument around; if photons are both wavelike and corpuscular, may not electrons also have wavelike as well as corpuscular properties? The conclusion that they do was not proved by experiment until 1926, but sated by earlier surprises, the research community cheerfully took de Broglie at his word. The real surprises came only in 1925 and 1926, when two German scientists independently put forward formal systems of mechanics designed to account for the strange assumptions underlying Bohr's account of the hydrogen atom. First, Werner Heisenberg at Göttingen (with Max Born and Pasqual Jordan) devised a system called "matrix mechanics" from which they were able to calculate the properties of quantized systems. The starting point for that enterprise was Heisenberg's proof that it is not possible simultaneously to measure the position and the speed of a particle such as an electron. This became known as his *uncertainty principle*. Almost at the same time, Erwin Schrödinger (an Austrian, but then at the University of Zurich) devised his "wave mechanics" which seemed to cast in the familiar language of mathematical physics the properties of quantum systems. It fell to Paul Dirac, then a young researcher at the University of Cambridge, to show that the two descriptions are equivalent to each other.

Those developments now constitute what is called quantum mechanics, the conceptual development that most markedly characterizes the twentieth century. It is a remarkably successful tool for calculating the properties of systems on an atomic scale, and also a system whose internal self-consistency is so compelling that it has been used successfully to predict that particles of matter have previously unexpected properties and even that particles of matter not previously known must somewhere or somehow exist. But in no sense is quantum mechanics a license for the belief that physical phenomena on a very small scale do not follow the principle that to each event (or happening), there is a cause. Rather, some causes may have several consequences whose likelihood can be calculated from the rules of quantum mechanics. Nor is everybody now content with the *meaning* of quantum mechanics. On the contrary, great attention has been given to this and related questions in recent years. The goals are practical as well as philosophical. Is it possible to avoid the limits imposed by Heisenberg's uncertainty principle? Is it possible to design computers that exploit the knowledge that a single cause may have several consequences? At what stage in the evolution of the contemporary computer industry will further progress be limited by the small size of the electrical components etched into the surfaces of pieces of silicon?

The emergence of quantum mechanics could not have been foreseen in 1900, and it would have required a perceptive author indeed to appreciate that gravitation would be as important an issue as it was in the twentieth century. There were clues, notably the failure of Newton's theory of gravitation to account for the details of the motion of some celestial objects—especially the rate of twisting of the elliptical orbit of Mercury about the Sun and the difficulty of making accurate predictions of the return of periodic comets, even well-known objects such as Halley's comet. There was also a current of theoretical speculation, triggered by the new knowledge of electromagnetism and typified by imaginative Fitzgerald's remark in 1894 that "gravity is probably due to a change in the structure of the ether produced by the presence of matter." In 1900, Lorentz read a paper on gravitation to the Amsterdam Academy of Sciences, starting a trend that occupied Europe's principal physicists for a decade and a half, until Einstein produced his general theory of relativity (which would have been better called a "relativistic theory of gravitation") in 1913. Fitzgerald's guess was vindicated: if "structure of the ether" is replaced by "curvature of space," we have the essence of Einstein's theory.

The general theory is equivalent to Newtonian gravitation provided that the concentration of mass (or, the same thing, of energy) is not too great; it is also naturally consistent with the special theory of relativity. Its effect, unique among physical theories, is to represent gravitational interactions geometrically, by the geometry of space-time. The gravitational field is not imposed on space-time as, say, Picasso applied paint to canvas: rather, it *is* space-time.[11] That is another illustration of how, whenever people dispense with the Newtonian notion of action as a distance, they are compelled to endow empty space with properties that Euclid never dreamed of. Einstein's theory of gravitation has survived all the tests it has been possible to make of it; that is the basis of the widely shared belief that Einstein's theory of gravitation is the outstanding achievement of human intellect and imagination of the twentieth century. Yet, as will be seen, there remains the horrendous unsolved problem of how to reconcile the theory of gravitation with quantum mechanics.[12]

The author of *What Remains to Be Discovered (in 1900)* would also, of course, have had to ask, What is life? There was much to say about how the living world came about due to the achievements of biologists in the nineteenth century. Pride of place would have gone to Darwin's theory of evolution by natural selection, published in 1858, which quickly became the guiding principle of biology in the closing decades of the nineteenth century. Nevertheless, the author of our book would have startled readers with news of the rediscovery of Gregor Mendel's observations in the 1850s of the patterns of inheritance in plants. By the turn of the century, Mendel's observations seemed a direct challenge to Darwin's notion that the inheritable variations that account for evolution are invariably *small* variations. That the implicit contradiction would lead to energetic research could have been foreseen in 1900; that it would lead first to the foundation of modern genetics (chiefly at Columbia University in New York) and then, in the 1930s, to a recognition that the apparent conflict is not a conflict at all would have been more difficult to predict.

Modern biology was being created. That cells are the essential units of living things had been recognized in the 1830s, when microscopes capable of making cells visible to an observer first became widely available. By the 1880s, the German physiologist August Weissman had shown that, from early in an embryo's development, the cells that are eventually responsible for reproduction (called *germ* cells) are physically distinct from ordinary body cells (called *somatic* cells). Weissman also recognized the structures in the cell nuclei that appear to be concerned with the transfer of inheritable characteristics from one generation of cells or even of whole organisms to their successors; these structures are called *chromosomes*. He also concluded that, in sexually reproducing organisms, germ cells [13] differ from somatic cells in having only half as many chromosomes. The group at Columbia University added detail to this picture. In quick succession came the proof that inheritable characteristics are determined by entities called *genes;* that the genes are arranged in a linear fashion along the chromosomes, much like beads on a string; and that different versions of the same gene are responsible for alternative versions of similar characteristics, say the color of a person's skin. Although the physical character of the genes was not known for half a century, the science of genetics (so named only in 1906) was given a solid foundation that is not yet superseded.

The eventual rapprochement between Darwinism and genetics could not have been anticipated in 1900. In retrospect, it is curious that so little attention was paid to the constitution of genes. What were they made of? The difficulty was partly technical in that there was no way the components of cells, and the chromosomes in particular, could be separated from each other and thus studied in isolation.[14] But in due course it became clear that chromosomes have two components: protein and a material called nucleic acid. In the decade after the Second World War, the crucial question was whether the stuff of inheritance consists of protein or of nucleic acid.

It was in 1900, as it happens, that Emil Fischer, Germany's towering genius of chemistry, was unraveling the molecular structure of both protein molecules and nucleic acids. Perhaps the author of *What Remains to Be Discovered (in 1900)* could have made a lucky guess at whether nucleic acids or proteins are the stuff of inheritance, but it took until the end of the Second World War to establish that the particular nucleic acid involved in chromosome structure is DNA;[15] chemically, it seemed unpromising material to provide the functions of genes because of its repetitive structure. DNA molecules are polymers of only four distinct chemical units differing very little from each other. How could such uniform and apparently featureless molecules generate the great variety with which the living world abounds?

The answer came in April 1953, when two young men at the University of Cambridge, J. D. Watson and Francis H. C. Crick, built a structural model of the DNA molecule whose details are in themselves sufficient to illustrate how these molecules function as repositories of genetic information. This structure of DNA was self-evidently also the repository of the recipe by means of which the cells in all organisms carry out the specific functions required of them. Indeed, its chemical structure even embodies the recipe by means of which a single fertilized egg in a sexually reproducing organism can develop into a fully functioning adult; what is called ontogeny had at last been unambiguously brought within the bounds of rational inquiry. That was the springboard for a detailed exploration of what has proved to be the universal biochemical machinery of living things, which continues still at breakneck pace.

New industries (under the general label of biotechnology) have been spawned, radically novel therapeutic techniques (such as gene therapy) await refinement and there is a prospect that the breeding of productive animals and plants will be enormously improved by the techniques now becoming available. The structure of DNA ranks with Copernicus's successful advocacy of the heliocentric hypothesis in importance. In 1900, a few brave spirits may have hoped that an understanding of life would be won during the century just beginning, but there cannot have been many of them.

. . . . . .

These observations of what has happened in science in this century illustrate two important truths. First, new understanding does indeed spring from current understanding, and usually from contradictions that have become apparent. Second, while it may be possible confidently to guess in which fields of science new understanding will be won, the nature of the discoveries that will deepen understanding of the world cannot be perfectly anticipated. This book does not pretend to describe discoveries yet to be made, but rather suggests which areas of science are ripe for discovery; that harvest of discoveries will be crucial both for the self-consistency of science itself and for its consequences in the wider world.

The text is divided into three parts: *Matter, Life* and *Our World*. As will be seen, contradictions abound, perhaps most conspicuously at the intersection of the apparently successful theories of particle physics and of the expanding universe. In the light of past experience, it is folly to believe that a so-called theory of everything is waiting to be formulated, but there may be what many physicists refer to as a "new physics," a physics regulated by principles not yet imagined. And in the life sciences, made exuberant by the understanding that has flowed from the discovery of the structure of DNA, problems abound: How did life begin? How will biology make comprehensible the vast amounts of data now being gathered? How the brain functions both in the everyday world and as the human attribute of mind is hardly clearer now than at the beginning of the century.

The river of discovery will continue to flow without cessation, deepening our understanding of the world and enhancing our capacity to forfend calamity and live congenial lives. As will be seen from the final chapter, there are crucial lessons in this tale for governments, the research profession and the rest of us.

PART ONE

Matter

Beginnings Without End

1

How did the universe come into being? Since about 1960, the answer has been clear in many people's minds: until about 10 or 20 billion years ago,* there was nothing, not even empty space. Then there sprang into being a tiny granule of space filled with such a huge amount of energy that it produced the 100 billion stars in the Milky Way, a comparable number of galaxies that lie beyond our own, the radiation that fills every corner of the universe and the momentum that even now sustains its expansion. That was the big bang.

This simple and beguiling picture of how the universe began is even consistent with the book of Genesis: in the beginning, there was the void, then suddenly there were the ingredients from which all of the world about us has evolved. Since the early 1960s, the big bang has been the generally accepted framework within which most studies of the

---

* The use of "billion" in this book follows U.S. practice: 1 billion = 1,000 million.

evolution of the universe have been carried through. As an organizing principle, it has been invaluable. And the respect accorded to the big bang has brought more than a quarter of a century of peace and quiet to cosmology, traditionally a field made quarrelsome by a chronic shortage of reliable facts.

That halcyon period is coming to an end. In the past few years, contradictions (as the Marxist philosophers used to say) between expectation and reality have begun appearing within the old framework. That framework cannot easily accommodate the questions now being raised about the origin of matter as well as of the physical universe, not to mention those about the origin of the rules that shape the behavior of the physical world, the laws of physics as they are called. The cosmological question is again an open question.

EVOLUTION BY EXPANSION

We forget how recent are the roots of present knowledge. The first doubts about the permanence of the universe were provoked at the beginning of this century by the failure to find a satisfactory answer to the question, "How long can the stars of the Milky Way keep shining?" The idea that thermonuclear energy could give them a realistically long life could not then have been guessed at. Not until 1917 was it established, after a long controversy, that the solar system is not at the center of the disk-shaped collection of 100 billion stars that make up the Milky Way.[1] By coincidence, that was just two years after Albert Einstein published his general theory of relativity, believing that it would stand or fall by the account it would give of the structure of the universe.

The idea that the universe is expanding was established only in 1929, by the U.S. astronomer Edwin T. Hubble. After nearly a decade's work with the brand-new 100-inch telescope on Mount Wilson near Los Angeles, he concluded that almost all the galaxies within range of the telescope are receding, and that the speed of recession increases with distance. That is now called Hubble's law. It ranks with identifying the structure of DNA in 1953 among the discoveries that have dramatically and permanently changed our view of the universe and our place in the world.

Some of the questions raised by Hubble's law are teasing. Thus: if all other galaxies are receding from our own, does that not imply that we are at the center of the universe, in conflict with the Copernican principle that it is a folly to suppose we have a privileged place? Not so. The galaxies are embedded in space, and space itself is being stretched, apparently uniformly in all directions. Or: does the universe have edges? Nobody can tell. The finite (although very large) velocity of light (just under 300,000 kilometers a second) means that it is possible to look only backward in time[2] and, even with the improvement of instruments in the past half-century, not yet to the beginning. What we can (at least in principle) see from where Earth happens to be is the "observable universe." What (if anything) lies outside it can only be inferred.

The practical value of Hubble's law is nevertheless huge. For one thing, it provides a yardstick for the universe. Simply measure the recession speed of a distant galaxy by measuring the degree to which the light it emits is reddened, or redshifted), and you can infer its distance from a knowledge of the Hubble flow.[3] The same principle is a guide to the age of the universe. Simply extrapolate the present motion of the galaxies backwards; the time that they were all together is when the universe began.

Hubble's own first estimate of the age of the universe was wide of the mark; he estimated something less than 500 million years, which would have made the whole universe younger than Earth (4,500 million years). Even in the 1950s, people were talking of a Hubble age of 2,000 million (2 billion) years. Since the 1960s, the preferred age has been 10 or 20 billion years or, more accurately, either 10 billion or 20 billion years.

The origin of this uncertainty is fully understood and generally accepted: the difficulty of making accurate measurements of distance in the universe. Not merely is every other galaxy a long way away from our own (the nearest one of comparable size, Andromeda, is at 2 million light-years), but obscuring gas and dust in our own galaxy makes distant galaxies seem fainter (and thus more distant) than they really are. Moreover, distant galaxies can be seen only at an earlier stage in their history, and for that reason may be intrinsically different from our galaxy now.

Relating the distance scale to the age of the universe raises another difficulty. Whether the universe has been expanding for 10 billion or for 20 billion years, there is no doubt that the speed of expansion has been declining all that time as the different galaxies in the universe have been pulled apart against their gravitational attraction for each other; in other words, the present expansion speed must be less than formerly. The difference will depend on the average density of matter in the universe: the more matter, the greater will have been the deceleration of the expansion speed. Numerically, the difference could be as much as a third; the true age of the universe may not be 10 or 20 billion years, but 6.6 or 13.2 billion years. The average density of the universe has become, in the 1990s, a crucial and contentious issue.

None of that belittles the enduring importance of Hubble's big idea. It remains a consistent picture of what is happening to the universe within the range accessible to the newest telescopes. An alternative explanation, that the universe is not really expanding, but that light becomes redder in color as it travels through the vast spaces is repeatedly advanced, but carries little conviction; there is no reason to believe that light behaves like that. So, since Hubble's time, there has been no escape from the haunting question of what will happen, as time passes, to the universe, our own galaxy and to us and our descendants.

BIG BANG

The other big idea underlying the cosmology of recent decades is due to George Gamow, a Russian emigré (from Odessa) to the United States. Gamow, a nuclear physicist, was an imaginative man, a lateral thinker among linearists.[4] What more natural than that it should fall to him to propose how the universe began? In 1947, he startled people with the idea that the universe began with the big bang.

To say just that would not have been to propose a scientific theory, but to make a statement, even a speculation. Yet Gamow's theory also offered an explanation of how the big bang would plausibly engender something like the universe we now see. In the first few instants, the universe would have been a tiny bubble of space no bigger than a football, or a baseball, or a golf ball, or a pinhead. . . . Outside the universe, there was nothing, not even empty space; inside, the temperature was huge, much greater than anything that has been seen in the past 10 or 20 billion years. That is all there was, for which reason it is wrong to describe the big bang as a "fireball"; there was nothing to burn, nor oxygen to burn it with.

How can even very hot empty space spawn the matter from which galaxies, stars and even people are made? Imaginative Gamow saw a way. At temperatures greater than the absolute zero of temperature (which is at minus 273 degrees Centigrade), all empty space is filled with radiation. That was what Planck had learned in 1900. More than that, the intensity of radiation increases very rapidly with the temperature. (Double the absolute temperature and the energy of the radiation increases sixteenfold.) But radiation also exerts a pressure on whatever contains it. So fill a tiny bubble of space with radiation at a great temperature, and it will expand. That is how Gamow's big bang embarked on the expansion that has continued ever since. All that was the physics of the nineteenth century.

But whence the matter now all around us? Gamow needed twentieth-century physics to account for that. The recognition that radiation can behave both as if it were a wavelike process and as if it were a collection of particles moving with the speed of light dates back to the first decade of this century. And the individual particles, called photons, can create matter out of empty space if they have enough energy. By 1947, everybody knew that a single photon, if energetic enough, can create a pair of electrons out of empty space: an ordinary electron carrying negative electric charge and a positive electron, called a positron, of the kind whose very existence had been predicted in 1928.[5] So why should not the exceedingly energetic radiation of the big bang universe create not only electrons, but the protons and neutrons that are the ingredients of nuclear matter?

Both the strength and weakness of Gamow's big bang is its simplicity. His universe expands naturally, at first under the pressure of radiation alone, so that it neatly accommodates Hubble's law. The theory also accounts for a distinctive feature of the present universe: After the ingredients of nuclear matter began to crystallize out of the furnace of the big bang, and as the infant universe was continually cooled by its expansion, there would have been a brief interval of time (a small fraction of a second) when both the temperature (between 10 and 100 billion degrees) and the density of protons and neutrons would have been great enough for protons and neutrons to combine to form helium and the nuclei of the heavy isotope of hydrogen called deuterium or "heavy hydrogen." Other materials (particularly the rare isotope of helium called helium-3, and possibly versions of nuclei of the elements lithium and boron) would have been formed in the same way at the same time. Certainly there is no way in which these materials could be formed in stars, as are more familiar elements such as carbon, oxygen and iron. Instead, they are consumed in stars.

The prediction about the light elements has been amply confirmed; the proportion of helium in the primordial gas is between 25 and 28 percent by weight. It is sobering that the many tons of heavy water with which some types of nuclear reactors are filled are literally a chemical relic of the early stages of the big bang.

Another feature of the real universe reproduced by the big bang was more surprising—and thus compelling. In 1966, two researchers at what were then the Bell Research Laboratories in New Jersey, investigating the causes of troublesome disturbances of radar signals, found that electromagnetic radiation in what is called the microwave region of the spectrum appeared to be reaching Earth from all directions with equal intensity at day and night and at different times of year. Those signs betoken a cosmic source of radiation. Daringly, Arno Penzias and Robert W. Wilson declared that their radiation was just that.

The finding was startling. The radiation was emitted by an object with a temperature of just 2.7 degrees above the absolute zero of temperature[6] (less than the boiling point of liquid helium, the coldest liquid on Earth). Their conclusion was that this cosmic microwave background radiation is a direct relic of the radiation that set the big bang universe on its expansion, and whose temperature has ever since been falling as the universe has expanded.

Hermann Alpher, a close colleague of Gamow in the 1940s and 1950s, insists that the existence of this radiation was predicted in the elaborations of the big bang view of how the universe began on which he and Gamow worked in the years after 1947.[7] Certainly the notion that the temperature of the universe would keep falling was clearly spelled out by Gamow and Alpher, but Penzias and Wilson (not to mention many of their contemporaries) were surprised by their discovery. That is probably a measure of how Gamow's account of the big bang was initially regarded by the research community: more as a model of how the universe might have begun than as a formal theory.[8] Penzias and Wilson gave it bite.

The cosmic microwave background radiation is not, however, a relic of the earliest instants of the big bang, but of a later time when all the matter in the present universe had already come into being in the form of hydrogen nuclei (mixed with small proportions of the nuclei of the light elements formed earlier). The temperature of the universe would have fallen, and with it the frequency of the predominant radiation, which would have been ultraviolet radiation of sufficient energy to prevent intact hydrogen atoms being formed. The electrons would have been free and would have prevented the transmission of all radiation; the universe would have been opaque. But as continuing expansion caused the temperature to fall, intact hydrogen atoms would have begun

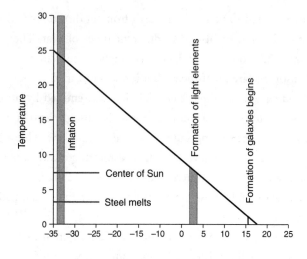

Figure 1.1 Milestones in the Big Bang. The temperature of the universe declined rapidly after the big bang. This chart has highly distorted scales (called logarithmic) for both temperature and time. The absolute temperature corresponding to a point on the vertical scale is obtained by multiplying 10 by itself as many times as indicated. Times on the horizontal scale are calculated in the same way, but negative scale numbers mean submultiples (by 10) of one second. (On each scale, the number "0" corresponds to a temperature of 1 K or a duration of 1 second, respectively.)

to form. Then, the bath of radiation would no longer have been coupled to the matter or, what comes to the same thing, the universe would have become transparent. The cosmic microwave background radiation is a relic of precisely that time, quite late in the evolution of the big bang, more than 100,000 years after the initial explosion.

A Steady Cosmology

For most of the past half-century, Hubble's law has been the organizing framework for novel information about the distant parts of the universe, while the big bang has been the handiest account of how the universe began. Hubble's law has not been seriously questioned, but the big bang

has not swept everything before it. In the event, two scientists at the University of Cambridge put forward a quite different proposal. They agreed, of course, that the universe is expanding, further increasing the vastness of the space between the galaxies. They asked whether this continual dilution of matter is compensated for by the appearance of new matter from which, in due course, new galaxies and stars will be formed.

That was called the "steady-state" cosmology, otherwise the "continuous creation" theory. Its authors were Hermann (now Sir Hermann) Bondi and Thomas Gold, now a professor at Cornell University. They were soon joined by Fred (now Sir Fred) Hoyle, who became the chief protagonist of, and eventually the whipping-boy for, the theory.[9] The steady-state theory made firm predictions about the numbers of galaxies of decreasing brightness visible in the sky. Philosophically, it was an attractive theory. It finessed awkward questions about the ultimate fate of the expanding universe by suggesting that they do not arise: is not the universe in a steady state? But in the absence of a mechanism by which energy or matter can appear from nowhere, the theory was no more explicit than the big bang about the reasons for creation of any kind, either once and for all or continuously.[10]

It is important that Gamow on the one hand and Bondi, Gold and Hoyle on the other asked a question about the world in which we live that had not previously been raised. Satisfied with Hubble's account of what the universe is like, they each sought to understand how it came to be like that. The steady-state theory is now largely of historical interest, but it is memorable for another reason: the animosity it created among cosmologists, especially at Cambridge. Although Hoyle's contributions to astronomy were great, notably as the principal author of the accepted theory of how most of the elements, from carbon to uranium, are synthesized in stars like the Sun, his espousal of the steady-state theory led to his being ostracized by his academic colleagues and to his almost unprecedented resignation of his Cambridge professorship. Until recently, he had dropped out of science. It is a tale that discredits

Cambridge, and a vivid illustration that cosmology, which touches religious among other chords, can be the cause of passionate dissension.

The big bang component of the standard cosmology, by contrast, has weathered well, at least until recently. But it is important that its significance should not be overstated and thus misunderstood. Neither Gamow nor his associates pretended to provide a detailed account of the evolution of the universe in the past 10 or 20 billion years; it is no shame on them, for example, that they did not predict the universe would turn out to contain stars made largely of the electrically neutral nuclear particles called neutrons. Nor, for that matter, did Darwin's theory of the origin of species by natural selection predict when, in the course of evolution, crocodiles or elephants would appear. The function of Hubble's law and the big bang in recent decades has been largely that of an organizing framework within which to fit newly gathered data.

THE DISTANCE PROBLEM

The bugbear of cosmology is the measurement of intergalactic distance and there is little hope that the difficulties will soon melt away. When distance is an essential elementary requirement for even describing the universe, that is a mildly shocking state of affairs, but not a scandal. The difficulties are immense.

The essence of the problem is that all information about extragalactic objects must be inferred from the radiation they emit. The degree to which the light from a visible galaxy is redshifted[11] is a direct measure of its recession speed. Hubble's law, which relates recession speed to distance, makes it possible to calculate the distance of a galaxy from its redshift. But how reliable is Hubble's law, based as it is on relatively nearby galaxies? The law says that the recession speed increases in proportion with the distance, but exactly how fast? That is one uncertainty. The bigger question—whether the expansion of the universe is nearly as uniform as Hubble suggests—appears to be settled (in the affirmative) by the uniformity of cosmic microwave background radiation, but this crucial point calls for independent confirmation. On both counts, there is a great need for an independent measure of the distance of an extragalactic galaxy. An obvious indicator of a galaxy's distance is its measured brightness (or, often more appropriately, faintness), but galaxies differ in their intrinsic output of light, so that there is no way of distinguishing bright galaxies a long way off and fainter galaxies lying closer to our own. To avoid such difficulties, there should be some identifiable feature of a distant galaxy whose intrinsic brightness is independently known. Borrowing a nineteenth-century term, cosmologists call these features "standard candles."

The workhorses of distance measurement in cosmology are the stars called Cepheid variables, first recognized in our own galaxy a century ago. They are stars essentially like the Sun but more massive, and in which a substantial part of the original thermonuclear fuel has been consumed. They are bright (at least 1,000 times brighter than the Sun) and unstable, and their output of light varies rhythmically over intervals ranging from a few days to a few months. The speed of their pulsation depends strictly on their average brightness; the brighter they are, the faster they pulsate. Once Hubble appreciated that the 100-inch telescope allowed him to distinguish individual stars in galaxies other than our own, he naturally looked for Cepheid variables and used them in the first measurements of distances to other galaxies.

This relationship was established only when Cepheid variables were seen in the dwarf galaxies called the Magellanic Clouds, which are visible in the southern skies and are apparently in orbit about our own galaxy at a distance of 100,000 light-years from its center. The dwarf galaxies provided a sample of Cepheids at roughly the same distance.

The only intergalactic distances so far measured with confidence and technical accuracy (with errors of a modest 10 percent) have relied upon Cepheid variables. There is an obvious snag. Ordinary galaxies put out many millions of times more light than the individual stars they contain, so that the more distant of them will be visible as sources of light even when individual stars cannot be picked out. Hubble's first survey was limited to galaxies less than about 10 million light-years away.

There is a second difficulty relating to Hubble's law that has been recognized as serious only in the past 20 years or so. Galaxies are not strictly isolated collections of stars, but are attracted gravitationally to their neighbors. As a consequence (and perhaps because of other influences) galaxies tend to be found in clusters. Our own galaxy, for example, belongs to what is called the local group of about a dozen galaxies, which in turn belongs to a larger structure called the local supercluster. This clustering process gives our galaxy a speed estimated at 200 kilometers a second towards the center of the local group, which is in turn supposed to be moving at roughly the same speed (but in a different direction) towards the center of the local supercluster.

None of that diminishes the value of the redshift as a way of telling the recession speed of a distant galaxy, but the local movement of our own galaxy (as well as the movement of our solar system within it, with an estimated speed of 220 kilometers a second) means that attempts to estimate the rate at which the recession speed increases with distance, the crux of Hubble's law, must be hazardous at best. One way of dealing with the problem is to correct the recession speeds estimated from the redshift to allow for the local motions; the drawback is that measurement errors are then increased. A better way would be to concentrate only on very distant galaxies, perhaps more than 300 million light-years away, when the local corrections would be comparatively less important, but the writ of the reliable Cepheid standard candle does not yet run so far away.

In 1929, Hubble knew nothing of these difficulties, which made it easier for him confidently to proclaim that the recession speed increases directly with the distance. The rate at which the present expansion speed increases with distance is a number now known as Hubble's constant and usually written as H_0. Typically, it might be represented as so many kilometers per second (of extra speed) for every million light-years (of extra intergalactic separation.[12] The larger H_0, the younger the universe.

The value of H_0 has been an embarrassment from the outset. Hubble's first estimates implied a universe younger than Earth, which was manifestly absurd. Latterly, as measurements have become more sophisticated, estimates of H_0 have tended to concentrate about two very different numbers—either about 15 kilometers a second of extra velocity for every million light-years of extra separation or about 25 kilometers a second for every million light-years. These estimates correspond to ages for the universe of 20 billion years and 11.5 billion years respectively.

Scientists faced with the need to estimate quantities that are prone to error are familiar with the problem. No measurement can be absolutely accurate. What usually happens is that first estimates cluster around an average value, which can then be taken as the provisional value of what is to be measured. But if the estimates cluster around two values, scientists smell a rat and ask (or should ask) whether their argument is flawed by some conceptual error.

The latest crop of Cepheid measurements has sharpened the dilemma. Beginning in the fall of 1994, three independent measurements of H_0, based on the identification of Cepheid variables in distant galaxies, have concluded that the value of H_0 lies close to 25 (kilometers a second per million light-years), not to 15. That implies that the universe is between 7.5 billion years and 11.5 billion years old. The publication of these measurements[13] caused a minor sensation for a simple reason: some stars in our galaxy are independently estimated to be more than 16 billion years old, with a possible error on either side of 2 billion years.[14] But how can the universe, supposedly the totality of everything, be younger than some of its parts?

These Cepheid measurements, which broke new ground in several ways, cannot be faulted technically. All of them are based on measurements of variable stars in particular (and different) galaxies of the Virgo

Cluster, a rich and compact collection of some hundreds of galaxies estimated to be between 50 and 75 million light-years away and toward which the local group of galaxies appears to be moving. Two of the measurements were made with instruments of the Hubble space telescope whose design specified the measurement of the true value of H_0 with an accuracy of 10 percent.

The conflict between the age of the universe estimated from the Hubble expansion and the ages of known stars has more recently been sharpened by a survey of white dwarf stars in our own galaxy. White dwarfs are stars comparable in mass with the Sun whose thermonuclear fuel has been exhausted, so that they have become compact objects radiating small amounts of energy (compared with the Sun) at high temperature (by the same standard). They are not easily found because of their faintness, but are easily recognized by their temperature. A research group based at the Florida Institute of Technology at Melbourne[15] has now carried out a more complete survey of white dwarf stars in the disk of our galaxy and has concluded that it is at least 9.5 billion years old.

Although only 50 white dwarfs have so far been measured (with a further 200 being studied), they are sufficiently near for their distances to be determined geometrically, so that uncertainties about their true

Figure 1.2 **Our Galaxy Edge-on.** Schematic diagram of our galaxy in which the dark central band represents the disk itself, the spherical region in the center is the "bulge" and the lightly shaded region is the "halo." The diameter of the galaxy is 100,000 light-years, while the disk is 400 light-years thick.

brightness (or faintness) do not arise. The measurements bear most directly on the evolution of our galaxy, especially on the relative ages of its components. The flattened disk with its spiral arms of stars and interstellar gas is the youngest part, perhaps 10 billion years old, the globular clusters in the galactic bulge are upwards of 12 billion years old and the stars in the more diffuse halo surrounding the bulge may be 16 billion years old or more. These measurements therefore support the belief that the correct value of H_0 is at the lower end of the range of uncertainty that has dogged cosmology for several years.

The conflict between the age of the universe determined from nearby Cepheid variables and that of the oldest stars in our galaxy may be a genuine contradiction. But it may also be the result of persisting uncertainty about the motions of the local galaxies. That will be an open question for many years. For the time being, the most prudent conclusion is that far too little is known about the positions, masses and motions of the million or so galaxies within, say, 200 million light-years of our own. Cosmology has a crying need of an accurate map of the nearby universe. When the general ambition is to understand how the universe began, is it not high time that we knew, with much greater certainty than we have at present, what the universe is like in our neighborhood?

That conclusion is confirmed by the announcement in February 1996 that a cluster of galaxies long known as Abell 3627 (after the catalog in which it was first listed) appears to be much more massive than had been thought. There are 600 visible galaxies in the cluster, whose total mass is between 1,000 and 10,000 times that of our own galaxy. That is 10 percent of the mass of the hypothetical structure identified in the mid-1980s and called the "great attractor," toward which most nearby galaxies appear to be moving. Why had the importance of this cluster not been recognized much earlier? Pure accident. Abell 3627 is partly obscured by the southern Milky Way, so that the brightness of its individual galaxies was hidden.

Do other standard candles than the Cepheid variables have more to offer in the measurement of distance? Several have been canvassed, and three appear to be helpful—the observation of the exploding stars called supernovae (of a particular type),[16] the brightness of the brightest galaxies in galaxy clusters, and the distant radio galaxies, some of which are too far away yet to be seen visually.

Supernovae can indeed be seen in galaxies far beyond the Virgo Cluster, but of necessity they are infrequent and sporadic. The exploding stars sought as standard candles are white dwarf stars that happen to be placed where they can accrete material from interstellar space. They may then acquire enough mass to reignite thermonuclear reactions in their outer layers, leading to a thermonuclear explosion. The expectation is that these supernova explosions evolve with time to be recognizable for what they are, while their intensity will be relatively constant. They are potentially reliable standard candles useful over greater distances than the Cepheid variables, but they are unpredictable so that gathering information about them is haphazard.

The arguments for believing that the brightest galaxies in rich galaxy clusters will always have the same intrinsic brightness are less persuasive, but their redshifts can be measured and the assumption that they are all equally bright does at least yield a good account of Hubble's law. Much the same is true of the distant radio galaxies, where the underlying assumption is that all of them are equally powerful emitters of radio waves, or at least that their apparent brightness as seen from Earth is, on the average, a good measure of their distance. One goal, for the near future, is to establish a region of space in which the nearby Cepheid scale overlaps the longer-range scales provided by supernovae explosions and galaxy clusters, both of which tend to yield the smaller estimates of H_0 (and thus the larger ages for the universe).

Opinions among cosmologists differ on the likely outcome of this dilemma. Even some of those who relish the idea that recent measurements are a setback for the standard cosmology suspect that, in the end, the smaller value of H_0 will be nearer the true mark. But a valid description of the expanding universe will not be possible until this question is cleared up. It is always possible, of course, that the older stars are relics of some earlier "universe," but nobody can yet conceive of a mechanism by which that has happened.

THE QUASAR PUZZLE

In 1963, astrophysics was startled by a quite unexpected discovery: there are objects in the sky whose images resemble stars, but whose emitted light is reddened to such a degree that they must be thousands of millions of light years away. Given that distance, the light they emit vastly exceeds that from any known stars and, instead, is comparable with the emission of light from whole galaxies. Yet they are pointlike objects: *quasar* stands for "quasi-stellar" objects.

The first discoveries were prompted by radioastronomers, who by the 1960s had found many powerful sources of radioemission in the sky and who were eager to find out whether there were optically visible counterparts. There was a strong suspicion that most of them were outside our galaxy; they were, for example, distributed more or less at random across the sky and not concentrated in the Milky Way. Some appeared to consist of two patches of radioemission close together, prompting suggestions that they were galaxies in collision. Then some of the most powerful sources of radiowaves were linked with bright starlike objects so small that their diameter could not be measured with optical telescopes.

Any belief that there are no surprises left should be dispelled by what followed. The optical spectra of quasars revealed redshifts greater than anybody had expected in the 1960s. The light from several quasars now known is reddened to such a degree that spectral lines normally appearing in the ultraviolet are shifted into the visible part of the spectrum. What could they be? Some held that they might be stars within or relatively near our galaxy that are so massive that the light leaving them is reddened by the gravitational forces at the surface, but that idea did not last for long.[17] Astrophysicists instead accepted that quasars do indeed lie at great distances from our galaxy—and then faced the problem of explaining how they can be such powerful sources of radiation.

Several thousand quasars have now been identified. Some of the most remarkable are radio galaxies which have flung off jets of matter in opposite directions, and with speeds that are a substantial fraction of the speed of light. What were thought (in the 1960s) to be colliding galaxies turn out to be the places where these jets of matter plough into intergalactic gas of some kind. A surprising feature of the central objects, which are often more than 1,000 times brighter than whole galaxies,[18] is that their output of radiation can vary very rapidly, in months, days or even hours. That is a sure sign that the source of the radiation is compact—a region whose dimensions are measured in light-months, light-days or even light-hours. The more rapidly fluctuating quasars must be so small that they would fit entirely within the orbit of Jupiter about the Sun. From the way in which quasar light carries telltale signs of having passed through cool material on its way to terrestrial telescopes, many quasars seem to be embedded within much larger structures, which are probably galaxies not obviously different from our own.

The explanation of this remarkable behavior? Provisionally, the engine at the heart of every quasar is supposed to be a *black hole,* a region of space where the density of matter was at one stage so great that it collapsed under its own weight, creating around itself a region of space from which radiation can no longer escape (whence "black") but still exerting its gravitational effect on its surroundings and thus sucking into itself whatever gas, dust or even stars there may be in the neighborhood (whence "hole").

Yet quasars are powerful sources of radiation, which is hardly what is expected of scavengers of anything massive in their neighborhoods. There is an explanation of that as well. Objects sucked into a black hole will not be stationary with respect to this massive center, but will move rapidly around the hole, probably in some kind of orbit. The result will be a disk-shaped structure around the black hole within which matter will be heated by collisions to a high temperature before disappearing into the void. Especially if the black hole is itself rotating, it may even generate magnetic fields strong enough to ensure that material is ejected in a narrow cone from the poles of rotation (thus accounting for the opposed jets of matter of the radio quasars).

Several questions arise. What, for example, is the role of quasars and other "active galactic nuclei" in the evolution of the universe? Being a long way off, quasars date from a time between 20 and 30 percent of the present age of the universe after the big bang.[19] Are quasars, then, an early stage in the evolution of all or most galaxies? And if so, what turns them off? Otherwise, what distinguishes a galaxy that becomes a quasar from one that does not? And how would we distinguish, among nearby galaxies, those that were once quasars? Presumably active black holes have become inactive, but there is no explanation of why they appear to have ceased sucking in material from their surroundings.

As will be seen, the concept of the black hole raises serious difficulties of a philosophical character. Yet quasars are but one of several kinds of unexpected phenomena in the universe to have emerged in recent decades, chiefly because of the improvement of observational techniques; neutron stars and X-ray sources have been found in our galaxy while the cosmic microwave background radiation is such a phenomenon on a cosmic scale. It seems imprudent now to settle on black holes as the engines of quasars when further observation may reveal that something much more interesting is going on. Significantly, astrophysicists and observers are less impatient than their model-building cosmologist colleagues and often refer to "putative black holes" only.

IS THERE ENOUGH MATTER?

One reason why the age of the universe is uncertain is that the age indicated by its present expansion speed is too great. The mutual self-attraction of the galaxies in the past, which affects the speed, demands a calculation based on the total amount of mass in the universe. How is it possible to weigh the universe, or even a single galaxy?

There are several ways. The output of, say, visible light from a galaxy is a rough measure of the number of stars it contains and, so, with assumptions about the average mass of stars, of its total mass. Difficulties crop up because light from individual stars may be obscured by interstellar gas and dust, which is one reason why astronomers have diligently classified galaxies into different types. Light output is a good measure of the relative mass of galaxies in the same category. Another way is to estimate the total mass of a galaxy from the systematic motion of individual stars. Because the stars are held together by their mutual gravitational attraction, they will orbit around the center, as do the planets around the Sun. From the speed of that motion, it is possible to infer how much mass lies inside the orbit. This procedure works well when it is possible to distinguish individual stars and to measure their speeds. The two ways of measuring the galactic mass typically give discordant results, and in a consistent direction: the mass estimated from light output is always less than that inferred from the rotational pattern. The difference is usually substantial. Even more disconcerting, the outlying stars are often moving more quickly than the visible mass of the outer stars allows.

The explanation? There must be material between the stars that does not emit radiation—that is "dark." Some obvious and several exotic candidate materials have been canvased to explain the discrepancies: Clouds of cool (and so invisible) gas within our own and other galaxies; some are known to be more than a million times as massive as the Sun.[20] Stars that have burned out may also be part of the explanation. So may be the objects called "brown dwarfs," made of primordial material but too small for their thermonuclear fuel to have ignited and to have set them shining; evidence of some such objects has indeed been uncovered in our own galaxy in the past few years.[21]

Yet so far there is no single case in which the discrepancy between the mass of a galaxy estimated from its light output and from the orbital motion of its stars has been adequately explained by conventional dark matter. Further observation may remove that difficulty, but meanwhile the speculation has arisen that galaxies, simply because they are concentrations of mass, may act as gravitational traps for particles of matter not bound into stars. That is part of a larger problem (see below), but for now the glaring discrepancy between the masses of galaxies estimated from their light output and from their rotation sticks out like a sore thumb.

Galaxy clusters present a similar but more serious problem. When the first Earth satellite equipped to record X-ray emission from the sky (a satellite named after Einstein) was launched in 1975, one of the surprises was the discovery of X-ray emission from the apparently empty spaces near the center of the Coma Cluster (six times further away than the Virgo Cluster); strikingly, some of the X rays carried the signature of iron atoms.[22] The implication was that distant intergalactic space is populated by material that has been recycled through stars, which are the only known means by which primordial hydrogen and helium can be converted into heavy elements such as iron. And the X rays emitted by the iron atoms imply either high temperature or energetic particles of some kind. Not all intergalactic space is devoid of matter.

The galaxy cluster problem is further sharpened by studies of how they form. Although the geometrical core of a cluster often includes the most massive galaxy, that mass is usually not enough to explain why other galaxies are attracted toward it. If there is a hidden stock of mass that drives the clustering, nobody knows what that is. Understanding of the clustering process is not yet good enough to tell whether it involves a small or a large fraction of the whole mass in the universe. Again, as within individual galaxies, exotic forms of hidden matter are candidate explanations.

Whatever the truth, the quantities of hidden mass seem not to be enough to decelerate the expansion of the universe to the point at which expansion would be halted, even reversed. The output of visible light from galaxies accounts for about 5 percent of the mass required to halt expansion. Generous allowance for the extra mass implied by the rotational pattern of stars within galaxies might double that percentage. Extra mass triggering cluster formation could bring the total to about 20 percent—one-fifth—of that required to halt the expansion. Yet for the past 15 years there has been an influential opinion among cosmologists that the mass of the universe or, more precisely, the density of the matter it contains, must be just enough eventually to bring the expansion to a halt, but not to reverse it. That *critical density* is equivalent to the mass equivalent of 7 hydrogen atoms for every cubic meter of space in the observable universe.[23]

In that view, the universe is balanced on a knife-edge. On one side lies continued expansion (at an ever decreasing rate because of gravitational self-attraction); that is called an "open" universe. On the other side of the knife-edge lies, to begin with, more rapidly decelerating expansion, an eventual halt and then contraction, so that the original stock of matter and energy is again concentrated together in a small region, in a "Big Crunch," perhaps then to start again. That universe is "closed." In between is the singular condition in which the universe expands at such a decelerating rate that the expansion almost, but not quite, stops.

The reasons for expecting the universe to be in a closed condition are theoretical. One of them is simply abhorrence of numerical coincidence, and is simply stated. If the density of matter in the universe at the outset were exactly the critical density, that condition would exactly persist now, 10 or 20 billion years later. On the other hand, any slight discrepancy at the outset would have been enormously exaggerated during the long expansion, and by many millions of times. Turning that argument around, if the density of the universe now is between 5 and 20 percent of the present critical density, the discrepancy between the density of the early universe and what would then have been the critical density (much greater than at present) would have been minuscule. But why should we suppose that the real universe differs from one balanced on the knife-edge between expansion and contraction only by such a small amount? Is it not more reasonable to suppose that the density at the beginning (and therefore now) was exactly on the knife-edge?

That argument, widely used or at least quoted, has little merit. Nor is it relevant that one of the earliest models of the universe in the language of Einstein's general theory of relativity (the "Einstein-de Sitter" universe[24]) expands so that the average density of matter is always equal to the critical density. There are many other solutions to Einstein's equations, describing universes of different kinds and different densities. Deciding between them should be a matter for observation, not presupposition.

The strongest support for the view that the actual density of the universe differs very little from the critical density is different, and rests on the first version of a theory (of the "inflationary universe") developed in 1981 to avoid some of the conceptual problems of the big bang (see below). Although later versions of the theory do not compel the belief that the present and the critical density should be equal, the issue has prompted a great variety of speculation and elaborate experiment designed to identify the ingredients of the "missing mass" required to "close" the universe, as the saying goes. It is true that galaxies and galaxy clusters are deficient in mass, and that the cause of these discrepancies has not yet been pinned down. Nevertheless, the continuing search for enough "missing mass" to close the universe, amounting to perhaps 80 percent of the total mass, is the modern equivalent of the fruitless search for the lumeniferous æther a century ago.

Why Galaxies?

Cosmologists and astrophysicists generally assume that the galaxies now in the sky were formed after, perhaps long after, the end of the big bang. Once the temperature of the universe had fallen below 3,000 K, so that matter and radiation were in relative isolation from each other, a human being would have seen a sky filled only with reddish sunlight from all directions. With continued expansion, the light would have become increasingly red in color until eventually, as infrared radiation, it would have been invisible. And then there would have been nothing to see in the sky: the universe would be dark.

What happened afterward is inference, but cosmologists believe that the universe would have remained dark for a billion years or so until atoms and molecules had clumped together into aggregations that were both stable and compact enough for stars to form and to begin shining. At that stage, the only matter in the universe would have been hydrogen mixed with the 25 percent of helium and the other trace elements formed during the big bang.

Which objects came first, stars or galaxies? Theoretical science offers no clear guidance, while observations of the distant (or early) universe do not yet go back far enough. The general (and fair) assumption is that the first objects in the sky would have sprung from naturally occurring variations, or fluctuations, of the density of matter in an otherwise uniform gaseous cloud. Places where the density of the matter is, by chance, greater than the average would sometimes grow by gravitational attraction at the expense of neighboring patches and then, having become still more massive, would attract further gas until they had assembled enough to make a star or, perhaps, a whole galaxy.

How likely is such aggregation to occur? Put that way, the question is intrinsically statistical; what is the chance that a clump once formed will grow and not shrink? Early in this century, essentially the same problem arose in connection with the formation of stars within our galaxy. James Jeans, a British astrophysicist, concluded that the formation of stars is not inevitable. The difficulty is that the material in an aggregating mass of gas is heated by the gravitational energy liberated during the collapse, so that many of the atoms and molecules in the aggregation acquire the energy with which to escape from it again.

Disaggregation would be accelerated if density and temperature at some point in a collapsing cloud of gas became great enough to ignite thermonuclear reactions. That is a principal reason why, in the past two decades or so, the general opinion has been that galaxies were the first to form, consisting at the outset simply of clouds of gas massive enough so that billions of stars would eventually be formed in them. The clouds would have continued to collapse at a majestic pace, one so slow that the gravitational energy could be successfully radiated away before further collapse was prevented by the ignition of thermonuclear reactions. There are further restraints on gravitational collapse if the contracting cloud is rotating as a whole or if there are magnetic fields buried within it.

Indeed, there are difficulties about the continuing process of star formation within our galaxy. Some argue that even in rich clouds of gas and dust such as the dramatic galactic nebula in the constellation of Cygnus, stars will not form unless there is some external trigger—the explosion of a nearby star, perhaps.[25] A study of rare isotopes of aluminum on Earth's surface suggests that the solar system includes some material scattered through nearby interstellar space by a supernova explosion less than 100 million years before the Sun formed, which supports the trigger hypothesis. But how could the first stars in a galaxy have been formed, before there were any stars to explode as supernovae?

After a century of close and successful observation of the stars of our galaxy, which has provided an apparently reliable understanding of the constitution of stars of different types and of their progressive evolution into one another, it is remarkable that their formation should be so much in doubt. The underlying problems are not conceptual, but there is every reason to regard the understanding of star formation as a challenge for the years ahead. So, for that matter, is the question of how the solar system formed; that will probably require not only more detailed exploration of the outer planets, but the observation of other planetary systems as well.

The formation of galaxies is another question. In recent years, there have been two conjectural arguments to suggest that there may be a bias in favor of galaxy formation. One argument is connected with the search for dark matter in the universe. Dark matter *might* consist of gravitationally bound aggregations of particles that act as invisible nuclei to catalyze the condensations of gas-forming primitive galaxies.[26] To qualify for this role, those particles must be relatively massive and so slowly moving that their kinetic energy does not allow them to escape. Nobody, as yet, has a convincing candidate particle, let alone a scrap of experimental evidence to support that choice. Indeed, the best argument in favor of cold dark matter as a catalyst of galaxy formation is that, if correct, it would also help to remove the suspicion that the measured density of the universe is less than it should be. Wishful thinking has a place in cosmology.

The second hypothetical aid to galaxy formation is also an open question. As will be seen, one way of dealing with some of the difficulties arising in the big bang is to suppose that the scale of the universe increased rapidly between the big bang itself and the first appearance of matter. In the process, fault lines (called "cosmic strings") may have appeared in the structure of space-time and could have been locations where particles of matter ("dark" or otherwise) would have congregated. That is also an open question.

Meanwhile, some telling observational evidence has come to light that supports an intermediate view of the first visible entities in the universe. In 1996, a research group from the University of Arizona[27] reported on a systematic search with the Hubble space telescope for blue galactic objects in a small patch of sky and found 18 of them, all smaller than our galaxy and other galaxies in our neighborhood. The inference is that these "galactic clumps" would afterward have merged to form galaxies more like our own. The objects' blue color is a measure of their distance: the light observed is really ultraviolet radiation from hydrogen atoms[28] which has been redshifted into the blue region of the visible spectrum by recession.

The authors of these observations infer that their galactic clumps are seen at about two-thirds of the way (in time) back to the big bang (if that is how the universe began) and that they contain a billion stars or so. Because objects such as these have not been seen in our more immediate neighborhood, it is reasonable to assume that they have grown larger as the billions of years have passed, perhaps by merging with galactic clumps like themselves eventually to form galaxies comparable in size with our own.

These observations point to a way in which the galaxies of the mature universe may have been formed from the "bottom up," by merger and accretion, but they are also a vivid reminder of how little is known of the early universe. The discovery of the galactic clumps is a technical triumph, pushing the Hubble telescope to its limits; they are as faint as the telescope can detect. Who knows what story will be told when the observations are extended to, say, the green region of the spectrum, which may reveal the origins of the galactic clumps themselves?

Other questions to be answered concern the distribution of galaxies across the sky. To a first approximation, they appear to be randomly distributed (which does not contradict the notion that many of them belong to clusters provided that the clusters are themselves randomly distributed). But there may be more than this to say. One telling observation is that even the most distant (and thus youngest) galaxies seem to include heavy elements such as iron that must have been formed in stars like the Sun. Another is that the clustering of galaxies seems to be commonplace on a large scale as well as on the small. On the small scale, our galaxy, for example, seems to be interacting with the half-dozen dwarf galaxies that surround it,[29] while a remarkable observation in 1995 revealed that a group of more than 1,000 stars in the direction of the constellation Sagittarius is moving as a group in a direction perpendicular to the plane of our galaxy as if it were a small galaxy that had already been captured and which had already made one transit through the disk.[30] Even well established galaxies like our own appear to be growing still.

On the large scale, observations in the past few years have suggested that galaxy clusters such as the Virgo cluster are moving as a whole in the direction of an unidentified gravitational influence called the "great attractor" and that there are also huge voids, or spaces empty of galaxies, in many nearby parts of the universe. There are even suggestions that the prominent galaxies in our neighborhood are arranged as if they were ornaments suspended from an invisible wall (which happens to be perpendicular to the plane of our galaxy and which is naturally called the "Great Wall"). These are mere hints that our part of the universe (and presumably all other parts of it) are more intricately constructed than previously supposed. The map that will eventually have to be constructed of the universe within the range of modern telescopes will be a map of discovery to match anything that Magellan could have produced.

Only when that map is complete will there be a rounded understanding of the natural history of galaxies. Meanwhile, it is important that the tendency of galaxies to be bound into clusters may point to a feature of the universe that could stand conventional cosmology on its head. If the galaxies of the Virgo Cluster (our own galaxy and Andromeda included) are destined eventually to coalesce, will Hubble's law still apply? The standard assumption that matter is (on the average) uniformly distributed in the universe could easily be vitiated. That would upset more than one applecart.

SMOOTHING THE BIG BANG

The difficulty about galaxy formation has come to light only since the 1970s, when a serious defect of the big bang explanation of the universe had also come to light. Imagine this experiment. Measure the temperature of the microwave background radiation reaching Earth from two diametrically opposite directions. That radiation can only have come from opposite poles of the universe when matter and radiation began to go their separate ways, more than 100,000 years after the big bang. The universe would then have been roughly the present size of our galaxy, but filled with all the matter that now survives as well as with the accompanying radiation.

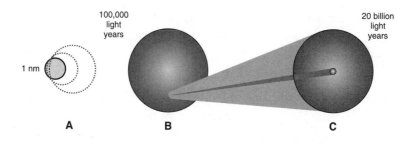

Figure 1.3 Inflation of the Universe. Time runs from left to right, but not uniformly. At **A**, the universe is one billionth of 1 meter across, when it embarks on inflation. At **B**, the universe is more than 100,000 years old and about the size of our galaxy; the small speck at the center of this frame is that which is supposed to have grown into our universe. The present state of the universe is represented at **C**. The narrow cone stretching back from **C** to **B** includes all of the universe we can now see.

The two opposite poles could not then have been in physical communication with each other; the most that can be expected is that the antipodean poles would each have been influenced by some earlier condition of the evolving universe, say 50,000 years after the big bang. That means that there could have been nothing to ensure that the antipodes would have the same temperature at 100,000 years. On the contrary, there is every likelihood that those temperatures would have been different. As a result, the temperatures in opposite directions now should also be different from each other, but that is emphatically not the case. So much had been clear from measurements in the 1970s, at least when allowance was made for the motion of the galaxy (and Earth with it) relative to the background radiation. In 1994, the COBE satellite* measured the temperature in all sky directions and found it to be the same in all directions to within one part in 100,000.[31] Gamow's universe should not be so uniform.

Not long after the importance of this contradiction sank in, it was ingeniously resolved—but at a price. In 1980, Alan H. Guth (now at the

---

* COBE = U.S. satellite Cosmic Background Explorer.

Massachusetts Institute of Technology) put forward what is called the theory of the "inflationary universe." He argued that almost immediately after the big bang, and before matter of any kind had appeared, the universe spontaneously expanded at huge speed and at an accelerating pace.[32] When the expansion had finished, all dimensions in the universe had been multiplied by a huge factor.[33] Yet that expansion, converting submicroscopic distances into cosmic distances, is reckoned to have been completed in a fleeting fraction of a second—one billionth of the time it takes for light to cross the diameter of an atomic nucleus.

What were the driving forces for this expansion? Guth's explanation is that the rapid and accelerating expansion has to do with a strictly non-Newtonian property of empty space, otherwise "the vacuum." That after all, is all there can have been at the beginning of Gamow's big bang. Since the beginning of the twentieth century, it has been known that the vacuum is not strictly empty; does it not embody Planck's radiation field? Now it is known that there are also other fields as well, many of them associated with tangible particles of matter possessing mass or substance. A decade ago, Guth's mechanism was a godsend to cosmology. Crucially, it was a way of smoothing out the temperature (and density) fluctuations expected in Gamow's big bang. It also explained the abrupt appearance of matter in the very early universe; the primeval vacuum spawned particles of matter spontaneously. Inflation, as Guth called the exponential expansion, flattened the universe, made it appear uniform and made its average density approximate to its critical density.

The philosophical cost of this innovation, however, has been huge and excessive. Inflation implies that the universe we can now see—the "observable universe"—began life as a tiny speck of the space-time in a cosmos still so young that matter had not yet appeared. So inflation inescapably requires that there must be a multitude of other universes alongside our own, each derived from a different speck of space-time and each evolving in parallel, but perhaps differently, depending on the properties of the space-time from which they sprang. Because the parallel universes are, for the time being, beyond our ken, there is no way in which evidence for or against their existence could be gathered.

Inflation also has the consequence that space-time in the present universe should not be curved as allowed for by Einstein's theory of gravitation, but flat, which is in agreement with observation. It is as if a toy balloon has been inflated to a huge degree (without bursting); a patch on the surface will originally have been curved, but after inflation will be virtually flat. The simplest versions of the inflationary theory, however, also compel the conclusion that the density of matter in the present universe should equal the critical density. The inflationary mechanism has therefore been one of the chief spurs in the search for "missing mass" in amounts sufficient to close the universe.[34]

It is a telling comment on the habits of the research community that its perpetual and healthy skepticism has not been lavished with the accustomed generosity on this daring and ingenious theory. The process that drives the inflation is the conversion of one form of empty space, the vacuum, into another—one that can accommodate particles of matter. This idea, called the "Higgs mechanism," plays a part in theories of how particles of matter are related to each other, but the validity of that theory cannot be tested for several years. Its invocation as the driver of the inflation of the big bang universe is something extra. Yet there is no direct evidence that the universe went through a period of inflation. To say that is not to deny the ingenuity of the theory or to suggest that empty space filled with high-temperature radiation would not have some of the characteristics required for inflation in the big bang universe, but neither is it surprising that the more pragmatic cosmologists have taken to affirming their adherence to the big bang "in the postinflationary period" only. Others may rightly ask, "But what is the big bang without that initial event?"

WHAT HAPPENS NEXT?

What, then, is the status of the big bang? It is not so much a theory as a model. It accommodates the conspicuous properties of the universe (the expansion, the primordial light elements and the cosmic microwave background radiation), but it is incomplete; why else would it have been necessary for Guth to introduce the notion of inflation? The unavoidable consequence of that innovation is that there must be several universes evolving in parallel. The unfamiliarity of that idea is not an objection, but the concept does raise difficulties to which the research community has not had time to accommodate.

In the many-universes picture, for example, it must be supposed that different universes, each derived from a different speck of the material created during a big bang, will differ in the density of the matter they contain. Some will resemble our universe in that they contain less matter than is required to form a closed universe, others will presumably have begun as supercritical and will have already gone through one or more cycles of expansion and contraction. That is not entirely beyond the bounds of imagination: imagine a group of toy balloons being inflated at different rates by pressurized gas from a set of central nozzles. But what would such a universe as the second one described above look like from somewhere else from which the boundary between a rapidly and a less rapidly expanding region of space might be visible? Little imagination has so far been spent on that question.

There is the further difficulty that the big bang, arguably the most significant event there has ever been, appears to be exempt from David Hume's dictum that to every event there is a cause. One way of exempting the big bang is to invoke supernatural causes: "God designed and made the universe," for example. Another is due to Guth and his colleague Paul Steinhardt, from the University of Pennsylvania, who argue that the vast energy of the big bang may well be an illusion, and that the great energy of the initial stock of radiation may be offset—perhaps even canceled—by the gravitational energy of the densely packed matter that appeared once the inflationary epoch was ended. In that view, the universe exists because of a quantum fluctuation 10 or 20 billion years ago, and is a transient once-and-for-all phenomenon, although one that lasts for, perhaps, 40 or 50 billion years.[35] More recently, Guth has claimed that in an inflationary universe, universes much like that we inhabit will begin to form (by means of big bangs) at random times and places in rapidly inflating space-time, leading to a so-called *fractal universe* with no beginning and no end.[36]

More recently, a different way of finessing this problem has come to light. In a book published in 1997, Lee Smolin, a professor at the University of Pennsylvania, put forward the idea that the universe had no unique beginning in time.[37] Instead, he argues, when an object such as a star collapses under its own weight to form a "black hole," the interior of the collapsed object may have the appearance of a big bang universe in its own right. From that, a whole new universe may grow, forever hidden from our sight. Just as the mechanism of Guth's inflationary universe rests on a theory of the behavior of particles of matter that cannot yet be tested, so Smolin's universe rests on a hunch about the way gravitational forces will be modified when proper account is taken of quantum mechanics. In the absence of solid ground for believing that black holes can engender whole universes within them, Smolin's scenario is no more persuasive than the account in *Genesis* of how the universe began.

The truth is that these questions cannot now be decided. It will be time enough to brood about the true origin of the universe when the gaps in present understanding of what the universe is like are less conspicuous than now. We forget that it is only in the past half-century that what is known about the universe has been augmented by information gathered by X rays, ultraviolet and infrared radiation and by radio waves as well as by telescopes for the first time freed from the difficulty of seeing through Earth's atmosphere—the Hubble space telescope will soon be followed into Earth orbit by even more sensitive and far-seeing instruments. There is much to find out about the real universe, and every expectation that the new knowledge will have an important bearing on how the universe began. That is why the prudent answer to the question "How did the universe begin?" is that we do not yet know. As will be seen, it is not the only challenge of that magnitude that faces science.

What, then, lies ahead? The most urgent need is to verify or, if necessary, modify Hubble's law. That will be possible a decade or so from now, when mapping projects now underway have been completed. Probably H_0 will turn out to be smaller rather than larger, meaning that the big bang universe is more like 20 billion years old than half as much. If, of course, the maps suggest that the clustering of galaxies is an inexorable process, Hubble's law and most previous cosmological speculation will be in serious trouble. But the same map making will provide a realistic natural history of galaxies; the likely conclusion is that galaxies like our own are the results of several stages in which smaller aggregations of matter have been merged together.

More particular questions cry out for answers. The nature of the hidden mass in apparently ordinary galaxies is a deepening embarrassment as the decades tick by. On the other hand, the search for "missing mass" to close the universe as a whole seems destined to be a wild-goose chase. The origin of the energetic particles in cosmic rays, a fashionable object of study three or four decades ago, remains enigmatic.[38] The role of magnetism in the structure of galaxies and perhaps of the universe at large is another issue that cosmology has put aside for the time being. The question of quasars—in particular whether the state of being a quasar is a stage in the evolution of all galaxies—will probably be answered in the affirmative, and by the rounded natural history of quasars to which we can look forward.

There is a further disappointment in the present state of affairs: Einstein's expectation that his general theory of relativity would stand or fall by the account it gave of the structure of the universe has not been fulfilled. That does not mean that there are serious doubts about the theory. Whenever it has been possible to put it to the test, as with the observation of the rate at which the orbits of pairs of compact stars bound gravitationally together change with the passage of time or when the images of distant quasars are split into several by the gravitational field of another galaxy intervening along the line of sight to the earth, its predictions have proved correct. But the observable universe is not a stringent test of the theory; from all appearances, space-time in our neighborhood is not noticeably curved, but flat. As will be seen, all that will change when it is possible to construct a bridge between the general theory of relativity and the other great conceptual innovation in the physical sciences this century—quantum mechanics.

Simplicity Buried in Complexity 2

The diversity of the substances in the world has been an intellectual challenge for at least 2,500 years: what can iron, wood and water have in common? Since antiquity, it has been known that certain rocks, when broken into small pieces and mixed with charcoal, yield metals such as lead, copper, zinc and even iron; now, common sand is similarly converted into silicon for the manufacture of computer chips. Chemists are adding tens of thousands of new chemicals a year (not all of them synthetic) to a catalog already stuffed with several millions of them.

In all of nature, diversity disguises underlying simplicity. All substances consist of very much smaller entities that are indivisible and embody the characteristics of the materials we see and touch. Twenty-five hundred years ago, Democritus used the word "atoms" for these unseen entities. This principle is both the foundation of the atomic theory developed in the first two decades of the nineteenth century and of the much more recent search, in the closing decades of this century, for the modern equivalents of what the Greeks believed to be the rudimentary ingredients of all matter: earth, air, fire and water. The present belief is that all matter is made of just two kinds of entities or particles, called hadrons and leptons,[1] usually in some kind of combination with each other.

There has also emerged a new and daring ambition. Many now argue that the particles from which tangible matter is constructed are not arbitrary facts of our life, but that their properties are consequences of the large-scale structure of the universe and of the manner in which it came into being. If, as is implicit even in the big bang, the first step is the emergence of the space in which the universe eventually evolves, and if the appearance of matter is a consequence of that initial event, the properties of the rudimentary ingredients of matter will be consequences of the manner in which the structure of space is determined at the outset. More generally, if our universe is but one example among many, may not the matter around us be peculiar to it and inappropriate to some other? Of course, we do not know whether the observable universe, which we call our own, is all there is, or whether it is a part of a larger structure or even one of many collections of material objects and radiation coexisting alongside our own.

This line of thinking has stimulated common interests between cosmologists and particle physicists typified by theories such as the inflationary universe described in the previous chapter. Taken to its extreme, the idea supposes that there will one day be a unified account of how the universe came into being and why the matter it contains has the properties with which we are familiar. The enormity of that goal, sometimes called "the theory of everything," is insufficiently appreciated. As a *goal*, it has powerfully stimulated much imaginative research. But the belief that such a theory is almost at hand is a distraction from important and identifiable problems not yet solved. The practical need is to understand the properties of real atoms in the real universe. What makes them what they are?

One Layer Under Another

The long search for more plausible atoms than those the Greeks proposed has been episodic, resembling the unpacking of a nest of Russian dolls. Unscrew Yeltsin and Gorbachev is revealed; unscrew him to find Andropov, Chernenko, Brezhnev, Khrushchev, Stalin, Lenin and Nicholas II. Repeatedly it has seemed that all the elementary atoms have finally been listed—whereupon new observations have proved the belief mistaken.

Dalton's atomic theory with which the nineteenth century began became (and remains) the intellectual foundation of the exuberant chemical industry that enlivened economic growth in the closing decades of that century. But then, another doll case had to be unscrewed. The discovery of the electron (in 1897) and of radioactivity (the previous year) suggested that Dalton's atoms are not indivisible. Electrons come from atoms and must be parts of them; radioactivity is the disintegration of atoms into fragments of themselves.

It fell to Rutherford, by coincidence also then at Manchester, to unscrew the Russian doll case carrying Dalton's portrait, showing that all atoms have electrons at the periphery; at the center, there is a much smaller positively charged nucleus which nevertheless carries more than 99.9 percent of the mass of each kind of atom.[2]

But even atomic nuclei cannot be indivisible—radioactivity shows that nuclei sometimes break apart. What are their ingredients? Answering that question is to unscrew the third doll-case. Protons, the nuclei of hydrogen atoms, were obvious candidate components; they each carry a positive electric charge identical in amount with the negative charge of an electron, and are nearly 2,000 times as massive. But must there not be something else as well? So much is clear from a comparison of hydrogen and helium atoms, which have one and two electrons respectively, but the helium nucleus is almost exactly four times (not twice) as massive as that of hydrogen (a proton). Where does the extra mass come from?

In a public lecture in 1920, Rutherford floated an interesting idea. Could there be novel particles (he called them "neutrons") comparable in mass with protons, but with no electric charge? Twelve years passed before James Chadwick, one of Rutherford's brigade of bright young men (then at Cambridge), took up the question. His stimulus was research in Paris by Frederic Joliot and Irène Curie (Marie Curie's daughter), who showed that puzzling and exceedingly penetrating (but invisible) radiation is given off when α-particles (or helium nuclei) from radioactive polonium collide with atoms of light elements such as boron and beryllium. Chadwick found that the radiation consists of electrically neutral particles with a mass virtually identical with that of the proton.[3] They were Rutherford's neutrons, the missing ingredients of mass in atomic nuclei.

The twelve-year gap between Rutherford's speculation in 1920 and Chadwick's discovery in 1932 is a vivid illustration of how the research community can live with contradictions—or suffer from collective blind spots. If Rutherford had energetically followed up his own proposal, he might well have made experiments similar to those of Joliot and Curie as well as the observation (by Enrico Fermi in 1932) that neutrons generate novel kinds of radioactive nuclei when they collide with uranium nuclei. That led directly to the discovery of nuclear fission (by Otto Hahn in 1939). Less than a decade after the discovery of the neutron, Chadwick was in Washington urging the United States to embark on the Manhattan Project.[4] It is a hair-raising speculation to ask how the course of recent history would have been changed if nuclear fission had been discovered a decade earlier.

Just as the atomic theory of the nineteenth century became the intellectual basis of modern chemistry, so the nuclear theories developed between 1910 and 1940 became (and remain) the foundation of what is now the nuclear industry. The chemical character of an atom is fixed by the positive electric charge of the nucleus, which determines how many electrons there will be in the periphery. And all the properties of an atomic nucleus can be inferred from the numbers of the protons and neutrons of which it is composed.[5,6]

In the 1930s, both in Europe and the United States, people began designing machines to accelerate particles of matter, both electrons and protons, using them to bombard the nuclei of atoms to induce nuclear transformations. Once it became possible to use neutrons for this purpose (after 1932), the discovery of nuclear fission was inevitable, as was the development of nuclear weapons and nuclear reactors. During this same period, it was also established (in 1938) that the energy put out by the Sun and other stars comes from thermonuclear fusion (the fusion of protons and neutrons into helium nuclei), whence the development of hydrogen bombs less than twenty years later, in 1952. A decade after that, by the early 1960s, understanding of the atomic nucleus was substantially complete.[7]

Unscrewing successive doll-cases soon became a technological matter. Rutherford had unraveled the structure of the atom by using α-particles from naturally occurring radioactive substances; he was able to infer the size and mass of atomic nuclei from the way in which the α-particles were scattered from a thin metallic film. Only occasional α-particles were scattered through large angles, suggesting that only a tiny fraction of the whole volume of an atom[8] contains material massive enough to deflect the projectile particles from their course; that is the nucleus of the atom. Most later discoveries about the structure of atoms and their nuclei have also used particles as projectiles—since the 1930s, usually particles given very large amounts of energy by machines designed for just that purpose. The first cyclotron (for accelerating electrons) was built at Berkeley, California, in 1930, while Rutherford arranged for his younger colleagues to build an electrostatic machine for giving protons enough energy to disintegrate some of the atomic nuclei with which they collided.

In the six decades since the first accelerators were built, the technology has become enormously more powerful. Machines have been designed to allow the head-on collision of streams of particles, which means that the full energy of both particles is expended when two of them collide. But the cost of these machines is very great. In 1995, the

U.S. Congress declined to provide the $9 billion required to complete construction of an accelerator in Texas, called the superconducting super-collider (SSC); for the time being, the initiative in the exploration of the structure of matter lies with a European machine (called the large hadron collider) now under construction at Geneva and likely to be in service early in the next century. Some participants in this enterprise believe the new machine will put a capstone on the present body of understanding by revealing the reality of particles of matter whose existence is now only predicted. Others hope the new machine will open up new paths of discovery. They have coined the phrase "new physics" to describe that hope.

Either way, the cancellation of the SSC has caused a crisis for particle physics everywhere. Since the end of the Second World War, the construction of ever more powerful accelerators has been sustained by international competitiveness between countries such as the United States, the Soviet Union (now Russia), various states of Western Europe and, latterly, Japan. Timely discoveries (and the Nobel prizes that seem routinely to follow) were widely regarded as counters in the Cold War. But chauvinism is now out of fashion (and not easily affordable). It seems improbable that there will be a successor to the European accelerator unless it is a genuinely international project—and even then, the research community will be lucky to see major new machines more often than one every 25 years. Yet despite recurrent speculation that a "theory of everything" will soon make the structure of matter and of the universe plain to all, the truth is that particle physics needs grounding in experiment and observation as much as any other field of science, and that it will be cruelly hampered after 2005, after the Geneva machine has yielded its first harvest of data.

RIVER OF DISCOVERY

The present belief that the true atoms of matter are entities called hadrons and leptons is part of the river of discovery let loose by the conceptual advances in 1925 and 1926—the development of quantum mechanics by the German scientists Werner Heisenberg, Max Born and Pascual Jordan in 1925 and by Erwin Schrödinger early in 1926.

Quantum mechanics is still spoken of in the tone of voice reserved for unintelligible intellectual abstractions such as "deconstructionalism" in literary criticism. Part of the trouble is that quantum mechanics is a teasing blend of the practical and the abstract. Many scientists call it a paradoxical way of describing the world; some write of the "quantum world" as if it were a foreign land. But that is emphatically not the case. Quantum mechanics is the only valid way of describing systems whose size is very small. What seem to be paradoxes are not paradoxes at all, but discoveries about the nature of matter. And quantum mechanics has become the most powerful tool in understanding the real world.

The illusion of paradox is born of the limitations of our senses; we cannot perceive events on an atomic scale, and so have no instinctive appreciation of them. So much was spelled out by Paul A. M. Dirac, one of the half-dozen young men who made this intellectual revolution, in 1930. In the introductory chapter of the first edition of his book, *The Principles of Quantum Mechanics* he listed several ways in which Newtonian mechanics could not answer contemporary questions in physics, and adapted in an interesting way the argument Democritus had used 2,500 years earlier to persuade himself that atoms are the ultimate constituents of matter. Newtonian mechanics, said Dirac, explains the motion of a large object in terms of the motion of small parts of it, and of those parts in terms of the motion of their smaller parts, and so on. But so long as

> *big* and *small* are relative terms, it is no help to explain the big in terms of the small. It is therefore necessary to modify classical ideas in such a way as to give an absolute meaning to size.

That, through Planck's constant, is what quantum mechanics does. If there are questions not yet answered about the meaning of quantum mechanics, they point not to inconsistencies in its formulation, but to discoveries yet to be made.

The obvious first goal of the new quantum mechanics was to understand how electrons are arranged in atoms, which certainly qualify as very small—the diameter of a hydrogen atom is a little more than a tenth of a millionth of a millimeter. The problem was first tackled in 1913 by the Danish physicist Niels Bohr, using a clever argument by analogy with the orbital motion of a planet about the Sun and the clue provided by Planck's quantum of radiation, showing that energy is transferred only in discrete amounts. Two important conclusions followed: there are states of a hydrogen atom (Bohr called them "stationary states") in which an electron does not radiate away its energy as Maxwell's nineteenth century theory predicted, while the frequency of all the spectral lines of hydrogen is exactly what would be expected from the difference between the energy of some pair of stationary states. Although the theory worked only poorly for atoms other than hydrogen, Bohr's contemporaries knew he was on the right track.

There followed two foreshocks of the earthquake to come. In 1924, Louis de Broglie asked why, if supposedly wavelike radiation had been given corpuscular properties by Planck, should not particles of matter such as electrons have wavelike properties as well as those of particles? He also correctly linked the purported wavelength of a particle with its momentum—the faster the motion, the smaller the wavelength. (Now, there are electron microscopes that exploit this principle.) Then, early in 1925, two graduate students at Leiden in the Netherlands, Sam Goudsmit and George Uhlenbeck, startled their elders by showing that anomalies in the spectral lines of some kinds of atoms in a magnetic field can be explained if all electrons are endowed with a certain fixed amount of angular momentum, or spinning inertia.

Wolfgang Pauli, an Austrian scientist then based at Zurich, took the lead in pouring cold water on this idea until he was persuaded otherwise,[9] whereupon he promptly became the prophet of the idea that all particles of matter carry a certain quantity of spinning inertia or *intrin-*

sic spin. Like the energy of Planck's radiation field, intrinsic spin is also quantized (or is invariably a multiple of some fixed amount); electrons, for example, have just half a unit of the universal quantum. Pauli then went a long way to explain the spectrum of the helium atom (which has two electrons) by postulating that if two electrons are in the same stationary state, their intrinsic spins must point in opposite directions.[10] As a bonus, this rule of thumb (called *Pauli's exclusion principle*) also accounted for the regularities of the properties of the chemical elements first described in the 1880s by the Russian scientist Dmitri Mendeleev. For example, the elements lithium, sodium and potassium all have similar properties because they have just one electron in a stationary state at the periphery of the atom, while all the other electrons are paired together by their opposite spins in states that are physically more compact.

The discovery of intrinsic spin, on the threshold of the climax in the development of quantum mechanics, typifies much that has happened more recently. There is nothing in Newtonian mechanics to suggest that particles of matter should have such a property. Inventing novel properties of particles to explain their observed behavior has been a striking feature of the past half century.

Werner Heisenberg seems to have been the first to appreciate how subversive were the new ideas with his advocacy in 1925 of what he called the "uncertainty principle."[11] If energy is exchanged only in discrete amounts, or quanta, he concluded that there must be limitations on the accuracy of physical measurements of an atomic system. For example, for an electron in the lowest possible energy state of a hydrogen atom, physical measurements can say nothing about its angular position relative to the nucleus at the center. A few months later, with Born and Jordan (both at Göttingen), Heisenberg produced a scheme (called "matrix mechanics") which is a comprehensive way of calculating the energy of the stationary states of atomic structures of any kind.

Then came a delicious irony. Erwin Schrödinger set out to construct "a theory which comprises both ordinary mechanical phenomena . . .

and typical quantum phenomena." His starting point was a set of ideas developed by the Irishman W. D. Hamilton in 1833 for describing in the same language rays of light and the classical movement of objects (such as planets about the Sun).[12] Schrödinger hoped to subsume the wavelike properties of objects such as electrons within a classical theory which, called "wave mechanics," was published early in 1926—but the valiant hope was quickly disappointed. Within a year, Paul Dirac, an electrical engineer turned physicist and a graduate student at the University of Cambridge at the time, had shown that wave mechanics and matrix mechanics differ only in the mathematical language in which they are described. Each is a comprehensive account of how systems on an atomic scale behave. There was no avoiding the subversive new ideas.

In what sense was this a revolution, as it has often been described?[13] It is just as correct to say that it was a series of exciting discoveries about the behavior of very small physical systems spread over the first quarter of this century and which still continues. Like the special theory of relativity, quantum mechanics is a correction of Newton's mechanics to allow for features of reality which are beyond the ken of the human senses. The most vivid proof that the river of discovery is unbroken is that Schrödinger's theory was founded on Hamilton's prescient generalization of Newton's mechanics a whole century earlier. Not only does Hamilton's mechanics remain a recurrent touchstone in what has been made of quantum mechanics since 1926, but what Hamilton knew as the "principle of least action"[14] became, in the 1950s, a powerful tool for dealing with the quantum interactions of electrons and other charged particles in the hands of Richard Feynman of the California Institute of Technology. The significance of 1926 is merely that it is the year in which systematic calculations of the properties of the stationary states of atomic systems first became possible.

The practical consequences of that development were profound. One was that people could begin to make good the deficiencies of Dalton's atomic theory—for example, ignorance of the reasons why some atoms combine with others to form molecules. [15] It was also quickly recognized that the reason why metals conduct electricity has to do with the existence within them of stationary states that allow electrons to travel the bulk of the solid; the semiconductor industry, which makes the essential components of electronic computers (among other things), depends on the knowledge of these stationary states accumulated in the past half-century as well as on inferences from Pauli's exclusion principle about the numbers of electrons they contain. It is no wonder that Dirac, one of the world's most modest men, wrote in *The Principles of Quantum Mechanics* in 1928 that "the whole of physics and chemistry have been reduced to applied mathematics." [16]

The conceptual upheaval was equally profound. Even the concept of the stationary state is at odds with everything that had gone before, and not simply because it is one in which an atom's electron does not radiate away its energy. Stationary states have no counterpart in classical physics, nor do they exist in the abstract. They are defined only in relation to some physical quantity—the energy of an atomic system or its angular momentum (spinning inertia), for example. They are the only states in which the value of the defining quantity is unambiguous. In all other states, the defining physical quantity (say the energy) does not have a unique value: measure it, and the result will invariably be one of the values corresponding to a stationary state, but just which value is strictly a matter of chance. Compared with classical mechanics, where all possible orbits of a planet around the Sun are in principle allowed, quantum mechanics entails restrictions on the possible motions of an electron around an atomic nucleus. That chimes in with the idea that the radiation field consists of quanta of energy: not everything is allowed on the atomic scale.

The uncertainty principle is another radical departure, foreshadowed though it was by de Broglie in 1924. Again the concept is not abstract but is related to particular physical quantities, which go together in pairs. [17] The position and the *momentum* (which is the mass multiplied by the speed) of a particle, say an electron, are such a pair. [18] What the principle declares is that nature allows one or the other to be known, but not both. To know something about both, it is necessary to accept that each will be in error by amounts ultimately determined by the value of Planck's constant. Indeed, the mutual uncertainties are reciprocally related to each other; if one is small, the other must be large.[19]

Heisenberg's principle sets the scale for the objects in the real world. Take a simple hydrogen atom, with a lone electron and a single proton as the nucleus. Where is the electron? Somewhere in a region surrounding the nucleus. And how big is that region? The closer to the nucleus, the faster the electron will be traveling, which means that its momentum ranges all the way from zero (at the periphery) to a value fixed by the energy binding the electron to the nucleus. The binding energy, in turn, is given by the observation that the electron can be detached from the atom by an electrical bias of 13.7 volts, which fixes the uncertainty of the momentum of the electron. Heisenberg's rule then fixes the uncertainty of the position of the electron and thus the size of the region it occupies. Simple arithmetic shows that the uncertainty in the position of the electron amounts to 3.3 tenths of a millionth of a millimeter, which is in the right ballpark.[20] Similar arithmetic can be applied to the case of the nucleus of deuterium, which consists of a proton and a neutron held together by what is called the "strong nuclear force," and where the binding energy is the equivalent of 2 million volts and where the masses of the particles are nearly 2,000 times those of an electron. The result is roughly one hundredth of a *billionth* of a millimeter. That agrees well with the experiments that show that atoms are typically 10,000 times larger than their nuclei.

Intrinsic spin was another conceptual upheaval. Its implication is that all particles of matter, by virtue of their mere existence, are endowed with some fixed amount of angular momentum or spinning inertia. The fixed amount is always a multiple of a universal quantum that does not depend on the mass of the particle, its electric charge or any other attribute. Pauli was the first to guess at an explanation: spin, he argued, exists because of the way the time dimension must be added to the three dimensions of ordinary space to provide a relativistic description of objects. Spin is nature's way of signaling the correctness of Einstein's theory of relativity.

By the end of the 1920s, there was more than this to say. There is a marked difference between particles whose spin is an integral number of angular momentum quanta (such as photons, for example) and those whose spin is a half-integral number of those units (such as electrons). The former are called *bosons*, the latter *fermions*. The distinction is important because Pauli's exclusion principle applies only to fermions. That explains why the stars which consist primarily of neutrons (which are fermions) do not collapse under their own weight and, in the last resort, why the ground on which we walk does not crumble under our feet.

Particles from the Head

The intellectual fashion of the decade from 1925 was one of daring that stops short of the foolhardy. Dirac set the trend. His project to reconcile the quantum mechanics of the electron with Einstein's special theory of relativity led him not only to mathematical support for Pauli's conjecture that intrinsic spin is necessary, but also to the conclusion that if negatively charged electrons exist (as they do), so must certain other particles carrying identical but positive electric charge. J. Robert Oppenheimer was among those who persuaded Dirac that the unknown antiparticle could not be the proton. Dirac then predicted[21] that there must exist a particle identical with the electron except that it carries positive, not negative, electric charge; that particle, now called the positron, was discovered in 1931.[22] Now we know that the same principle applies to all other fermions. And if electrons are particles of matter, positrons (and the other antiparticles) are particles of antimatter: one will annihilate the other, producing photons.

The fashion for theorizing the existence of particles caught on, perhaps bolstered by the fruition of Rutherford's offhand prediction more than a decade earlier that neutrons would exist. By 1931, Pauli was brooding on the phenomenon of β-decay, in which an unstable atomic nucleus sheds an electron (gaining one unit of positive electric charge in the process). To make the energy, momentum and intrinsic spin of the process balance, he postulated the existence of a particle of matter called the neutrino, with no mass and no electric charge, but with intrinsic spin, momentum and thus energy. Particles with those properties are virtually undetectable, but people worked happily in the belief that neutrinos exist until the first direct proofs in the 1960s.

There followed a further startling prediction, and from an unexpected source. In 1934, Hideki Yukawa from the University of Kyoto in Japan attempted to explain the forces that hold atomic nuclei together. Compared with electromagnetic and gravitational forces, the forces in the nucleus are effective only over the small dimensions of atomic nuclei, but they are also very strong: the binding energy of the proton and the neutron in a deuterium nucleus is 100,000 times greater than that of the electron in a hydrogen atom. Following the model of the radiation field, Yukawa's idea was that nuclear forces may be similarly mediated —by an analog of the photon that would show up as a novel particle endowed with mass and electric charge.

Yukawa's venture started from the picture then built up of how the radiation field generates the forces between moving electrically charged particles. The idea that radiation is both wavelike and corpuscular was by then familiar and the word "photon" had entered the scientific vocabulary: the energy of a photon is a quantum whose size is proportional to its frequency. Planck's radiation field became simply a region of space in which the numbers of photons of different frequencies are those required to keep the system in some kind of statistical equilibrium with itself. But the equilibrium is not static. Photons are repeatedly appearing from the vacuum and then disappearing again, but in such a way that on the average, the radiation law is not violated.

Any electrically charged particle passing through empty space may stimulate the omnipresent radiation field to generate a photon of some frequency. Where does the energy for that come from? The obvious source is the moving charged particle, but charged particles moving through empty space do not generate showers of photons—flashes of light.[23] The reality, unreal though it may seem, is that the energy required for the appearance of the photon comes from nowhere: the uncertainty principle has it that the energy of any system will be uncertain over short periods of time. Photons generated in this way, of course, cannot last for longer than the uncertainty principle allows, for which reason they are called "virtual."

Then the generation of forces between *two* charged particles in empty space is easily described. Each generates virtual photons of all frequencies as it moves through space. Occasionally, some of these will interact with the second particle, perhaps being absorbed by it. But photons carry not only energy but momentum, and so will deflect the second particle from its original course, just as if it had been acted upon by some external force—the electrical repulsion or attraction of the other particle. That is how the interaction—gravitational, electromagnetic or nuclear—between particles of all kinds is now conceptualized.

Superficially, this account of how particles interact may seem a gimcrack mechanism, dependent on the chance creation of virtual particles at the expense of Heisenberg's uncertainty principle. Yet it works superbly. That knowledge meant that when the world's physicists turned their attention to the demands of the Second World War, they had every confidence that they would afterward soon identify the essential ingredients of matter. Instead, they were unscrewing yet another dollcase.

Surprises Galore

There were two awkward surprises in 1947. One was the confirmation of Yukawa's prewar prediction that there would be a particle (called the π-meson) that mediates the strong nuclear force between protons and neutrons, just as impermanent photons mediate electromagnetic forces.

The particles (found in cosmic rays) proved to be 273 times as massive as electrons, to carry either positive or negative electric charge and to decay in a fiftieth of a microsecond on the average. And the surprise? The π-mesons were by no means the most common particles of comparable mass in cosmic rays; far more common are those now called μ-mesons, which are 206.8 times as massive as an electron and which also carry electric charge (either negative or positive). Except that they are also radioactively unstable (with an average lifetime of a few microseconds), μ-mesons have now been shown to be identical with electrons, into which they decay in an average time of one hundredth of a microsecond. The question of why nature needs a second electron had hardly been formulated precisely before a third version (called the τ-electron or tauon) was found in 1975. (That object is nearly 3,500 times as massive as an electron, or nearly twice as massive as a proton.)

The other surprise sprang out of cosmic ray studies at the University of Manchester. Again in 1947, two photographs of the interior of a cloud chamber, the device then used routinely for observing particles in cosmic rays, were each found to contain a V-shaped track. What could the neutral object be? There followed four heroic years of hauling a powerful electromagnet up the Pic du Midi in the Pyrenees and another up to the Jungfraujoch in Switzerland to escape as much of the atmosphere as possible; many more V-shaped tracks were found, and also a second group of tracks with the shape of a letter K, suggesting that a charged particle had been transformed into one charged and one neutral particle.

Understanding the new particles was an even more daunting task. The V-particles proved to be 19 percent more massive than protons, to be spontaneously unstable (as if they were radioactive) but to survive for longer (if only for a fraction of a billionth of a second) than would be expected for particles of that mass.[24] The new particles were christened "strange," with the understanding that they constituted a novel kind of nuclear matter, with kinship to protons and neutrons (then collectively known as nucleons) rather than to electronic matter.[25] Fuller understanding of strange matter came only a decade afterwards, by which time doubts had grown about the status of protons and neutrons as indivisible particles.

Immediately after the Second World War, the world's theorists knew there was a crucial problem to be solved: how to reconcile the idea that electromagnetic forces are brought about by the exchange of virtual photons between electrically charged particles with the special theory of relativity. This was the goal of what Dirac called "quantum electrodynamics," shortened to QED by his younger colleagues. At this time, three independent solutions of the problem were found; the most imaginative was that with which Richard P. Feynman put his stamp on the physics of the past half-century.[26] His version of QED proved to be both a physical theory of how quantum systems are transformed with the passage of time and also a powerful means of calculation. Feynman's reputation is in no way diminished by the recognition that part of his inspiration was the work by Hamilton in the nineteenth century that set Schrödinger off.

In Feynman's version of QED, a charged particle such as an electron does not travel from one place to another along a single path, but by all possible paths in parallel. Feynman then used the quantum rules developed in the 1920s for telling what would happen at the destination. Because particles are also waves, the influences of unrealistic paths between, say, A and B, tend to cancel each other out. And the interaction between one charged particle, say an electron, and another like it? That is a matter of working out the chance that a virtual photon emitted along the track of one electron will be absorbed by the other, so transferring momentum between the two. And more complicated happenings, say the conversion of a photon into an electron pair—an ordinary electron and a positron? Just as the vacuum can generate virtual photons, it is also a source of virtual electron pairs. For such a virtual pair to turn into real particles, several conditions must be satisfied, the chief of which is that the photon must have enough energy to create the mass of the two electrons (which is the equivalent of just over one million volts). Again the chance that such a pair will be created is calculated by working out the chance of the separate events.

The success of QED has surprised even its practitioners. Many of the calculations of how electrically charged particles behave in interaction with photons and with the vacuum turn out to agree with experiment to within a few parts in 100 million.[27] Now, there is even a laboratory-scale measurement that confirms the reality of the role of the vacuum in processes like this: take two flat metal plates separated by a small distance and mount them in a larger chamber from which the air has been pumped out. The vacuum as a whole can support all possible wavelengths of radiation, but the part of it between the two plates cannot sustain electromagnetic vibrations whose wavelength is not geometrically related to the dimensions of the gap. The result is that the pressure of the surrounding vacuum exceeds that of the space between the plates. So there should be an attractive force between the plates. Astonishingly, that has now been measured.[28] The idea that the vacuum—supposedly empty space—exerts a pressure on objects within it is a startling innovation.

But even Feynman's incisive treatment of QED, like the alternative schemes, has a persistent flaw: attempts to calculate real physical quantities, such as many of the quantities arising in the theory turn out to be infinite. Ways of subtracting these infinite quantities in a self-consistent fashion have been developed, but nobody is entirely content with that state of affairs. The general expectation is that these infinite quantities will melt away when there is a more comprehensive theory of the particles of matter covering not just the electromagnetic interactions, but others as well. As will be seen, that goal is still some way off.

The problem of the new "strange" particles, important in itself, typifies how later discoveries have been handled. The question was tackled in the 1950s by Murray Gell-Mann, also at the California Institute of Technology. He put forward the startling hypothesis that nuclear matter (then only the nucleons, Yukawa's meson and the newly discovered strange particles) is really constituted from three particles called quarks

—a name taken from James Joyce's *Finnegan's Wake* and intended to convey the whimsy that entered particle physics then. True to the fashion, the three quarks were given nonsense names: *up*, *down* and *strange*; each was supposed to carry half a quantum of intrinsic spin. (There are also anti-*up*, anti-*down* and anti-*strange* quarks.) Unlike all previously discovered particles, the then-hypothetical quarks were supposed to have electrical charges which are not simple multiples of the positive charge on a proton. Instead, the *up* quark was assigned a charge of two-*thirds* of the proton charge, the *down* and *strange* quarks *minus* one-third of that amount.

This development was a signal for people (Gell-Mann in the lead) to begin building models of composite particles. The proton consists of three quarks, mesons of two (one of which is invariably an antiquark).[29] The model building was spurred not only by curiosity, but by the hope of making predictions about the real world: what clinched the notion that quarks are real was the prediction by Gell-Mann (in 1964) that there should be a composite particle made from three strange quarks which he called *omega-minus* (written Ω^-); in due course, that particle was found. This is a latter-day echo of the way Mendeleev's periodic table carried conviction only when the three gaps in the table that the author had identified were filled in the first decade of this century.[30] In the same way, quarks have been part of our reality since the discovery of the Ω^-.

Now there are six quarks in total. The three for which evidence has come to light in the past two decades rejoice in the names of *charm*, *bottom* and *top*. The discovery of each of them was preceded by its prediction, lending conviction to what the experimentalist eventually discovered.[31] One unexpected feature of quarks is that their masses span a range from about the equivalent of 70 million volts (for *up* and *down*) to 175,000 million volts (for *top*); that is comparable with the range spanned by the masses of the three electrons (the ordinary electron, μ and τ), which is no doubt no accident. Composites of the six quarks, with or without their antiparticles, constitute the hadrons, the self-standing particles of nuclear matter. Among these composite particles, there is an important distinction to be made: particles with three quarks (which include the nucleons) are called "baryons," those with two are the mesons.[32] With the exception of the proton, all of them are unstable.

What about single quarks as such? In the ordinary world, they do not exist. It is a fact of our life that quarks occur only in pairs or as triplets. How does that square with the expectation that the vacuum, which is potentially a source of all possible particles and their antiparticles in pairs, should be capable of yielding any quark together with its antiquark? Such pairs are indeed produced, in energetic collisions between protons and antiprotons for example, but their existence is only fleeting. Invariably, the quarks conjure other particles out of the vacuum —the more massive of them are converted into narrow bundles of hadrons traveling more or less in the same direction and called hadron jets.

The belief that bare quarks do not exist in the real world is called "quark confinement." It remains to be seen whether the interdiction against the existence of free quarks is absolute, while there is as yet no compelling physical explanation why it should be so. Whatever the reason, it must somehow be linked with the strong nuclear force that hold quarks together in composite hadrons. Where does this force come from? From another quantum field, of course. In the established spirit of whimsy, the field concerned is called a *gluon* field. The idea is that quarks are glued together by the mutual exchange of novel particles called gluons, which stand in relation to the strong nuclear force as do photons to electromagnetic forces.

There is, however, a complication; indeed, there are several. Quarks carry ordinary electric charge (either one-third or two-thirds of the charge of the proton), which means that a π^+ meson consists of two mutually repelling charged quarks (an *up* and an anti-*down*). The strong nuclear force must at least be powerful enough to overcome that repulsion if mesons are to exist even briefly. What then, for the strong nuclear force, stands in for the electric charges that determine electrostatic and electromagnetic interactions? The answer is that the attraction and repulsion of quarks is determined by a novel attribute called (whimsy again!) "color." To make sense of the great variety of quark combinations discovered at the particle accelerators in the 1960s and 1970s, it has been concluded that "color" has three "values" (by contrast with the two values of an electric charge, positive and negative), which are conventionally called *red, green* and *blue.* (There are also anti-*red,* anti-*green* and anti-*blue,* related to *red, green* and *blue* as the positive electric charge of a positron is related to the negative charge of an ordinary electron.) As with electric charge, quarks with the same color charge repel each other (which means, for example, that the two *downs* in a neutron must be differently colored, and that the accompanying *up* carries the third color). This property leads naturally to an empirical rule for telling which, among all possible hadrons, will exist in the real world: overall, they are color-neutral. Baryons (made of three quarks) are colorless if the color charges are *red, green* and *blue* or the same combination of the anticolors. Mesons (with two quarks) are colorless if they are a combination of a particular color-charge with the corresponding anticolor (say *red* and anti-*red*).[33]

This scheme of things is the basis of the now-standard way of treating the strong nuclear forces that hold together the quarks, called *quantum chromodynamics,* or QCD, in obvious analogy with QED. But QCD has proved a less successful theory. The multiplicity of charges and quarks is not so much the difficulty. The real complication arises because the gluons that mediate the strong nuclear force must themselves carry color charge, so that the gluons interact with each other as well as with the constituent quarks.

That complicates the use of QCD for calculating the properties of particular hadrons and their interactions with each other, but there are conceptual difficulties as well. Inside a hadron, say at π^+ meson, both of the quarks (the *up* and the anti-*down*) will be in motion relative to each other, as will the gluons repeatedly interchanged between them. Together, these entities will carry substantial amounts of momentum or kinetic energy. Because the π^+ meson exists, if only briefly, its energy and thus its mass must be less than the energy embodied in the mass of its two quarks if they could exist separately. But the internal motions represent a positive amount of energy which also shows up in the measured mass of the meson. The difficulty is that the quantity of kinetic energy involved can be calculated only inaccurately. (It can also be measured by collision with fast electrons, for example, but again imprecisely). What that means is that the accuracy of knowledge about the masses of the quarks is as yet only poor. Yet these quantities are essential ingredients of QCD.[34]

Conceivably, the way around difficulties of this kind will come from the program to unify all the known forces between pairs of particles, of which there are four: the strong nuclear force, the much weaker nuclear force that links leptons to hadrons (as in β-radioactivity), electromagnetism and also gravitation. However different the four interactions appear, there is an influential school of thought that they are all manifestations of one single underlying interaction between material particles whose identity will become apparent only at very great energy.

This ambition has been encouraged by the second great intellectual triumph of fundamental physics since the Second World War—the successful unification of GED and the theory of the weak nuclear interaction (by Steven Weinberg and Sheldon Glashow at Harvard University and, independently, by Abdus Salam at the University of London). The idea that two such different interactions between particles, one of them essentially infinite in range and the other confined to dimensions no bigger than an atomic nucleus, could be successfully described in the same mathematical language is in itself a daring notion.

Nevertheless, by the early 1970s, the originators of the electroweak theory were ready with a prediction: that there would exist three new particles, called W^+, W^- and Z^0. The two W particles (the + and − refer to ordinary units of electrical charge) mediate (by exchange) the weak nuclear interactions in much the same way that Yukawa's mesons were supposed to hold nuclei together, or gluons to hold quarks together. The Z^0, on the other hand, was offered as a massive equivalent of the photon, the quantum of the radiation field. The originators of this clever theory knew that these particles would be massive, but not that they would be nearly 100 times as massive as a proton. But that is what emerged when these particles were discovered at the European Laboratory for Particle Physics at Geneva in 1985, putting a triumphant capstone on the most remarkable of all the quantum theories of matter yet devised.

One measure of the confidence this electroweak theory enjoys is that, in combination with QCD, it is known as the "standard model." As will be seen, that implies a sense of finality that may well be premature.

Gaps in Understanding

There are several missing ingredients in this conceptual edifice, one of which is tangible: there should be yet another particle called (in advance of its discovery) the "Higgs particle."[35] The role of this entity in the scheme of things, and many of its properties, have been amply anticipated. For example, it has no electrical charge. What remains uncertain is the mass of the Higgs particle, which could be 150 times greater than that of the proton and possibly much greater still. That is a crucial matter, because it will determine whether the large hadron collider being built at Geneva will be energetic enough to create a supply of individual Higgs particles through the collision of protons with antiprotons.

The importance of the Higgs particle cannot be exaggerated. It is exclusively an invention of the standard model. Like all the other particles that go to make up matter, it is also an all-pervading field that can manifest itself in the form of particles. Its function is to account for the demonstrated existence of the W^+, W^- and Z^0 particles in a consistent fashion. If the accelerator experiments fail to uncover the Higgs particle, the standard model will be in serious trouble, which is why most of those working in the field, buoyed up by the genuine successes of the past few years, rather take the view that it must exist.

The physical meaning of the Higgs field is recondite. Thus the value or "strength" of the field is supposed not to be zero even in empty space. Is that nonsense? Not necessarily, say the theorists. What if the energy of a Higgs field differing from zero is *less* than that of one that vanishes everywhere? That is a very strange notion; with an ordinary electric or magnetic field, for example, the energy is known to beginning students to increase markedly as the strength of the field increases. The standard model supposes the opposite to be true for the Higgs field. The hope is that when the mass and other properties of the Higgs particle have been measured, the relationship between the other material particles of the real world, notably the leptons and the quarks, will be naturally defined. That, it will be noted, would go a substantial way toward accounting for the properties of particles as well as for their presence in the world. By the same test, the standard model will be in serious trouble if the Higgs particle does not appear on cue.

The mechanism that holds the quarks together is a different matter. Not only is quantum chromodynamics (QCD) a difficult recipe for calculation, but it does not provide a natural explanation why quarks ordinarily exist only as composites. The rule that only "color-neutral" hadrons exist (however fleetingly) in the real world presumably reflects some property of underlying space and time, but nobody knows what that may be. The situation is much like that in 1925, when Pauli's exclusion principle was first announced. The difference is that Pauli's principle made sense of a wide range of physical phenomena other than atomic spectra.

Nor is there much to boast of in the attempts that have been made to extend the successful unification of the electromagnetic and the weak nuclear interactions to encompass the strong nuclear interaction as well. Soon after the virtues of the electroweak theory were recognized, Glashow and Howard Giorgi (also at Harvard University) set out to build what they called a grand unified theory, or GUT, to describe by a single set of equations the electromagnetic, the weak nuclear and the strong nuclear interactions. One achievement of the scheme is that it accounts for one of the most remarkable features of the natural world—that the electric charge carried by all the particles capable of independent existence have either the negative electric charge of the electron or its exact opposite, the positive charge of the antielectron. Understandably, Giorgi has described the equations at which he and Glashow arrived as "beautiful"; they are to the cognoscenti.

There are now several versions of GUTs, all of which describe the three interactions by one set of equations. But no version of the theory has yet secured the prize of predicting the properties of the particles of matter as well as a description of how they interact. On the contrary, they make predictions that have not been confirmed. First, all GUTs predict the existence of what are called magnetic monopoles, the "north" or "south" poles of a magnet detached from each other; attempts to find such entities have failed. Second, all GUTs suggest that the simplest nucleon, the proton, should be radioactively unstable, very occasionally disintegrating into electronic matter. Attempts to verify that prediction have also failed.

Even within the standard model, there are difficulties. Why, for example, are the mass-less charge-less particles called neutrinos always *left-handed?* That is a curious business. Neutrinos were conceptually invented to ensure that energy, momentum and intrinsic spin are not lost in the disintegration of atoms by β-radioactivity: they are momentum and intrinsic spin (half a unit of it). Lacking mass, Einstein's special theory of relativity requires that they travel with the speed of light. Relativity then requires that the spin should point either in the direction in which the neutrino is traveling or in the opposite direction. What the experiments show is that true neutrinos invariably have one orientation, that in which the spin points backward along the direction of travel; they are like left-handed corkscrews. Antineutrinos, by contrast, are invariably like right-handed corkscrews. The nonexistence of right-handed neutrinos points to a fundamental asymmetry in the structure of matter. The electroweak theory is not compromised, but the asymmetry must be included there "by hand," as the mathematicians say. Even so, an important question is left unanswered: what is it about our world that caters only to left-handed neutrinos?

There is a related difficulty about the symmetry of the world we live in that calls for an explanation also not yet forthcoming. In Newton's mechanics, it is possible to reverse the direction of time, or to make time run backward, and still have a description of a plausible motion so long as energy is not dissipated by extraneous influences such as friction. The equations of mechanics have just that form. So, too, do the equations of quantum mechanics.

Two other symmetry properties affect particles of matter and their composites. Suppose that you replace every electrically charged particle by its antiparticle and vice versa. The result is to turn a positively charged π^+-meson into a π^--meson, for example, or a proton into an antiproton. The symmetry idea is that the relationships between particles should remain unchanged by the transformation.

The third natural symmetry is geometrical. All theories of mechanics, in Newton's time and latterly, use algebraic language for referring the positions of objects to some arbitrary reference point. If that reference point is used as a kind of pointlike mirror through which rays of light pass from all points of an object to form an image an equal distance on the other side, that operation—strictly a geometrical device—is called inversion. This procedure (unlike that of reflecting an object in a plane mirror) does not change the relative orientation of the different parts of the object as they appear in the image; a right hand has an image which is a right hand, but which is upside down. The object can be exactly superimposed on the image by rotation through half a circle. This geometrical operation nevertheless plays a distinctive role in quantum mechanics because the mathematical quantities describing the state of a system may change their algebraic sign (from plus to minus or vice versa) when they are inverted. That leads to the definition of an attribute of the state of a particle (or of several of them) called *parity*, which is either plus or minus and which has no counterpart in classical mechanics.

Until 1956, it was generally expected that parity would be unchanged when particles are spontaneously transformed into one another. Then came a shock. It was discovered that parity does change in some cases of β-radioactivity. Since then there have come to light cases of particle transformation in which the combinations of charge conjugation and parity are altered. The classic case is that of the unstable strange meson called K^0, whose stability differs markedly from that of its antiparticle, and in such a way as to suggest that even symmetry with respect to time is violated.

These difficulties seem only to affect the transformation of particles brought about by the weak nuclear interaction. They do not compromise the usefulness of the standard model in making calculations, but —like the left-handedness of neutrinos—they are another indication that there are features of the structure of space and time, and of its symmetry, that are not yet catered for by the theories. Bluntly, their origin is not yet understood.

Neutrinos raise a further difficulty. The Sun is a copious source of them because of the thermonuclear reactions that sustain it. In contrast with the radiation generated in the core of the Sun, which takes many millions of years to reach the surface because of its encounters with free electrons on the way, the neutrinos escape relatively unimpeded, and can be detected on Earth's surface through their rare encounters with atomic nuclei (which can change one nucleus into another). Measuring these effects requires that huge amounts of material should be assembled, usually at the bottoms of deep mineshafts to avoid contamination by particles.

The first experiment of this kind, begun in the United States, involves the conversion by neutrinos of chlorine atoms into atoms of an isotope of argon which is radioactive (allowing the number of them to be counted). The surprising result, consistent over a quarter of a century, is that the Sun appears to emit between one and two-thirds fewer neutrinos than expected. The most plausible simple explanation appears to be that, on their brief journey from the Sun's core to Earth, neutrinos associated with electrons are randomly converted into those associated with μ and τ electrons. There is no reason, within the framework of the standard model,[36] why that should happen.

There is yet another question mark over the role of neutrinos in the scheme of things: the possibility that they may have some of the properties required to provide the "missing mass" for which the cosmologists have been searching for many decades. But are not neutrinos devoid of mass by definition? That is no more than an assumption and may not be exactly correct. But if one or more of the three neutrinos has mass, the symmetry of the equations that represent the standard model would be spoiled. That would be another sign that the understanding of the structure of matter is still incomplete. The goal of specifying just why the particles in the real world are what they are is still a long way off.

Experiments to Be Done

The understanding of matter won in recent decades often seems to have been an abstract business almost entirely dependent on theorists weaving theories. That is an illusion. The theories are intellectual triumphs of the first order, but none of them would command confidence without the crucial clues provided by experimenters. The idea that hadrons consist of quarks, for example, stems directly from the discovery of strange particles in 1947. While the discovery of the Higgs particle is now widely expected, it may not be the last piece of information bearing on the structure of matter.

In keeping with past experience, further investigation may also show that the quarks themselves are composites, much as hadrons were found to be made of quarks. In the early weeks of 1997, there were suggestions that even electrons may have an internal structure, or be composite.[37] If observations such as these eventually prove substantial, there would be another doll case to unscrew. Given the recurrent discovery that seemingly indivisible particles are composite after all, who would be surprised?

It is also true that the ambitions to find a unifying language for all four of the interactions between particles of matter is unfulfilled. That will remain so while there is no clear idea of how the ubiquitous gravitational interaction between particles is to be reconciled with quantum physics. Many have been driven by the frustration of their ambitions in this respect to entertain entirely novel ideas of how particles, hitherto regarded as pointlike specks of matter, may be projections into our own three-dimensional world of much more complicated structures (see Chapter 3).

So how can particle physics, whose extraordinary success since the early 1930s has depended almost entirely on scraps of information gathered from cosmic rays and accelerating machines, make progress in circumstances like that? By taking to the skies. Already, the connections between the structure of matter and of the universe are close, as the sequence of events in the first few instants after the big bang makes plain. But there are other opportunities. There are stars in our own galaxy whose interiors may consist of quarks and gluons held together as a kind of liquid. The distant galaxies that are exceptional sources of energy are other potential test beds of the behavior of particles of matter on a small scale and high energy. Learning about the particles of matter from the study of astrophysical objects is not new; between 1920 and 1930, the cosmic rays were the only source of energetic particles on the surface of the Earth and were still of great value until the 1950s.

So what are the fundamental particles of matter? For the time being, hadrons and leptons. But if quarks and electrons should prove to have internal structure, the chase that began in 1947 to understand strange matter will begin all over again. And the goal to account for the properties of all these particles by some natural law? That remains in limbo, and will remain so while there is such a meager understanding of the reasons why the GUTs are relatively unsuccessful (even though they are every bit as ingenious as their originators say), why gravity stands out from the other interactions like a sore thumb and why, for that matter, neutrinos are left-handed. Particle physics, which has recently made common cause with cosmology, will soon be wrestling with the meaning of space and time—in theory and in the laboratory.

Everything at Once 3

The ambition of those eager to understand the origin of matter is to explain not only why the known particles consist of hadrons and leptons (quarks and electrons) but also why they have the particular properties attributed to them by observation. It is a reasonable and commendable ambition. To the extent that it links the origin of matter with that of the universe itself, it has greatly stimulated both particle physics and cosmology in the past 20 years. Is it not one of the goals of fundamental science to illuminate the nature of disparate phenomena by accounting for them all together?

In its extreme form, this same ambition has prompted the search for what is called the "theory of everything," or the TOE—an explanation for both the nature of the universe and what it contains. Einstein set the trend, long before quarks were ever heard of, with his long and fruitless search for a universal field theory. The volume of discovery in the past century has provoked such a sense that this ultimate triumph is almost within sight that people (including many who should know better[1]) believe it is a matter of years only before they will be writing textbooks on the subject. They have reckoned without the simple truth that the recent pell-mell pace of discovery has left them intellectually breathless, without a true understanding of what they are talking about.

Such harsh words require exculpation. In all fields of science, the enthusiasm of investigators is sustained by knowing that there are always novel problems to be tackled and, sometimes, solved. So it is now in the understanding of matter, the universe and where they both came from. There are still novel experimental data to predict, and known equations to be modified to see whether they then make better sense of what is already known. As will be seen, there are also novel ideas, going beyond the standard model of particle physics and big bang cosmology, to be explored and further elaborated; those working in these fields keep themselves busy. Yet as the years pass and these models of reality are neither convincingly confirmed nor denied by the scant tests that can be made of them, it is inevitable that attention should turn to the conceptual foundations of the nature of space, time and matter. Ridding fundamental science of the obvious conundrums that now abound is the best recipe for escaping from the present dilemma.

What Is Space-Time?

Newton, who first codified the laws of mechanics and of gravitation, had a simple picture: space is the background in which events take place —in which billiard balls collide and flywheels keep turning—while time is like a river running always in the same direction, serving as a means of cataloging events in an order that makes sense. A billiard cue, for example, strikes the cue ball, which in turn impacts on a colored target billiard that may (or may not) then roll into the pocket for which it is intended. In that orderly description, events are always preceded by their causes; there is no danger that the target billiard will tumble into a pocket before the cue ball has been struck. That is the principle of causality, which is automatically satisfied in Newton's mechanics.

Einstein's special relativity upset that applecart by stripping time of its simple meaning. Most graphically, the rate at which time passes depends on how quickly you are moving relative to the timekeeping device—whether a clock, an atomic clock or even a heartbeat. A person able to measure the passage of time on a moving object (say an Earth satellite) will find that it runs slow compared with his or her own clocks, and to a degree determined by the ratio of the speed of the moving object to the speed of light.[2] There are disconcerting consequences, not least that events that seem to happen at the same time on, say, an Earth satellite, may seem not to be simultaneous when seen from elsewhere. Even simultaneity is a relative concept, made so by the fact that nothing, not even information, can travel faster than light.

By Einstein's own account, he came to that radical idea (at the age of 16) by asking himself what would happen if he were to "pursue a beam of light with the velocity c" (the conventional symbol for the speed of light in empty space). Would he then see a stationary pattern of oscillating electrical and magnetic fields? There "seems to be no such thing, whether on the basis of experience or according to Maxwell's equations." Hot pursuit of a light beam generates merely another light beam.[3]

Special relativity also requires that the dimensions of tracts of space (or of objects occupying them) should shrink when they are moving at a substantial fraction of the speed of light and, for good measure, that the mass of a quickly moving object should increase indefinitely as its speed approaches that of light. Now there are even more radical things to say about empty space, the vacuum. Since Planck's time, it has been plain that empty space is filled with the universal radiation field, whose energy increases rapidly with the temperature. Since the 1930s, it has also been clear that pairs of electrons can be conjured out of the vacuum. Now we must believe that the vacuum is pervaded by all the other fields corresponding to the particles discovered in the past half century

—the lepton (three kinds of electron and their neutrinos) and hadrons (three pairs of quarks) together with the fields that hold them together, gluons for the strong nuclear force and the heavy bosons (Z^0, W^+ and W^-) for the weak force. There is also the hypothetical Higgs field, potentially capable of making matter appear from nowhere. The vacuum, commonly called empty space, is far from empty.

That is not the whole of what has been learned about the structure of space and time from the study of quantum particles. The case of intrinsic spin is instructive. It seems that all particles, by virtue of their existence, embody definite amounts of angular momentum[4] or of *spinning inertia.* That is a curious business. In the everyday world, angular momentum is associated with massive objects, flywheels in rotation for example. The quantum particles that carry intrinsic spin are, for all practical purposes, pointlike objects without dimension, yet their intrinsic spin behaves as if it is a real attribute of the particle; the spin of an electron in an atom, for example, is added to, or subtracted from, its angular momentum around the nucleus to give the total angular momentum. It is as if space has a previously unsuspected property that requires any pointlike object such as a particle to carry a certain amount of angular momentum.[5] And what is it about space that prevents two particles being in the same quantum state at the same time (the exclusion principle), or that makes it impossible to measure accurately both the position and the momentum of an object (the uncertainty principle)?

There are further difficulties. There is, for example, the problem of nonlocality, dramatized by Einstein (with his colleagues Podolsky and Rosen) in 1935, when his skepticism of quantum mechanics had firmly taken hold. Named after its three authors, the "EPR paradox" began life as a thought-experiment—as well as a challenge to the interpretation of quantum mechanics worked out by Niels Bohr's Copenhagen School. Imagine some unstable quantum particle that decays by emitting two identical particles in opposite directions. The emitted particles might, for example, be photons, which are quanta of radiation whose intrinsic spin is one universal unit[6]; the spin is manifested by the direction in which the photons are polarized. If no net angular momentum is generated in the original decay, the pair of photons must be polarized in directions at right angles to each other.[7] But in which particular directions?

There is nothing in the decay of a single particle into two photons to suggest that one direction should be preferred over others. The conventional statement of quantum mechanics is that each of the two photons is in a "mixed" state comprising all possible directions of polarization. What then if the polarization of one photon is measured, say with a polarizing filter? Photons, being particles of light, are indivisible; either they are transmitted through the filter or they are not. But then a remarkable property of quantum particles comes to light. If, by chance, a photon successfully traverses such a filter, the chance that it will get through a second filter with exactly the same orientation will be 100 percent. What has changed? The state of the photon. It is no longer in a mixed state, but in one whose polarization is defined. The mixed state has "collapsed" into a pure state, in the language of the trade.

So what can be said about the second photon? As stated, the direction of polarization must be at right angles to that of the first, so that without making a second measurement, it could be proclaimed with absolute confidence that the second photon is polarized in the opposite direction to the first. The essence of Einstein's challenge is the question, "How can the second electron instantaneously 'know' what its polarization must be once the polarization of the first is measured?" Is the information gathered about the first photon transmitted to the second faster than the speed of light, and how, even though the two photons may then be several centimeters or even meters apart?[8]

The conventional answer, originally due to Niels Bohr, is that it makes no sense to regard the two photons as separate objects. They are irretrievably linked together by their common origin in the decay of a single particle: they are in what Schrödinger first called an "entangled" state. But not everybody is satisfied with that explanation, notably the physicist John S. Bell, who for thirty years before his death in 1991 was the principal analyst of the EPR problem and its ramifications.[9]

The reality of entangled states is not now seriously disputed. Beginning in 1994, they have even been advocated as the basis of a new breed of electronic computers working on quantum principles. But entangled states are extended objects, certainly bigger than pointlike particles, thereby suggesting that quantum mechanics is not concerned exclusively with the very small. It remains to be seen (as some have suggested) whether they play a part in the communication between nerve cells[10] or between relatively distant parts of biologically active molecules.

The collapse of the mixed state of polarization into a pure state following the simple measurement of the polarization is not merely another curious feature of quantum mechanics, but is also the nexus of still unresolved controversies about its meaning. Before a measurement, all states of polarization are possible, but afterward the photon is in the condition that the measurement uncovers. What exactly does this mean?

Heisenberg in 1926 was the first to argue that systems with atomic dimensions (such as photons) are bound to be changed in some way when measurements are made of them; a mixed state may collapse into a pure state, for example. That, indeed, is the origin of the popular mantra that, in quantum mechanics, "the observer influences the event observed." But this is not the license for the introduction into fundamental science of personal subjectivity it is sometimes claimed to be. For one thing, the measuring devices that provoke the collapse of a mixed state are not people, but mechanical devices. For another, there is no great mystery involved: the polarizing filter that causes the collapse of the mixed photon state is merely a device with built-in directionality that transmits only photons matched to that direction.

The more substantial controversy centers on the link between the phenomena on an atomic scale which are the proper subject of quantum mechanics and the full-sized world in which we live. In principle, there should be no difficulty in combining together the quantum descriptions of atoms to give the equivalents for molecules or for, say, crystals of solid substances in which atoms of different kinds are laid out in regular arrays. That is precisely the field in which quantum mechanics has been brilliantly successful in the past half century; now, predicting the properties of arrangements such as these is limited only by the availability of computer power.

So why not construct a quantum description of, say, a motor car, or of Earth, or even of the whole universe? Questions like that prompted Schrödinger to construct his well-known "cat paradox," which amounts to asking whether there can be a mixed quantum state consisting of a live cat and the same animal dead.[11] The orthodox (and inescapable) answer is that the questions are wrongly posed and make no sense; the transition from an atomic to the macroscopic scale is not just a matter of careful mathematics but also entails some consideration for what is called the "arrow of time." But precisely how the transition is to be made remains unclear.

Einstein's theory of gravitation, which mixes together space and time more intimately than the special theory, raises other important questions about the character of space-time. There are, for example, difficulties about causality: it is not always self-evident that events are preceded in something akin to time by the events that cause them. More immediately, the theory allows that there should be regions of space that are permanently shielded from our ken by "horizons"—barriers impenetrable even by beams of light. The most familiar examples are the objects called black holes, supposedly formed by the collapse of stars literally under their own weight; each is supposedly surrounded by a surface forming just such an horizon. The status of these objects will not be settled until there has been substantial progress in making a bridge between Einstein's theory of gravitation and the rules and regulations of quantum mechanics.

GRAVITATION AND THE QUANTUM

If the evolution of the big bang is ever to be treated by rational analysis, there will have to be a way of describing the behavior of dense agglomerations of matter, including the bizarre objects called black holes. But there are other reasons why a theory of what is called "quantum gravity" has emerged as the most clamant need in fundamental physics: the daring ambition to unify the known interactions between particles of matter has so far foundered on the failure to bring gravitation within a framework that accommodates the electromagnetic and the strong and weak nuclear interactions.

The place to start is with the big bang. What was the universe like at the very beginning? It was simply a tiny bubble of space-time packed with energy—the energy of the fields from which particles of matter later materialized, and notably the Higgs fields (whose energy is greatest when they have zero value). But energy is equivalent to mass, which means that space-time in the nascent universe would have been tightly curved, following Einstein's theory of gravitation. How quickly would the universe have been growing? Einstein's relativity, from which is derived the principle that nothing within the universe can travel faster than light, does not restrict the speed at which a bubble of space-time can itself expand. (That is the reason why Guth's inflationary universe, discussed earlier, is not in conflict with general relativity.) The result is that, one microsecond after the big bang, the observable universe would have been just 300 meters in radius—just over five times bigger than a standard football field in diameter—but would have been embedded in a much larger structure, the parts of the larger universe that are permanently beyond our ken. At that stage, the temperature would have *fallen* to 10,000 billion degrees and the density of the particles of matter would have been comparable with that in atomic nuclei or in neutron stars. In other words, the full rigors of the general theory of relativity would apply. Space-time would have been tightly curved.

These circumstances would also have required the rules of quantum mechanics to apply. In what would have resembled a gigantic atomic nucleus, particles of the same kind would have been indistinguishable from each other. Pauli's exclusion principle, requiring that no two particles of the same kind can be in the same state,[12] gives the internal constitution of the meganucleus properties now only guessed at.

The intense radiation is a ferocious complication. At a temperature of 10,000 billion degrees, the predominant photons are the energetic X rays called γ-rays, individually more energetic than the particles produced by the world's most powerful accelerators. Only this radiation could have prevented the early big bang universe from collapsing under its own weight, and instead set it on the path of expansion that is still its outstanding characteristic. Quantum mechanics and gravitation would then have been inextricable.

The plan to find a uniform description of all the interactions between particles of matter is relevant, but is also a distraction. The project goes beyond the less-than-triumphant grand unified theories (GUTs) with which people have sought to unify the strong and weak nuclear forces with electromagnetism, seeking as it does to bring in gravitation as well. The assumption (or, perhaps, the hope) is that if the energy of individual particles is sufficiently high, all four interactions will appear to be four different aspects of the same phenomenon. There is experimental evidence that something of this kind does happen: the short-range nuclear forces become less restricted in their range of effectiveness, as the energy increases.

There is also a counterargument. The gravitational force seems to differ in character from the others. The strong nuclear force, which holds quarks together (with the help of gluons) in composite objects such as mesons and the nucleons, becomes a strong repulsive force at small distances; only that property gives these objects their characteristic dimensions. The gravitational force between two objects, on the other hand, is always attractive and becomes ever stronger as the separation of two masses is decreased. Moreover, there is no known way in which the gravitational influence of massive objects, however distant, can be prevented from affecting all other masses in the universe. In other words, antigravity shields have not yet been invented. By contrast, the similarly long-range electrostatic forces between electrically charged particles, electrons for example, are much stronger than the gravitational attraction. For example, the electrical attraction between the electron and the proton in a hydrogen atom is greater than their gravitational attraction by a factor of 10^{38}, which means 10 multiplied by itself 38 times. That by itself explains why electrons, carrying the same negative electrical charge, are never prevented from repelling each other by their mutual gravitational attraction. Considerations such as these suggest that gravitation may defy the grand unification program.

How great would the energy have to be for all four interparticle forces to seem aspects of the same phenomenon? The energy should be the equivalent of the mass of roughly one billion billion protons or neutrons. Then, the argument goes, there would be a single quantum field from which all four interactions would be derived. As always, there would be a particle associated with the field; that "Planck particle" would have a mass of a microgram or thereabouts.[13] Until the feasibility of this grander unification scheme is established, the status of the Planck particles is strictly hypothetical. They are the embodiment of a belief, even a hope.

The more urgent need is to find some way of reconciling the theory of gravitation with quantum mechanics. Nobody has yet succeeded, but not for want of trying. The search for a viable quantum gravity has been under way since the late 1960s. So far, people have set about the task by starting with Einstein's theory of gravitation (otherwise the general theory of relativity). One approach, pioneered by the U.S. physicist Bryce DeWitt in the late 1960s, was to treat Einstein's equations much as Maxwell's equations were dealt with in the 1950s in the successful foundation of QED.

Something has been learned. Just as the interaction between electrically charged particles is mediated by the quanta of the radiation field, or photons, so the gravitational field must be mediated by particles, which are called "gravitons." Like the photon, the graviton has no mass, which is appropriate for particles mediating long-range forces such as electromagnetism and gravitation. The surprise is that gravitons must have an intrinsic spin of two units. As yet, nobody has observed a graviton, but only the effects of their existence. Did not Newton's apocryphal apple fall to the ground, after all? Gravitons (like neutrinos) may be all around us, but we sense them only through the gravitational attraction they cause.

The program to reconcile gravitation and quantum mechanics by applying to Einstein's equations the techniques that worked well in QED is nevertheless now in the doldrums. The obvious difficulty is technical: whereas the development of quantum electrodynamics was plagued by calculations yielding infinite quantities, which then had to be rationalized away, quantum gravity yields even more horrendous infinities; so far, no self-consistent way of turning them into finite numbers has been found.

Most probably, the mathematical problems reflect physical difficulties, of which there are several. The most immediate is that, in Einstein's gravitation, the elements of the gravitational field (of which there are ten) are themselves indicators of where energy or mass is to be found, which means that elements of the field at one place must interact with elements of the field at all other places. (Technically, the equations are nonlinear, which is a recipe for mathematical complexity—as will be seen.)

There is a deeper anxiety: the failure of the familiar procedures to cope with Einstein's gravitation may stem from an underlying assumption that is false. One possibility is that Einstein's equations, paragons of rectitude though they seem, are incomplete; from time to time, there are investigations of that possibility.[14] It is more probable that underlying (and implicit) assumptions about the character of space or time are faulty. Whether the omnipresence of intrinsic spin, for example, among the attributes of quantum particles is a characteristic of particles or of the underlying space may be a matter of semantics, but there is at least a possibility that it may be the latter. If so, it may be necessary to abandon, in this context, the algebraic description of space devised by Descartes in the seventeenth century and to replace it with something else. But with what?

So far, there have been some ingenious, but unconvincing, essays in that direction. What if space has a foamlike structure, for example? In the early 1950s, Homi J. Bhabha, later the father of the Indian atomic energy program (and of India's capability for causing nuclear explosions) was toying with the idea that points in space are intrinsically ill-defined. His research articles on the subject amounted to an attempt to build Heisenberg's uncertainty principle into all estimates of the distance between two objects. Bhabha readily persuaded his colleagues that he was a clever man, but failed to convince them that his recipe was the right one.

Whatever the explanation, there is a sense that the reason why the quantum gravity project is becalmed is not simply mathematical, but that the problem to be solved is not yet fully understood. If "new physics" springs from anywhere, it will be from this formidable challenge.

NEW PICTURES OF PARTICLES

People have not been sitting on their hands. Faced with an impasse, as in the search for quantum gravity, it is a fair strategy to pose the problem differently. What if the general opinion of what constitutes a particle of matter is conceptually mistaken? Or what if the ambition to unify the four interactions between particles is misconceived, perhaps by relying too literally on present notions of what these interactions are? Instead of aiming directly for fourfold unification, might it be more prudent first to unify something else?

That is part of the inspiration of the most elegant approach to quantum gravity yet dreamed of: supersymmetry. Remember the three generations each of hadrons and leptons—the quarks on the one hand

and the electrons and neutrinos on the other, all with a half-integer intrinsic spin? If quantum gravity is ever attained, it will have to describe both kinds of particles in the same language, so why not do that from the start? But what then of the particles whose function is to mediate one or other of the four interactions—particles such as the photon, the W^+, W^- and Z^0, the gluons and the graviton? Tackle them as well, of course.

Quarks and electrons are so different that the task is like designing a language that will be equally well understood by those who speak Chinese and those who speak French. At least formally, however, the common language can be described. Abdus Salam, who with Weinberg and Glashow was one of the independent co-inventors of the electroweak theory, has written of supersymmetry that it is "an incredibly beautiful theory—a compelling theory if ever there was one."[15] One snag is that it compels the existence of extra particles; for example, the existence of the still hypothetical graviton compels the existence of an object called the *gravitino* whose distinguishing mark is that its intrinsic spin is 3/2 (1.5) of a unit of angular momentum. The more obvious snag is that the estimated masses of these extra particles are comfortably outside the range of the particle accelerators likely to be built in the foreseeable future. But echoes of this theory will recur in the years ahead.

Meanwhile, most of those who seek to capture quantum gravity by stealth, avoiding a direct assault on the seemingly invulnerable redoubt, have embraced the notion that the particles of matter are not like very small billiards, each with the distinctive properties of quarks, electrons and so on, but are different states of vibration of entities called strings.[16] Like piano wires under tension, they vibrate at predetermined frequencies. Because each string is a quantum system, its vibrational energy comes in quanta, and each vibrational quantum corresponds to a mass. So why not arrange that the vibrational quanta correspond to the masses of the three versions of electrons, or of the three generations of quarks, and then see if it is possible to work out rules that will allow vibrating strings to simulate the particles of the real world.

This radical idea has its roots in the discovery in the early 1960s that ordinary nucleons, the proton and the neutron, have an internal structure. Strings were briefly advocated as a means of explaining the presumed internal states, but the innovation seemed to raise more problems than it solved: certainly the explanation in terms of quarks seemed more familiar. But string theory, as this field is known, never quite disappeared from the research community's consciousness. One attraction is that it allows several related particles to be embodied in a single object, but by 1974 strings began to seem as if they might also be a natural way of describing the gravitational interaction between masses. By 1996,[17] there were even suggestions that the description of the particles of matter as vibrating strings requires that there should be a gravitational force between all particles of matter; from being the most obdurate of problems, gravitation becomes part of the solution of all others.

Only the suspicion that strings may be a natural way of describing gravitons, not to mention the other particles linked with them, has kept the theory alive. The hope is that the apparently insuperable difficulties in making an accommodation between quantum mechanics and gravitation would be reduced and, perhaps, might even melt away. Since the mid-1970s, successive gusts of enthusiasm have been succeeded by waves of disappointment because new promises have been repeatedly unfulfilled. For example, there was great excitement in 1984 when John Schwartz from the California Institute of Technology and Martin Green from Queen Mary College, London, hit on a way of using string theory to account naturally for the left-handedness of all the neutrinos in the real world. But then there followed a decade when the calculations appearing in the scientific journals seemed even more complicated than they had been before; there seemed little chance that people would come leaping from their baths crying "Eureka!"[18]

Part of the reason is the feature of string theory that most challenges the imagination: the strings whose vibrations are supposed to represent particles exist not in ordinary space-time, but in ten-dimensional space-time. How, then, can strings represent particles in the real world? Space-time, after all, has only four dimensions; what happens to the six superfluous dimensions of the strings representing particles?

The standard answer is that they have been "compactified," which means that they have each been shrunk, or curled up, into a space so small that its dimensions compare with what is called the Planck length. This length is as small a fraction of the size of an atomic nucleus, itself the smallest tangible entity in the real world, as that is as a fraction of a grown person's height.

These are strange notions almost calculated to excite skepticism, even disbelief, yet those working in the field are persuaded that string theory will eventually yield the rapprochement between quantum mechanics and gravitation that has eluded other approaches. Time (and a lot more effort) will tell whether they are right. It seems already to have been established that the string theories are in many ways equivalent to the more conventional theories (QCD and the GUTs constructed to deal with the known particles of matter), although nobody has yet been able to write down a prescription of what the world is like in the language of string theory.

Not only are the mathematical problems unfamiliar, but the conceptual basis of the theory is unclear. What exactly does compactification entail, for example? Do the six hidden dimensions have the properties of regular space-time or, possibly instead, those of the surface of a torus (which, in ordinary three-dimensional space, is the familiar figure of a doughnut)? May the hidden dimensions be physically accessible without the intervention of processes whose energy is that of the Planck mass? What exactly is the significance in the ordinary world of the scale on which compactification happens? Is that somehow related to the uncertainties inseparable from quantum mechanics? Or, more radically, is compactification a sign that our concept of space is a construct of the human mind devised for the description of large-scale phenomena that makes no sense in the description of particles or their equivalents, the quantum strings?

Meanwhile, the concept of a particle as the state of a vibrating one-dimensional string has been generalized. Why stick with one-dimensional objects, after all? Why not represent particles as the vibrations of small pieces of two-dimensional membrane, which is what the surface of a drum-head consists of? Or, more generally, why not think of them as multidimensional surfaces capable of vibration which are somehow embedded in ten-dimensional space? Those who lavish their intellectual energy on these so-called N-branes do so in the hope that some version of such a theory will yield such a neat account of all the particles that sheer elegance will prove persuasive.

How has it come about that such an elaborate network of theories has been constructed while basic questions such as the meaning of the missing six dimensions remain unanswered? That is to ask for too much. String theory is an elaborate metaphor for physical reality. Founded on the simple notion that a string vibrating in a direction at right angles to its length may serve as a means of representing groups of particles (originally the quarks), it has been carried by the requirements of mathematical consistency into lines of argument that, for the time being, mean little to the research community at large. The enterprise engages the energy of enthusiastic groups of talented people. Their conviction that they are on the way to a unified description of the world is infectious. But none of that ensures that they are correct.

Eventually, the theory will stand or fall by its success in explaining the real world while avoiding the infinite quantities that have plagued earlier efforts. If the enterprise succeeds, it will be time enough to consider what the implications are for our general understanding of space and time, but there is every hope that we shall all be wiser. And even if the enterprise fails, there will be a better understanding of the nature of the difficulty presented by the universal gravitational interaction.

ARE BLACK HOLES REAL?

Meanwhile, there is a more practical question linked with gravitation: the status of the concept of a black hole. The idea is simple. In due course, all stars will collapse. The Sun, now burning hydrogen as thermonuclear fuel, will switch to helium fuel in about 5 billion years, will then become more luminous (and on that account much larger), will shed some of the material in its outer layers through explosive instability and eventually will be incapable of generating energy at its core. The future of what is left of the star will then be shaped by the gravitational attraction of all parts of the star to all others, which will make the spent Sun shrink at a pace determined by the rate at which heat liberated by the continuing collapse can be radiated into interstellar space from the shrinking surface. The Sun is destined to end as an object called a "white dwarf." More massive stars have potentially more interesting fates.

What is matter like inside a white dwarf? At the center, to begin with, the temperature exceeds 20 million kelvin and the pressure is great enough to separate all but the heaviest atoms into their components—electrons and atomic nuclei. (This is the kind of hot electrified plasma that engineers are trying to create for the production of thermonuclear power.) But white dwarfs are no longer producing energy, so that the temperature declines and the pressure increases, perhaps only modestly,[19] but sufficiently to prevent the recombination of the electrons with the nuclei from which they have been separated. The material is a metal of a kind, either a fluid or a more familiar solid. (Hydrogen made metallic by pressure is now believed to be a substantial part of the interior of the planet Jupiter.) The density also becomes very large—perhaps 1 million tonnes of matter for every cubic meter (a volume not much bigger than that of an upright piano).

How can such material resist the inward pressure of the outer envelope, which steadily increases as the temperature declines? Quantum mechanics has the answer. In a white dwarf, the electrons are free to move through the whole electrified volume, but only within the constraints of Pauli's exclusion principle, which prevents two electrons being in the same state. The result is that further compression of the material of the star requires that electrons occupy states of ever greater energy, which translates into ever greater resistance to gravitational pressure. For the burned-out Sun, calculations of this "degeneracy pressure" suggest that the resulting white dwarf will be stable for the rest of time.[20]

More massive stars may go further along the road to collapse. If the density of the material in the interior of a star increases 100,000-fold above a million tonnes for every cubic meter, the nuclear particles will themselves be unstable. In particular, electrons and protons will combine together into neutrons (the reverse of the normal radioactive decay of a neutron into a proton) until the neutrons themselves (also subject to the exclusion principle) yield a degeneracy pressure resisting further collapse. In that state, peculiar things may happen: neutrons under sufficient pressure may be turned into their components (forming agglomerations of quarks and gluons). That is one of the places particle physicists will be looking if governments decline to pay for more energetic particle accelerators than those now being built.

One of the triumphs of astrophysics in the past few decades is the identification of stars in our own galaxy that represent all the successive platforms of stability in the process of collapse. White dwarfs are recognizable by their comparatively low output of light and high surface temperature (both conditioned by the small surface area of objects perhaps less than 1,000 kilometers across).

Neutron stars are even more intriguing objects. The first two were recognized in 1967 at the University of Cambridge by pulses of radio waves emitted every few seconds or so.[21] The pulsation comes about because the stars are surrounded by intense magnetic fields, up to 100 million times as strong as Earth's magnetic field; interaction with electrons near the surface of the star generates a pattern of radiation which sweeps across the sky like a lighthouse beam. The pulsation rate, which is also the spinning rate, is a sign of how compact are neutron stars. Thousands are now known, with pulsation rates between the lower limit of detection (about a thousandth of a second) and several seconds; inevitably, perhaps, they were called "pulsars" before it was realized that they are neutron stars. Most (perhaps all) pulsars are the remnants of the cores of stars more massive than the Sun that have run through their evolution, expelling their envelopes explosively, and which are recognizable in our galaxy and others in a supernova explosion.

Pulsars have enormously enlivened the astronomy of the past three decades. Their number is a clue to the rate at which stars in the galaxy have already run their course, so pointing to the amounts of heavy elements formed from primeval hydrogen and helium. Several pulsars have also been found to be members of binary star systems, held by gravitation to a companion star, which has made them invaluable diagnostic aids in the understanding of X-ray emission from binary star systems. A remarkable double pulsar, consisting of a pair of neutron stars in orbit about each other, promises to be the proving ground for the instruments now being built for the detection of gravitational waves from elsewhere in the universe; the more rapidly spinning pulsars are more accurate timekeepers than the atomic clocks now used for setting international standards of time.

Neutron stars are not black holes, but are gateways to them. Just as the internal degeneracy pressure of a sufficiently massive white dwarf may not be enough to prevent its collapse to a neutron star, so a sufficiently massive neutron star will continue to collapse—to a black hole.

The concept goes back earlier, to 1915 when the first version of Einstein's theory of gravitation was published. The German physicist Karl Schwartzschild then found a prescient solution of Einstein's equations, corresponding to a mass of material located at a fixed point in space. He also recognized that, whatever the amount of mass involved, there would be a sphere around the center within which space-time would be so tightly curved that even radiation could not escape from within it.[22] The surface of the sphere is called an *event horizon* because it marks the boundary between two regions of space that can no longer communicate with each other. An object that has shrunk within its own event horizon (about 3 kilometers for a star with the Sun's mass) becomes invisible, although it continues to exert its gravitational effect on its surroundings, notably by capturing material. But once matter has been captured into a black hole, nothing can be learned of its fate.

Until his death in 1955, Einstein vigorously resisted this use of his theory of gravitation on the grounds that it is a "metaphysical" notion to suppose that there may be regions of space (within the event horizons surrounding black holes) of which nothing can be learned. That objection may well be overturned when there is a successful quantum gravity. Meanwhile, it is striking that observational astronomers appear to share some of Einstein's skepticism, habitually referring to "putative" or "candidate" black holes.

Observational evidence for black holes goes back to the discovery in the early 1960s of quasars, starlike or quasi-stellar objects, which are compact and powerful sources of radiation at a great distance in the universe. Despite the distance, the output of radiation from a quasar may flicker rapidly, in a matter of hours, which implies that their size must be measured in light-hours—the distance from the Sun to Saturn, say.[23]

Are quasars really black holes? If so, how can a black hole, whose gravitational attraction for everything is supposed to be so strong that even light cannot get away, be an exceptionally powerful source of radiation? Because matter in the course of being sucked into a black hole collides with other matter and dissipates the energy thus liberated as radiation. If a black hole happens to be spinning, the radiation of energy is reckoned to be especially efficient; matter awaiting engorgement accumulates in a disk lying outside the event horizon and at right angles to the axis of rotation, much as Saturn's rings are arranged around the planet. There it is continually accelerated by the attraction of the central black hole. On some estimates, as much as 80 percent of the mass of the material gathered into an accretion disk will be converted into radiation.

A process like that could easily account for the compactness and the huge output of radiation from the quasars.

Many other kinds of astrophysical objects have now been suggested as provisional black holes, both in the far reaches of the universe and in our galaxy. The extragalactic objects are called, generically, "active galactic nuclei"; they include the spectacular objects charted by radioastronomers and called giant radio galaxies, which appear to be propelling jets of matter outwards from their cores.

Further support for the idea that black holes are real come from objects within our galaxy. One such object, discovered in the 1950s, is the source of X rays called Cygnus X-1.[24] The source consists of two stars held together by gravitation. One, which is optically visible, may be 20 times as massive as the Sun; the other, which is invisible, is believed to be a black hole on the grounds of its mass, estimated as more than three times the mass of the Sun and thus above the limit of stability for a neutron star.[25] What can it be but a black hole?

Other "putative black holes" have been supposed to be responsible for the distribution of the velocities of stars in otherwise unremarkable galaxies. The principle is that individual stars travel about the center of the galaxy with speeds that increase with the total amount of mass lying within their orbits. The general observation in galaxies that stars are traveling faster than can be accounted for by the mass of the visible stars prompts the question whether black holes may account for some of the discrepancy. In our own galaxy, the motion of the stars near the galactic center suggests that there is a black hole there, with a mass equivalent to that of 2,600,000 stars as massive as the sun.

There are two things to say. First, large though these estimates are, they are only a tiny fraction of the visible mass of a typical galaxy (perhaps 100 billion times as massive as the Sun), so that they are not a solution to the problem of the missing mass on the scale of the universe as a whole. Second, if black holes are commonplace at the centers of nearby galaxies, it is remarkable that they seem not to be generating radiation as the presumed engines of the quasars do so spectacularly. An attempt to finesse this problem in respect to the putative black hole near our galactic center raises many other problems. It is known that the central light-year of the galaxy is filled with a hot ionized gas made from material thrown off from stars in the region. There is enough gas to suggest that the black hole would be engorging the mass of Jupiter every year, which should make it a powerful source of X rays if the local black hole radiated like a quasar. Now Ramesh Narayan and a group of colleagues have proposed that the gas near the black hole, which is hot enough to be ionized into electrons and atomic nuclei, behaves in such a way that the two components require different temperatures, with the nuclei at the higher temperature. The result of that is that the electrons radiate much less energy than would be expected of them.[26] The obvious difficulty is why that does not apply to quasars; are they perhaps powered by black holes colliding with each other?

This evidence, although mostly circumstantial, seems powerful support for the reality of black holes. Many astrophysicists have cheerfully accepted it at its face value; there is even a flourishing (if small) academic industry known as "black hole physics." The habit of others in referring to black holes as "putative" seems to imply a collective uneasiness about the concept.

How can this issue be resolved? Again the long-sought rapprochement between gravitation and quantum mechanics will be important. If one outcome were that Einstein's equations should somehow reflect quantum uncertainties along the road of collapse, that might blur the line of discontinuity at the event horizons. Such a development would rob black holes of their most puzzling feature: that what happens inside them is beyond our ken despite their continuing gravitational effect on their surroundings. That is the chief reason why these remarkable objects remain in limbo, suspended in many people's minds between belief and disbelief. That state of affairs lends urgency to the search for a realistic quantum gravity. Until that search succeeds, the most remarkable objects in the universe will have only a dubious status.

INVENTING TIME

The need to construct a bridge between gravitation and quantum mechanics may be the most urgent challenge in the physical sciences, but it is not the only one. There is an influential opinion that there should also be an explanation of the sense of time that we all share. Where should we look for that?

Our sense that time passes inexorably from past to future seems to be predominantly a matter of psychology. By definition, the past is what we can in principle remember; the future, which has not yet arrived, can only be imagined. Thus our sense of time seems to have a logical foundation. Moreover, the second law of thermodynamics provides a definition of the direction of time with its statement that an isolated system will adjust itself, as time passes, so that the quantity called entropy is as great as possible. Put two liquids (say, saltwater and alcohol) into a vessel and they will eventually be intimately mixed together (unless, like oil and water, they are chemically prevented from doing so). An ordered state spontaneously becomes disordered, or the entropy is spontaneously increased. The reality of the process is beyond dispute; mixing two liquids as described releases heat.[27] Does not behavior of that kind provide an objective sense of the direction in which time increases that is independent of sensations in our heads?

This is the classical problem of "the arrow of time," made respectable by the Viennese Ludwig Boltzmann in the 1890s. Boltzmann set himself the task of showing that the tendency for the entropy of an isolated system to increase is simply a statistical consequence of the properties of atoms and molecules capable of colliding with each other as, for example, they do in a gas or a liquid. In the course of his work, Boltzmann laid the foundations of what is now called statistical mechanics, and was even able to construct a mathematical quantity (defined in terms of the positions and speeds of its atoms or molecules) that could not increase in the course of time, and which he naturally identified with the entropy.[28] In the event, it turned out his supposed proof of the second law of thermodynamics was logically faulty; Boltzmann had implicitly ignored improbable outcomes of the interactions between atoms and molecules.

The origin of the arrow of time is nevertheless an important question because there is nothing in either Newton's mechanics or Einstein's to suggest that inanimate systems should have a built-in direction of time. On the contrary, both systems are formally indifferent to the direction in which time runs: if it runs backwards, velocities simply change direction, with the consequence that Earth would move backward in its orbit about the Sun, or that an apple falling to the ground would move upward instead, as if it had been thrown.[29] There is no arrow of time in the basic laws of mechanics.

Where else may its roots lie? In quantum mechanics, perhaps? There is certainly one point at which the standard view of quantum mechanics leads to descriptions that are not reversible with time: the process of measurement. Merely measuring the properties of a particle in a mixed state, as in the EPR experiment discussed earlier, causes it to collapse into a pure state. So the act of measurement has brought about a change in the system.

Roger Penrose is one of those who suspects that collapsing wave functions may conceal the origin of the arrow of time,[30] but the argument is far from clear-cut. Ilya Prigogine, the Belgian physicist who has spent a lifetime exploring the properties of systems not in internal equilibrium, would deal with the problem by describing even classical systems in the language of probability, in part to allow for the phenomenon of chaos, but also on the grounds that this is a more appropriate language, and so would build the arrow of time into the elementary description of the objects of the physical world; unfortunately it is far from clear whether such a prescription can be carried through the whole of physics, and whether it would meet the need for an explanation of the arrow of time.

Nevertheless, our perception of the world implies that the arrow of time is real. So much is famously illustrated by the twists and turns that followed Maxwell's objection to Boltzmann's claim that he had found a mathematical proof of the second law of thermodynamics. Seeking to show that the enterprise must be fruitless, Maxwell asked his readers to imagine a container filled with gas and with a partition dividing it roughly into two halves. The partition has a hole which can be covered with a small door, at whose handle is a very tiny person able to see which molecules are likely to traverse the hole in the partition. The small person, called "Maxwell's demon," is instructed to close the door if exceptionally fast atoms or molecules are approaching the hole from his side of the partition, thereby ensuring that they remain on his side of it. The result is that, after the energy of the molecules has been redistributed by collisions among themselves, the temperature on the demon's side of the partition has been increased. If the door closing the hole can be operated without the use of energy, this quaint thought experiment is meant to show that microscopic processes can lead to the apparently spontaneous redistribution of heat in a counterintuitive fashion. Entropy has been decreased, not increased. To look for a mathematical basis for a process bound to run in the other direction was, by inference, fruitless.

Not until the arrival of quantum mechanics in the 1920s was the issue resolved by the question "How would Maxwell's demon measure the velocities of molecules so as to separate the fast from the slow?" The most convenient instrument would be a laser linked with a computer for closing the door. To measure the speed of atoms, the laser would have to operate at X-ray wavelengths, which means that its photons would carry substantial amounts of energy. But then, because the photons also carry momentum, the demon's activities would substantially alter the velocity of the molecules of the gas. Maxwell's demons, the argument goes, could not carry out their instructions (and do not in any case exist).

The issue is not whether the second law of thermodynamics is valid in the ordinary world; nobody doubts that. Even the notorious variety of statements of the law are a stimulus, especially the way in which the definition of entropy can be related either to the exchange of heat between an object and its surroundings or to the degree of internal disorder. The question is simply whether the arrow of time is implied by the equations that embody the basic rules of mechanics, Newton-Einstein on the one hand and quantum mechanics on the other.

John Archibald Wheeler, one of the most versatile relativists of Dirac's generation who has been at Princeton University and the University of Texas at Austin, is quoted[31] as having written that "time is defined so that motion looks simple." Is the arrow of time all in the mind?

THE ANTHROPIC TRAP

That is an illustration of the working of what is called the anthropic principle, first put forward (in the mid-1960s) in a modern context by Robert H. Dicke, then a professor of physics and an astronomer at Princeton University. The idea is that the laws of physics obtaining in the real universe of which we are a part must be consistent with the eventual emergence of living things, and in particular of people, on Earth.[32] If the universe were younger than 4 billion years (the length of time so far occupied by the evolution of life forms on Earth), there would have been no people to build telescopes with which to observe the sky, and thus no cosmology. Similarly, if the stars of our galaxy had not been able to manufacture carbon atoms from primeval hydrogen and helium, even the raw materials for the evolution of life would have been lacking.

The anthropic principle is most directly provoked by the now-fashionable questions whether the first appearance of the universe (by a big bang or some other means) is itself a physical process susceptible to analysis and whether there might have been, or even may be, "other" universes. These questions are innovations of some importance and break with the tradition since Newton's time. But at least since before the time of Plato, there has been a school of opinion holding that, once order had been created out of chaos, the universe ran itself.[33] Newton, a deeply religious man, took essentially the same view: the orderliness of the universe and the elegance of the laws of physics are a sufficient proof of the hand of God. A little later, however, the German mathematician Wolfgang Leibnitz (whose supporters held him to be the true inventor of the differential calculus) coined the term "the best of all possible worlds" to distinguish our universe from others there might be (or might have been). In the eighteenth century, the notion of several parallel universes cropped up more often, notably in the head of the French military engineer de Maupertius.[34]

Now, of course, we know that the observable universe cannot be all there is, at least in the big bang view of how it began. The big bang by itself would have generated a universe far more lumpy than observations show it to be. The only device so far found to smooth out the big bang universe is that of inflation, whose effect is to ensure that the observable universe derives from only a tiny part of the hot matter created after the big bang. And while we can never learn directly what happened to the rest of the material, except that it must have been used in the construction of other universes, we can be certain that if our universe began with a big bang, it can only have been one of a collection of universes. Since the time of Josiah Willard Gibbs, professor of chemistry at Yale University at the turn of the century, that has been something to be studied: an ensemble of universes.

This view may even help to resolve some of the difficulties that have come to light from the study of our universe. For example, the discrepancy between the known density of matter in the universe and the expected critical density may simply be a consequence of the properties of the small speck of preinflationary space-time from which the observable universe derives. Other universes derived from other specks of space-time may have differed from the universe we can see in having a density greater than the critical density, with the result that they may already have reached the limit of their expansion determined by their enhanced density, and may now be collapsing again.

The idea that our universe is but one of several whose potential for evolution simultaneously came into being is strange, but it seems to be an inescapable feature of a big bang in which inflation plays a part. It is no wonder that the big bang itself compels less than full-hearted conviction. What can be said about the laws of physics in these parallel universes? According to the inflationary model of the big bang, inflation is itself driven by the properties of the Higgs fields that feature in all versions of GUTs, implying that the laws of physics have been fixed from the first appearance of the original bubble of space-time. That in turn implies that all the parallel universes evolve by the same rules as ours has done—by the same laws of physics—and with the same values of the fundamental constants.

The laws of physics in this connection denote not only the general principles that determine the behavior of objects in the physical world —the gravitational interaction between all massive objects, for example, or Heisenberg's uncertainty principle. Just as important are the numbers that determine the size of these effects—the actual strength of gravitational forces, the degree of uncertainty in simultaneous measurements and the like.

If Newton's gravitational constant (usually written G) were substantially greater than the value measured on Earth's surface, for example, the world would seem a very different place. Evolution would no doubt have given us thicker legs to sustain our greater weight. The fate of the Sun would also be different. The value of the constant formally determines the gravitational force between two objects of specified mass at a certain distance. The greater the magnitude of the force between different parts of the Sun's fluid envelope, the greater would be its density and, as a consequence, the higher would be the temperature at the center. That in turn would mean that the thermonuclear reactions in the Sun's core would be more vigorous and that its output of energy would be increased. The Sun would run its course much more quickly than its estimated time span of 10 billion years (of which half has already gone).

The universe also would seem very different. The density of matter in the universe at present might well exceed the critical density, which would be reduced in the same ratio as G is increased.[35] That could well mean that the expansion of the universe was halted a long time ago, perhaps in less time since the beginning than it has taken for life to evolve on Earth. Then there would have been no people like us to make the observations which are the basis for modern cosmology.

The anthropic principle is both entertaining and stimulating. All kinds of questions about the world we live in can be answered by reference to it. For example, why does the world appear to have exactly three space dimensions? That topic has been seriously discussed at least since the time of the philosopher Immanuel Kant in the late eighteenth century. He was impressed that Newton's gravitation, involving as it does a force that decreases with the square of the distance, is ideally suited to a three-dimensional world. By the early years of this century (and notably because of inquiries by the Dutch physicist Paul Ehrenfest), it had been recognized that the orbits of the planets around the Sun would be unstable if there were other than three dimensions. The same is true for the states of an electron in an atom predicted by the quantum theory.

So what? The anthropic principle in this connection amounts simply to the observation that people such as ourselves and the objects that surround us would not exist if there were, say, two or four space dimensions. In other words, of all the universes and the accompanying laws of physics that might have made their appearance 10 or 20 billion years ago, that which we inhabit is a variant that is conducive to our own existence. It is a matter of temperament and personal idiosyncrasy to decide whether that statement is meaningful. Empirically minded people take the view that the statement is a monumental banality. But there is no doubt it lends support to religious opinions that the universe we inhabit embodies divine design.

The version of the anthropic principle that holds that the properties of the universe must satisfy the condition that life exists is called the "weak" anthropic principle. Its role in contemporary science is marginal at best; it has not proved to be a predictive tool, for example. There is also a "strong" anthropic principle which goes much further: "The universe must have those properties which allow life to develop within it at some stage in its history." That statement[36] may superficially appear no different from the weak anthropic principle, although it presupposes that life "at some stage" is a qualifying feature of the universe, which therefore is not easily distinguished from a universe that accommodates living things by design. Those who take this form of the anthropic principle seriously say that it is self-evident that we should not now have our present understanding of what the universe is like if it were not that there have been people to observe it, but in doing so they mistake the description for the reality.

Towards Fuller Understanding

The chief present obstacle to a better understanding of the physical world is the lack of an accommodation between gravitation and quantum mechanics. Both the description of the universe and the understanding of where matter came from are hamstrung as a consequence. There is a chance that the still nascent theory of particles as strings may be a way of avoiding this impasse, but, after a quarter of a century of effort, the chance of success declines with the passing years.

The days have long since gone when a little scribbling on the back of an envelope or two could yield a worthwhile theory of anything, let alone a theory of everything. It is not just that the research community has become more exacting in its requirements of innovators, but that the present crisis in fundamental science arises from the now evident need for a deeper understanding of what is meant by space, time and matter. Is intrinsic spin a property of space, an attribute of particles of matter or an amalgam of the two, for example? The need to confront basic questions like this is not a novel one. Did not Newton face a similar task in the seventeenth century?

As always, the lack of data against which new ideas can be tested is an impediment. The pace of the development and deployment of new instruments has become exceedingly slow—the Hubble space telescope, which has been making entirely novel observations of the distant universe since 1992, was designed in the early 1970s with technology now surpassed. The instrument being built in California for the detection of gravitational waves,[37] which could (among other things) help explain where black holes fit into the scheme of things, is unlikely to be in service for some time to come, more than 40 years after the first abortive attempts to detect these phenomena. The large hadron collider being built at the European Particle Physics Laboratory (CERN) at Geneva will be, for several years at least, the only particle accelerator powerful enough possibly to put a capstone on the standard model of particle physics by the discovery and study of the Higgs particle.[38]

The lack of data, while serious, is not as black as it is sometimes painted. The Hubble space telescope, antiquated though its design may be, has already transformed the exploration of the distant reaches of the universe. In Europe, Japan and the United States, more specialized (and much cheaper) telescopes are energetically being built for launching into orbits about Earth; early next century, there will be a rich harvest of new data about the early universe to make sense of.

The lack of data is more serious in particle physics, but is not calamitous. Existing machines are still providing important information (on questions such as the relative importance of quarks and gluons as carriers of angular momentum in composite hadrons, for example), the free parameters of the standard model are much in need of more precision, the constitution of neutron stars deserves more attention than it has so far been given while the whole field should have a vivid self-interest in explaining itself to the rest of the research community more clearly than it has done so far.

So what lies ahead? What follows is simply speculation, but is seriously intended. The idea that the laws of physics are somehow embodied in the structure of what we call empty space, the vacuum, has come to stay, but whether string theory or its elaborations is the explanation remains an open question. If, eventually, a picture of quantum gravity commands attention, we shall find that it is possible to know more about a black hole than its mass and angular momentum; Einstein's hunch that the classical concept of a black hole is metaphysical (and thus intolerable) will be proved correct.

By extension, the idea that the universe began in a single event, the big bang, will be found false. For most of the five centuries since Galileo first saw the moons of Jupiter through a telescope, observers of the heavens have been like kidnap victims seeking to learn where they are from the chinks of light that reach them through imperfect blindfolds; half a century from now, cosmologists will have a much better idea of what kind of universe they are expected to explain. The once-and-for-all universe of Genesis, or of Guth's equivalent, is an improbable outcome.

It will be time enough to talk about a theory of everything when we know what everything is.

PART TWO

Life

The Likelihood of Life

We now know *when* life appeared on Earth's surface, but we do not yet know *how* it began. Already serious attempts are being made to identify planets in orbit around stars other than the Sun that may harbor living things of some kind. How will people know when they have found a planet capable of supporting life if there is such general ignorance of how life emerged spontaneously on the primeval surface of our own planet?

The word "spontaneous" begs an important question. Many hold that the appearance of living things was an act of divine creation, as in *Genesis.* For them, the first appearance of life-forms is not so much a problem to be solved as an explanation of why Earth seems a "special" place, the Copernican principle notwithstanding.

Yet there are important differences even among those who believe that we (and all animals and plants) owe our lives to an act of creation. At one extreme are those who deny that the fossil record, and in particular its antiquity, represents Darwinian evolution by natural selection. Instead, they claim that both the fossil record and life itself are products of an act of divine creation just a few thousand years ago. They are properly known as creationists but they often call themselves "scientific creationists." Logically, their position is self-consistent, but it is illogically and mischievously at odds with the rules of evidence in science.

Most of those who hold to divine creation for the origin of life accept that the fossil record means what it says and suppose that the very first life-forms were specially created, since when elaborations of them have followed much as Darwin outlined in *The Origin of Species by Means of Natural Selection* (1858). For them,[1] it is in no sense sacrilegious to try to identify the first forms of living things. How else, many religious people ask, is it possible to come to a better understanding of the divine purpose?

Darwin originally took that position. It is inevitable that the one who startled the world with the notion that the past succession of the species had been determined by changes in the environment and by the genetic capacity of existing life-forms to make advantageous adaptations to them (the modern statement of what "natural selection" means) was repeatedly asked how life began. In a letter to Joseph Hooker, then the secretary of the Royal Society, Darwin said in 1863: "It is mere rubbish at the present to think of the origin of life; one might as well think of the origin of matter." A few years later, he had changed his mind. In 1871, he wrote again to Hooker to say:

> "It is often said that all the conditions for the first production of living organisms are now present, which could ever have been present. But . . . if we could conceive in some warm little pond, with all sorts of ammonia and phosphoric salts, heat, electricity, etc., present, that a protein compound was chemically formed ready to undergo still more complex changes . . ."

Darwin clearly acknowledged that there might be circumstances in which life could arise spontaneously from inorganic sources.

In France at about the same time, Louis Pasteur underwent a similar conversion. By about 1860, he had shown how to prevent the putrefaction of foodstuffs and even of human flesh by avoiding contamination by germs, often by simple heating (whence "pasteurization" as the technique for preserving perishable foodstuffs). In defending his experiments on sterilization, Pasteur went so far as to claim that only living things could beget other living things. He went on to claim that the "doctrine of spontaneous generation" would never recover from the "mortal blow" he believed that his work had delivered, yet by 1878 he was able to write: "I have been looking for it [spontaneous generation] for 20 years but I have not yet found it, although I do not believe that it is an impossibility."

By the middle of the nineteenth century, the doctrine of vitalism—that living matter is intrinsically different from inanimate—had all but been abandoned. Symbolically, it ended in 1828, when the German chemist Friedrich Wöhler prepared urea by the simple device of heating the inorganic material ammonium cyanate. His cleverness was to identify the product as urea. Urea is unambiguously a natural product, a constituent of human urine and one of the chief means by which mammals excrete excess nitrogen. But even before the publication of *The Origin of Species,* enough was known of the chemistry of living things for people to appreciate that they differ from the inanimate in the complexity of the chemicals they contain, not in some undefined quality.[2]

The willingness of Darwin and Pasteur to contemplate the use of the word "spontaneous" may have been due to the influence of the German biologist Ernst Haeckel,[3] who published in 1868 a series of lectures he had earlier given at the University of Jena in celebration of *The Origin of Species.* Among other things, he suggested that the first living things may have emerged on the primitive Earth by the spontaneous assembly of suitable chemicals into primitive organisms. From that point, Haeckel argued, natural selection would have taken over, leading to the life-forms now extant.

That is the current agenda of the search for the origin of life on Earth. The goals have not much changed since 1924, when the Russian A. I. Oparin published a book called *The Origin of Life* which used then-modern knowledge to dress Haeckel's ideas in more plausible language.[4] Remarkably, a further 30 years went by before experiments were carried out to test the speculations of Oparin and of those who followed him.

The starting point for Oparin's argument was that the atmosphere of the young Earth was very different from what it is now. In chemical language, it was a *reducing* atmosphere, virtually devoid of the free oxygen that now sustains the life of animals, ourselves included. What oxygen there was would have been locked up in its chemical compounds with hydrogen (which makes water, or H_2O) and carbon (carbon dioxide, or CO_2). There would also have been some carbon monoxide (CO), while other carbon would also have been present in the atmosphere in the form of its compound with hydrogen called methane (CH_4). Nitrogen, now the predominant component of the atmosphere (to the tune of 80 percent), would also have been present as ammonia (NH_3) in the atmosphere of 4,000 million years ago. These are the materials now present in the atmosphere of the planet Jupiter.[5]

Oparin guessed that simple organic compounds could be formed from these components of the primeval atmosphere by chemical reactions, perhaps induced by ultraviolet light from the Sun. He went on to argue that the products could have accumulated in the oceans, there to be assembled into life-forms. It fell to Stanley W. Miller, a graduate student at the University of Chicago in 1956, to pass a high-voltage electrical discharge through a mixture of gases of that kind and to recover from the bottom of the reaction vessel a mixture of chemicals including several of those called amino acids, the building blocks of proteins.

Technically, Miller's experiment was a great success. From different mixtures of methane, nitrogen and water with small traces of ammonia he produced ten of the twenty amino acids that occur naturally in modern protein molecules. The idea was to simulate the effects of lightning flashes. The more common products, the amino acids glycine and alanine, appeared in surprisingly large amounts; some 3 percent of the amount of methane consumed from the reaction vessel was converted into alanine. Miller has calculated that if the Sun's ultraviolet radiation were used with the same efficiency for the production of amino acids on primeval Earth, more than 100,000 tons of alanine would have been produced each year.

Yet the most remarkable feature of that experiment, still lauded as a milestone in studies of the origin of life, is not the outcome (which Oparin had predicted), but that 30 years should have passed before somebody put the prediction to the test. Miller has told how his graduate-thesis supervisor, the venerable Harold Urey,[6] urged him not to waste time on such a purposeless exercise. What Urey meant is that research in this field would be regarded as speculative, and inappropriate for those aiming at tenure-track positions at universities.

Urey's unease is not surprising. Even now, the problem that remains to be solved is daunting. It goes without saying that the ultimate test of success will be the replication in some laboratory of processes that might plausibly have led to the formation of living things. Given present ignorance, it is impossible to guess whether that will take several years or several decades.

Interestingly—and disconcertingly for some—the key assumption underlying Oparin's argument, that the primeval atmosphere was a chemically reducing atmosphere,[7] is not now so readily accepted as in 1924. The counterargument rests on the comparison of Earth's atmosphere with that of similarly placed Venus and on the attempts made in recent decades to reconstruct the original composition of the nebula of gas and dust from which the solar system is supposed to have formed about 5,000 million years ago. In this view, ammonia in particular would not have been conspicuous in the early atmosphere. This controversy has an obvious bearing on the availability in the primeval seas of the chemicals (whatever they were) from which the first living things were constructed. Researchers in this field, familiar as they are with the Oparin tradition, tend to resist the new arguments. Instead they should welcome them because they may well lead to a better account of what the early atmosphere was like.

Meanwhile, the time *when* life began on Earth has been pinned down. Earth itself is 4,500 million years old. Rocks recovered from the surface of the Moon by the U.S. Apollo missions between 1968 and 1972 revealed that the cratering of the Moon's surface was intense between 4,200 million and 4,000 million years ago, whereupon it almost abruptly ceased. It is unthinkable that Earth, then even closer to the Moon than now, would have escaped the bombardment that so pockmarked the Moon, and equally unthinkable that conditions then would have permitted life to emerge.

So life began more recently than 4,000 million years ago. The other end of the time frame comes from dating the earliest plausible fossils found in Earth's surface rocks. Early fossils are necessarily scarce, because much of the early continental crust has already been recycled back into Earth's interior. But the first living things would not have had the bones and skulls that help preserve more recent fossils—those of dinosaurs, for example. Nevertheless, there are not many of them, but there are structures with the size and shape of modern bacteria in sedimentary rocks dating from 3,800 million years ago. A little later in the fossil record, there are structures called stromatolites that appear to be relics of huge agglomerations of single-celled organisms, either bacteria or algae, comparable to the huge bacterial "mats" found floating in modern oceans. Structures of this kind have been found in Australian sedimentary rocks dating from 3,500 million years ago. It seems fair to suppose that life was well established by then and possibly by 3,800 million years ago. By the yardsticks of geological time, that is a narrow window—perhaps no more than 200 million years, and 500 million years at most.

Science as History

The origin of life on the surface of the Earth is a unique historical event whose character cannot be established by experiments in contemporary laboratories: that statement (which equally applies to the evolution of a particular species such as *Homo sapiens* or, say, to the fact that the geology of central Asia has been shaped, in the past 35 million years, by the collision of the Indian subcontinent with preexisting Asia) has often been used to argue that the origin of life is not and cannot be a proper part of science. For how can we hope to reconstruct the singular circumstances leading to what may have been a unique event? History in general is fraught with the same difficulties: however sophisticated may be historians' understanding of human nature and of public affairs, what theory of history could have predicted that the American War of Independence would have begun as it did?

Many scientists have taken this position on the origin of life. Jacques Monod, the distinguished French molecular biologist, said as much in 1970 in his elegant book *Chance and Necessity*. There is no way, he argued, that an event as improbable as the emergence of life on Earth could be analyzed by science, which is able to deal only "with events that form a class." A decade later, Francis H. C. Crick, co-originator of the structure of DNA, put the argument more specifically: the chances that the long polymer molecules that vitally sustain all living things, both proteins and DNA, could have been assembled by random processes from the chemical units of which they are made are so small as to be negligible, prompting the question whether the surface of the Earth was fertilized from elsewhere, perhaps from interstellar space. "Panspermia" is the name for that.

The chance of assembling from its component parts one of the large molecules that now plays a vital part in life processes must indeed be exceedingly small. Protein molecules that appear in living things, both as structural elements (muscle fibers, for example) and as enzymes that stimulate vital chemical transformations, are made from just 20 of the many small-molecule chemicals called amino acids linked together in a chemically specific way. Enzyme molecules (which may consist of 100 or many more amino acid units linked together) act as catalysts by accelerating chemical transformations. The effectiveness of an enzyme is usually exquisitely sensitive to the precise arrangement of the amino acid units along the length of the molecule.

Mere arithmetic shows that the chance that such a molecule will be assembled from its elementary components by random processes is so small as to be virtually zero. The number of differently arranged protein molecules with 100 units is easily stated (but the arithmetic is not so simple[8]): multiply 20 by itself 100 times. The result is unimaginably great—very much greater than the number of particles of matter in the whole of the observable universe. The combined mass of a single copy of each of the possible protein molecules 46 amino acids long would be almost twice the mass of the observable universe.[9]

Monod outlines a particularly fiendish version of his argument in relation to the genetic material DNA, whose molecules are constructed from four particular chemical units called nucleotides strung together.[10] The simplest organisms, bacteria for example, may have several million nucleotides arranged in a precise sequence. (The number of different structures of that size is found by multiplying 4 by itself several millions of times, giving an even larger number than that of all possible 100-unit protein molecules.) But when bacterial cells divide, the DNA molecules are also replicated, which requires the intervention of the enzymes whose structures are themselves determined by the arrangement of particular stretches of the original DNA, now identified with genes. How much less likely must it be that DNA molecules embodying that capacity to replicate will emerge from random assembly of the nucleotide units?

There are two ways in which these arguments can be countered. The more traditional is to observe that even the improbable will happen if enough time passes. If Oparin's primeval sea contained amino acids in solution, for example, and *if* there were a mechanism for linking them together in random order, all possible protein molecules would appear after a sufficient length of time. Those with special properties, especially the capacity to stimulate their own replication, would become predominant and might eventually become the raw materials for assembling true life-forms. But the probabilities are still outrageously too small. Making a viable organism by random assembly requires, after all, that several different long molecules should be formed at the same time at the same place.

The argument that time will accomplish everything has been further undermined as the antiquity of the fossil record has been enlarged. As recently as the early 1970s, it seemed that the first traces of life appeared in the fossil record about 800 million years ago, allowing the previous 3,000 million years for the random assembly of molecules capable of replication, and from which living things might be formed. Now the window of opportunity has shrunk to at most 500 million years and perhaps to a mere 200 million years.

But the first living things did not have to wait on the random assembly of molecules comparable in complexity with those now found in modern organisms. Life did, after all, emerge nearly 4,000 million years ago. And there is no reason why the first life-forms should have depended for their survival on molecules with all the complexity of the chemicals in modern cells, the proteins and the DNA in particular. On the contrary, the starting point for life need not have been an organism in the familiar sense, but simply a molecule or molecules with the property of being able to catalyze their own formation from raw materials in the environment of the times. That is the *prebiotic* phase of the history of terrestrial life.

A world accumulating similar molecules is not necessarily a living world, of course. If the self-replicating molecules were, for example, molecules of starch (which are branched chains of simple glucose mole-

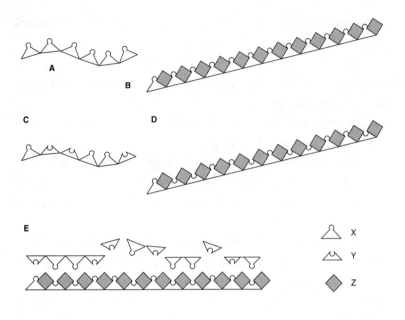

Figure 4.1 A hypothetical prebiotic replicator. The graphic symbols represent the chemical units X, Y and Z (see here at lower right) whose properties are given in the text, where the successive stages are also defined.

cules chemically strung together), they would simply accumulate in the oceans (making the water sticky) until their concentration reached that at which the rate of their accumulation was matched by the rate of their chemical destruction by water.

Models of a more interesting prebiotic world are easily constructed. What follows is a *strictly hypothetical model* of what the prebiotic molecular replicators may have been like (see Figure 4.1). It serves two purposes only: to illustrate that relatively simple chemicals might yield complicated structures capable of replication in suitable circumstances —and to draw attention to what those circumstances might be.

Suppose that the primitive ocean contains a chemical X with the property that several molecules of it will link together to form a chain-like molecule, called a polymer, say poly-X. To what length will these molecules grow? New X units can be added only at either end, so that all molecules grow longer at a fixed rate. But each molecule may be broken into two at any time by the chemical action of the seawater; that happens more often to longer molecules because there are more links to break.

So the primeval sea will contain poly-X molecules of various lengths (panel A in Figure 4.1). The average length will be decided by the competition between the tendency of water molecules to break polymer molecules into two and the tendency of the molecules to grow by the addition of fresh X molecules at the ends. If there is plenty of X, the number of poly-X molecules may increase continually, but the average length will stay the same.

This system has one of the properties of a living system: self-replication. Under a microscope—it would have to be an electron microscope—the number of molecules would continually increase, but the longer molecules would seem to snap into two from time to time. What is that but replication?

It is even possible to imagine ways in which the average length of the poly-X molecules could be increased. If, for example, the primeval sea contains a common chemical (call it Z) that tends to form a water-repellent sheath around poly-X molecules, the links between the X units will then, on the average, break less often, the competition between growth and disintegration will be shifted to some degree towards growth and the average length of poly-X molecules will increase (see panel B in Figure 4.1).

But mere growth will not qualify poly-X as alive. The only attribute of a poly-X molecule that seems to matter is its length and, by that yardstick, poly-X does not remotely breed true; most molecules will be shorter than the average length, a smaller proportion will be longer. One of the essentials of life is that individual molecules must be able to impress their own characteristics on their own offspring. How can that be done?

Evidently not by the random snapping process, for then one of the "offspring" will usually be less than half the length of the parent. Something else is required for inheritance. So suppose that poly-X molecules can form offspring not only by being snapped in two by the action of the water but also by directly influencing the polymerization of X units into fresh molecules of poly-X. Imagine, for example, that molecules of X line up alongside the X units of an existing poly-X molecule and are then polymerized into a separate parallel molecule. The longer the original poly-X, the longer will be the offspring molecule produced by this mechanism. But so long as molecules of poly-X can be broken into two pieces by the action of water, the result will still be a mixture of molecules of different length: the system of molecules is not yet alive.

One obvious deficiency is that there is no influence in favor of uniformity. While long and short molecules coexist—and long molecules are still broken into two by the action of water—the collection of molecules seems a long way from being a species of even the most rudimentary kind. The missing ingredient of this *strictly hypothetical* model is a means of discriminating between long and short molecules. Otherwise, there is no selective advantage that will favor one over the other.

How might such an influence arise? One possibility is that the water-repellent sheath of Z molecules becomes more effective as the length of a poly-X molecule increases. There are some precedents for such behavior: clusters of a few atoms or molecules capable of forming well-ordered crystals nevertheless have a disordered structure, but as further atoms or molecules are added to the cluster, the long-range order characteristic of a crystal spontaneously emerges. If increasing the length of poly-X increases the degree of order in the Z sheath which, because it is then well-ordered, substantially reduces the chance that molecules of poly-X will be broken by the action of water in their central regions,[11] the result could be a dramatic increase in the average length of poly-X.

But even then, the poly-X would not be "alive." Both long and short molecules of poly-X would presumably have similar chemical properties, and so would have no distinctive biochemical characteristics.

To survive as a class, polymer molecules must have some function that will enhance the chances of their own perpetuation. That calls for a further complication of this strictly hypothetical model.

Suppose, then, that two similar chemicals, Y as well as X, can participate in building polymer molecules (see panel C in Figure 4.1). Instead of identical molecules of poly-X such as ...X-X-X-X..., there will then be poly-(X,Y) molecules such as ...X-Y-Y-X-Y... in which the X and Y units are jumbled together and which, for that reason, will differ from each other chemically. As with simple poly-X, there must be a way in which a molecule of poly-(X,Y) can perpetuate itself, perhaps by inducing X and Y units to line up alongside the corresponding units of an original polymer molecule and to be polymerized there into a separate molecule.

Now the possibility will arise that some particular arrangement of X and Y units, perhaps at one end or the other of the polymer molecule, will provoke or catalyze a chemical operation that assists the polymerization process. Anything that would make the Z sheath less permeable to water molecules would help (see panel D in Figure 4.1). So would a process of energizing X and Y units to allow them to form part of new polymer molecules or an improvement of the efficiency of the primitive replication process.

This is the essence of molecular evolution: any molecule that makes the formation of others like itself more efficient, either by processing them more efficiently or by enhancing their stability, will be preferred over others.

If the average length of the molecules poly-(X,Y) has grown to, say, 20 units, the number of possibly different polymer chemicals with exactly that length will exceed one million, many of which will coexist at any time, and all of which will crop up at some time or another. What if one particular version of poly-(X,Y), say the molecule poly*, with a particular arrangement of the units X and Y, has a decisive effect either on the efficiency of the underlying polymerization process or its outcome? An improvement of efficiency would make no essential change to the molecular mixture; it would simply ensure that greater proportions of X and Y are polymerized. But if poly* is relatively much more stable than other versions of poly-(X,Y), there will be an increase of the proportion of poly* in the mixed population of molecules. In the extreme case, if poly* is entirely immune from attack by water, the whole system would be one for converting X and Y units into molecules of poly*.

That system illustrates how prebiotic replicators could have arisen spontaneously from simple chemicals, but it does not qualify as a living system because it has no potential for further development; it is simply a (hypothetical) way of synthesizing a particular chemical from simpler starting materials. (The idea that poly* may be immune from degradation by water is not as artificial as it may seem; the fatty-acid molecules that constitute modern cell membranes, which spontaneously line up alongside each other, may owe their durability to their uniformity of length and structure.)

This is not necessarily the most efficient way in which even this synthetic process may be accomplished. If energy is required for the addition of X and Y units to polymer chains, repeated polymerization and degradation of these molecules will be wasteful of energy. But what if poly* has the property of inducing X and Y units to line up alongside its own structure in precise positions, with X against Y and vice versa (as in panel C, Figure 4.1), there to polymerize into a parallel molecule (see panel D)? The outcome of that operation will be not one molecule but two, the original poly* together with a *complementary* molecule, with the same length but a chemical structure that is the inverse of the original. (It will be seen that this is the relationship between molecules of DNA and those of the related chemical called RNA into which genetic information is transcribed in real cells.) The directed synthesis of poly* (and its complementary molecule) is likely to supercede the process of random assembly if it requires less energy.

Even with that elaboration, the prebiotic replicator poly* and its complement are not in any sense "alive." Two further elaborations are required for that. One is that poly* and its complement should have some chemical function other than that of being the most stable polymer in their class. There are several possibilities. If the reason for the extra stability of poly* is that it is sequestered into a kind of cell membrane, for example, a useful extra function for poly* would be that it has the facility for expelling water through the membrane, the effect of which would be to enhance the stability of the whole structure.

The second elaboration of the model consists in recognizing that even though poly* is supposed to be the most stable molecule in the poly-(X,Y) class, there may be others nearly but not quite as stable which can also be included in an accumulating membrane without fatally undermining its stability; some of these may even be more efficient than poly* at expelling water across the membrane. Then, with the passage of time, there will be a trade-off between the defining characteristics of poly*—its stability and its second chemical function. In the end, the dominant version of poly-(X,Y) will be that representing the compromise between the two functions that best matches the external environment, which is the source of X and Y units and of such energy as this hypothetical model requires for its functioning.

This model is not intended as a serious suggestion of what the prebiotic replicators on Earth were like. Its purpose is simply to illustrate that random chemical processes, the polymerization of X and Y units, can lead to a situation in which particular molecules are favored over others and that extra chemical functions (such as the expulsion of water) are also likely to be favored if they enhance the stability and growth of the system as a whole. The second property, which depends on the assumption that there are at least a few molecules with properties similar to those of poly*, is the origin of genetic diversity in this hypothetical model. That is the diversity upon which natural selection acts, molding the properties of an evolving entity to the external requirements of the environment.

Three features of this model stand out. First, it is entirely a matter of chance whether, in the process of becoming longer, a particular poly-(X,Y) would acquire the capacity to catalyze the formation of molecules in its own image. (The other side of that coin is that there is no reason why there should not be repeated essays in this direction.) Second, there is no predetermined design for the successful molecule: the sole criterion for the dominance of the preferred form is that it is the most efficient replicator. Third, the question of whether the preferred form would be capable of further elaboration would depend on whether the replication mechanism (which must be faithful enough to ensure a degree of uniformity) is sufficiently error-prone to allow for further chemical innovations to occur (again by chance). If so, life in the proper sense of the term would have begun.

This is a strictly hypothetical illustration of what prebiotic life may have been like. The template mechanism is obviously fashioned by analogy with the ways that DNA and RNA function as genetic molecules,[12] but the first organisms might well have been built around other materials. And nobody has yet been able to specify a set of molecules that functions in this way in a laboratory test tube.

Nevertheless, the illustration points to several likely features of prebiotic life. First, the polymerization would require energy; in living things as now evolved, one of the most common ways of effecting chemical reactions is that enzymes called *kinases* add energetic chemical groups called *phosphate groups* to one or another of the components of a chemical reaction. On prebiotic Earth, activated chemical components of long molecules might have been produced inorganically. Second, even the acquisition of a water-impermeable sheath would involve an energetic cost, for such a sheath would necessarily have a degree of order that the same molecules in solution would have lacked, so that there will be a loss of entropy. But there is now laboratory evidence that hydrophobic molecules, especially the materials known as fatty acids, will readily assemble into membrane-like structures (called micelles) around water-repellent materials such as drops of oil

The most conspicuous feature of the hypothetical prebiotic organisms is that they have no purpose and even, at the outset, no function external to themselves. Only when some of the molecules by chance acquire functions that contribute to their own efficiency can they become conspicuous among the presumably rich and diverse array of chemicals in the Oparin sea, but even the most successful classes of molecules will not be assured of immortality. A successful class of molecules may persist for a long time, but the individual members of the class will be at risk from destructive processes in the environment and will be repeatedly dismembered into their components. It is the same, of course, with the cycle of birth and death that marks the fate of the members of persistent and apparently successful species, *Homo sapiens* included.

This strictly hypothetical model also illustrates the importance of fidelity of replication: that is how a particular molecule becomes distinguishable from the statistical mishmash that surrounds it. But too much fidelity is the death of further development. Unless there is variation, there is no room for improvement. That is what Darwin was saying a century and a half ago. The survival of a life-form, however primitive, requires a delicate balance between fidelity of replication and intergenerational variation. Too much of either, and extermination follows.

Thus the emergence of life on Earth may truly have been an historical accident, but a more probable accident than the appearance in the primeval sea of the long polymer molecules that are now the vital components of living things. It is also possible that there may have been several alternative routes to self-replication, with the result that we shall never be able to decide precisely what the first replicators were like.

That is a further reason why the claim that life on Earth did emerge spontaneously must eventually be sustained by a laboratory demonstration. It will have to be shown that a particular collection of molecules will replicate autonomously in circumstances plausibly similar to those on the surface of the primeval Earth. It is impossible to guess when the demonstration will take place, but we may all be pleasantly or (according to temperament) otherwise surprised.

The same is true of other proposals for the founding of terrestrial life. One, for example, is a scheme advocated by Stuart Kauffman of the

Sante Fe Institute in New Mexico, who proposes that the techniques of molecular biology should be used to generate a great many chemicals, perhaps millions of them, of kinds that may have played a part in the first life-forms. The idea is that tiny quantities, perhaps only a few molecules, of these distinct but related chemicals should be mixed together so as to tell which will act as catalysts (or enzymes) to effect biochemical changes in other components of the mixture.

The technique, called *combinatorial chemistry*, in whose development Kauffman has played an important part,[13] is being used by commercial companies in the search for new drugs. It amounts to the simulation of molecular evolution in the laboratory; even inefficient catalysts would enjoy a selective advantage over other molecules in such an artificial soup, and would at the same time evolve into more efficient forms. Computer simulations have shown that the technique should yield the results predicted, but here, again, only a practical demonstration of what can be achieved in a real laboratory will carry conviction.

Panspermia, or the idea that Earth was populated at an early stage by organic chemicals or even living creatures from elsewhere in the solar system, the galaxy or beyond, is in a different case. Reports in August 1996 that there are traces of life-forms more primitive than terrestrial bacteria in meteorites originating on Mars have made the subject fashionable again.[14] It is certainly true that carbon is plentiful in the solar system[15]; meteorites of the kind called carbonaceous chondrites contain a high proportion of carbon and even organic chemicals including amino acids. It is similarly the case that the giant molecular clouds in interstellar space within our galaxy and others include quite complex organic chemicals—acetic acid is one of those most recently found by radioastronomers.[16] But this evidence can be interpreted in two ways: *either* the organic chemicals so apparently plentiful elsewhere were ingredients for the first replicators on Earth *or* their common occurrence elsewhere is a sign of how easily they may also have been formed on Earth.

Most of the controversy about panspermia (and there is a lot of it) centers on the hazards that life-forms from elsewhere would encounter during their transit to Earth—the initial and presumably explosive event that sent them on their way, the long exposure to cosmic rays en route and the fiery entry through Earth's atmosphere.[17] But the serious objections are not mechanical but biological. The chance that extraterrestrial life-forms or their components, say bacterial spores or naked genetic material, would reach an environment on Earth conducive to their replication and survival must be exceedingly small, no bigger than the chance that long protein molecules would be assembled by random processes.

Logically, the case for panspermia is shaky. It does not explain the origin of life from inorganic materials, but merely displaces it elsewhere, and somewhere less accessible. What reason is there to believe that the surface of some other planet, let alone the surface of a meteorite or the interior of an interstellar cloud, would be more hospitable than the surface of the Earth, which has nurtured the descendants of prebiotic terrestrial life for the best part of 4 billion years? Is it not more prudent to regard the presence of carbon chemicals, including amino acids in, say, meteorites as a sign that the formation of organic molecules from inorganic materials is more probable than has been thought?

WHAT IS LIFE?[18]

It would be convenient if there were a definition of life, but there is none. The best that can be done is to specify criteria that living things must satisfy—and then it turns out that almost every organism is an exception. For example, it is reasonable to ask that living things must be able to reproduce themselves faithfully, but "faithfully" can have different meanings. Too much fidelity (or too little Darwinian variation) and the organism and its progeny may not be able to survive quite small changes in the environment.

Or should living things be self-contained, able to sustain themselves only with inorganic chemicals scavenged from the environment? Viruses are usually excluded from the company of living things on the grounds that they are not competent to reproduce themselves without the aid of intact cells. That exclusion is, however, inconsistent with the inclusion of the microorganisms called mycobacteria, which are parasites of mammals—and just as deficient as viruses in their capacity to fend for themselves. (They are responsible for several human infections, including listeriosis.) In reality, all the organisms now on Earth depend, perhaps less conspicuously, on other organisms. People need vitamins, trees need soil bacteria to scavenge nitrogen from the air and most flowering plants need insects for the transfer of pollen. That is why the complex of living things, the biosphere, is sometimes called the "web of life." Among the organisms now alive, some bacteria and single-celled marine algae may meet the two criteria of fidelity of reproduction and self-sufficiency, but no others.

Even so, all things now living are organized on the same principles. Although there is no reason why several different schemes for the organization of life should not have arisen in prebiotic times and have persisted after the emergence of true organisms, in the inevitable competition for resources, the more efficient modes of organization would have emerged as dominant, obliterating traces of the less successful forms.

The clinching proof that all modern forms of life are organized on the same principles has come with the development of molecular biology since 1953. For example, the *genetic code* specifies how the sequence of nucleotides in a DNA molecule determines the sequence of the amino acids in the proteins manufactured in all cells: with a few significant exceptions, the same code is used in every organism.[19]

There is a related and surprising discovery whose bearing on the question of how life began could well be profound, but which remains for the time being a conundrum. In 1995, a group of French scientists announced the discovery of bacteria at the bottom of an oil well[20] at a depth of 1,500 meters in the Paris Basin (in continental France). Now, it seems, nearly identical bacteria have been found in oil wells in Cameroon, in Africa, some thousands of kilometers away.[21] Both sets of organisms live in water at a temperature of 70 degrees Centigrade, using both sulphate and hydrocarbons as food. Contamination of the oil reservoirs by seawater and underground aquifers is unlikely; most probably the bacteria are living fossils from the time when the sedimentary oil-bearing rocks were originally laid down, more than 100 million years ago. Inevitably they use the same genetic code as the gut bacterium *E. coli* (not to mention the mammals in whose digestive tracts those bacteria live).

So life is not only universal, but ubiquitous. In the 1970s, there was great surprise at the discovery of bacteria in apparently inhospitable regions of the Antarctic, often in pools of water appearing only in brief Antarctic summer. In semitropical regions, bacteria (called *halobacteria*) have been found in ponds and lakes in which the concentration of salt is so great that more familiar bacteria could not possibly survive there, but would be dehydrated. The more recent discovery of bacteria and even invertebrate animals at depths of thousands of meters in the world's oceans is a sign that life-forms, and microorganisms in particular, seem to have colonized all the niches of Earth's surface that could conceivably support life.

These exceptional bacteria now alive have been recognized for what they are only in the past quarter century. Of necessity, their occupation of strange niches means that only a small proportion of them have so far been identified and described. Called generically *Archaeabacteria* or *Archaea,* at first it seemed that they might be more primitive (in the sense of preexisting) forms of life than ordinary bacteria, but that question appears to have been settled (negatively) during 1996, most decisively by a remarkable study of the complete genome of a particular species.[22] Now, as will be seen, there is evidence that the Archaea diverged from the less specialized bacteria about 2,000 million years ago (see below).

The divergence of the Archaea from other bacteria is a fateful point in the history of life, for it marks the point at which the biochemical machinery that all modern life-forms share with bacteria had been clearly established. Whatever may have been the repositories of genetic information in the prebiotic world and in the succeeding millennia, by 2,000 million years ago DNA had become the genetic material in all but a few organisms (viruses such as HIV, which causes AIDS, are an exception).

Also established at that point must have been the complicated means by which cells, both ancient and modern, manufacture specific protein molecules with vital functions by the intervention of structures called *ribosomes*, themselves made mostly of molecules of the material called RNA (chemically similar to DNA and also capable of carrying genetic information[23]) together with particular protein molecules. They perform the vital function of assembling protein molecules from amino acids in the precise sequence required for their effective functioning, but the manner in which they accomplish this is only now being worked out.[24]

As will be seen, there are many other features of the biochemical machinery used by organisms now extant that have this appearance of great antiquity. There is every reason to expect that the fuller understanding of these processes will throw light on the mechanisms by which they have arisen, thus helping to specify the nature of preexisting life-forms. The program to understand the origin of life on Earth is thus neither more nor less than an effort to write the natural history of the biochemistry that sustains all life even now.

The best way of meeting Monod's objection that the origin of life was an historical accident that cannot be reconstructed will be to show that there is indeed a set of "events that form a class" from each of which living things could have emerged. That will be done if it proves that there is more than one system of interacting molecules that can yield prebiotic replicators capable of further molecular evolution.

In any case, what has happened here on Earth may well be mirrored on the surfaces of other planets. With the recent development of sensitive telescopes and other astronomical techniques, it is only a matter of time before there are hundreds of candidate planets to investigate.[25] If, as seems likely, the odds against the emergence of life on Earth are not nearly as great as previously supposed, it will only be a matter of time before living things are found elsewhere in the galaxy. And that will transform our understanding of our place in nature more profoundly than the Copernican revolution 500 years ago.

LIFE IS AN ABERRATION?

Even if the first living things were not the result of an improbable event, their persistence requires explanation. The economist John Maynard Keynes is often best known for his dictum, "In the long run, we are all dead." He might have added, ". . . but even in the short run, we *should* be dead." For the most stable state of the materials of which we are constructed—carbon, oxygen, hydrogen, nitrogen and phosphorus together with mineral elements such as calcium, iron and cobalt—is not the state in which they occur in living things. Carbon is most stable as carbon dioxide or as carbonate rock. Hydrogen and oxygen are more stable locked together as water molecules. In present circumstances, mineral elements such as iron are more stable in combination with oxygen as components of solid rock. How do living things avoid the degradation of their tissues into gas or rock?

Chiefly by the continual expenditure of energy. Appearances notwithstanding, all living things are well-ordered collections of molecules. The literally vital importance of chemical order is most vividly illustrated by the arrangements in most modern cells for ensuring that DNA molecules retain along their length the original and precise arrangement of the four nucleotides from which they are constructed. That is not a simple matter. Even in inactive stretches of DNA, chemical changes occur more or less at random in the chemical units of which they are built[26]; errors are more common in active stretches of DNA.

How can such errors be made good? Famously, the genetic information in each cell is encoded in the sequence of the nucleotides along the length of not just one, but two, chemically complementary DNA molecules, twisted together in the now familiar double helix. If one molecular strand is broken at some point, there are enzyme systems ensuring that the break will be repaired. If several successive nucleotides are excised from one strand, perhaps because of a malfunction of the processes by which genetic information is extracted from the DNA or by the effects of external radiation, repair enzymes will usually bridge the gap. Only because the same information is encoded in the complementary strand of DNA can the repair mechanism ensure that the repaired strand has the "correct" nucleotides in the "correct" order.[27]

This sophisticated mechanism (the details of which are still poorly known) is a considerable burden for the biochemical economy of a cell. The several protein molecules produced to function as enzymes in the repair of DNA (at least 50 of them in human cells[28]) require the expenditure of energy to secure the survival of the cell concerned and ultimately the accurate transfer of its genetic information to future generations.[29] Yet this is only one of several ways in which cells expend energy on the maintenance of internal order in the face of the natural tendency for their components to revert to the most stable chemical forms.

Energy expenditure is necessary but not sufficient to counteract this tendency. It is not an accident that all autonomous living things, from bacteria to the cells that make up the tissues of complex organisms, are surrounded by membranes which, if ruptured beyond repair, allow the vital contents of the cells to react with materials (usually water) in their environment. Membranes do not entirely prevent these homogenizing chemical reactions, but they make them slower and therefore unimportant in the short term, perhaps the lifetime of the cell concerned. But keeping the membranes intact is a further burden on the energy budget of a cell. The question of where the membranes came from in the first place, and when, remains open.[30]

Where does the energy come from? The most conspicuous of Earth's organisms, the grasses and the trees, are sustained by the output of energy from the Sun, whose surface temperature exceeds 6,000 K. Although sunlight is modified to some degree in passage through Earth's upper atmosphere, notably by the removal of ultraviolet radiation, a green leaf of a plant growing on the surface of the Earth will sense that it is bathed in radiation from a hot object (at 6,000 K) at a great distance (100 million miles, or just under 150 million kilometers). In other words, the green leaf is irradiated by photons (*quanta* of radiation) each carrying much more energy than photons corresponding to the temperature of the environment in which the plant is growing. With the intervention of the green-colored material called chlorophyll, the result is to turn carbon dioxide from the atmosphere into simple chemicals based on carbon, including sugar molecules such as glucose. Oxygen is a by-product of this process, called *photosynthesis*. Herbivorous animals eat plants and so, like plants themselves, depend on photosynthesis directly for their survival; carnivores eat herbivores, and so depend on it indirectly.

That is how things are now, on Earth, but it would be mistaken to suppose that photosynthesis had been the energetic basis of life since the beginning. Indeed, there is circumstantial evidence to the contrary. Because photosynthesis converts carbon dioxide into chemicals based on carbon and releases oxygen into the atmosphere in so doing, the emergence of photosynthesis as a major source of vital energy would have been accompanied by an increase of the proportion of free oxygen in the atmosphere. There is evidence of just such a transition in the history of the Earth's atmosphere, but surprisingly late. Some of the oldest deposits of iron ore based on oxides of iron have a characteristic banded appearance, suggesting a seasonal pattern of their deposition as sediments; strikingly, several of the oldest of these structures date from roughly 2,700 million years ago.[31] The inference is that the photosynthetic apparatus that sustains modern plants embodied a striking improvement of efficiency about 2,700 million years ago. That is another landmark in the evolution of modern life.

Many organisms depend on photosynthesis only in the most indirect fashion. The organisms of the luxuriant bacterial fauna of the human gut never see the Sun. Their energy is derived exclusively from undigested chemicals in the human food supply (some of which may be vegetable) or of the unwanted products of digestion. The same is dramatically true of the communities of bacteria and small invertebrate animals, including shrimplike creatures, clams and tubeworms, now recognized on the deep ocean floor. (There can be no true plants there, because there is no sunlight.) The more luxuriant communities are found near the deep-sea vents, which are fissures in the sea floor through which gush large quantities of hot water laden with inorganic chemicals of all kinds—sulphates, nitrates and phosphates, together with toxic materials such as the compound of hydrogen and sulphur or hydrogen sulphide (H_2S) and carbon dioxide in solution as well as methane gas. The bacteria are the primary energy converters; the invertebrates graze on them. The ultimate source of energy is the heat locked up in the Earth's crust, which raises the water temperature and provides the rich solution of chemicals on which the communities survive. It remains a question for the future to know how these communities became established, especially because the deep-ocean vents are likely to be short-lived by geological standards.

There are obvious attractions in the idea that some of the chemicals essential for the first prebiotic replicators were produced by chemical reactions on or beneath the Earth's surface, not only in the atmosphere. Chemical reactions between water and hot rocks are potentially prolific sources of energetic chemicals, which can be produced in bulk, and in high concentration.[32] Strictly geochemical life should in principle be possible.

The possibility of bacteria in near-surface rocks bears on the chance there may be (or may have been) life elsewhere in the solar system. All the conditions believed necessary for the support of terrestrial life—ample supplies of water and carbon-based chemicals, an atmosphere (to assist the exchange of gases and even to protect organisms on the surface from ultraviolet radiation) and an external supply of energy—may not be essential everywhere. Among the four inner planets, only Earth now meets all the conditions, although Mars and Venus may both have met them in some past stage of their evolution. Jupiter and Saturn are usually dismissed as planets capable of supporting living creatures because they have no solid surface and are much further from the Sun, but who is to say that exclusively airborne life is impossible? Much the same goes for Jupiter's moon Europa, which appears to have ample water and methane, and which probably has an internal store of heat comparable with that of Earth. That there may be living things somewhere on Europa cannot be ruled out because of its great distance from the Sun. But nobody can guess what such creatures, if they exist, would be like.

In what sense, then, is life an aberration? The occupation by living things of all conceivable niches on the surface of the Earth where there is a source of energy and appropriate nutrients suggests the opposite. Yet technically, all living things are out of equilibrium with their surroundings; without a constant supply of chemicals from their environment and a means of producing energy from them, they would revert to the "dust and ashes" of the Anglican *Book of Common Prayer*. All living things owe their existence to their status as intermediaries in the conversion of high-temperature energy (solar radiation or geochemical energy as the case may be) into low-temperature energy, that radiated from Earth's surface. Only this flux of energy has assured the continued existence of the biosphere (the totality of all living things).

The effects of a flux of energy can be dramatic. The discovery that the massive clouds of gas and dust within our own galaxy contain surprisingly large amounts of simple organic molecules is one sign of how materials such as hydrogen and carbon can be converted into more complex chemicals by a flux of radiation, in that case from the embedded stars. Those chemicals will revert to their constituents when the stars exhaust their energy.

Earth's atmosphere is sustained in the same way. The energy entering the atmosphere is high-temperature radiation from the Sun, but that which leaves it, similar in total quantity, is infrared radiation. The present layered atmosphere, with its lower troposphere and the higher stratosphere, is kept like that only because the atmosphere is involved in converting high-temperature to low-temperature radiation. If the Sun went out, the boundary between layers would be one of the first casualties.

It seems to be a general rule that where there is a flux of external energy that keeps some complex system in a *steady state*, the system acquires a structure, or a degree of order, it would lack in isolation.

CLUES TO PRIMEVAL LIFE

The starting-point for reconstructing the origin of life is the biochemical mechanisms that now sustain the rich diversity of animals, plant and microorganisms now on Earth and the fossil record that links living species to more primitive organisms. The search has been made realistic only by the revolution in biology centered on the molecular structure of DNA.

The broad stages of the evolution of life are plain. First, there was a period lasting somewhere between 200 million and 500 million years —the prebiotic phase—during which molecular evolution threw up molecules capable of replication that were potentially the founders of the first truly living things. There followed a period of between 2,000 and 3,000 million years during which all organisms were essentially single-celled creatures.[33] Multicellular organisms appeared later, probably more than 1,000 million years ago,[34] and were abundant by the Cambrian Explosion about 550 million years ago.[35]

What can be learned from present life-forms about still earlier forms of life, even the self-replicating entities of prebiotic times? As recently as 20 years ago, the standard reply would have been "almost nothing!" Now there is a great deal to say—and the prospect of much more to come. What follows is a listing, roughly by order of evolutionary antiquity, of some of the biochemical landmarks in the evolution of life whose further study promises to yield important information about the origin of life.

The emergence of organisms consisting of specialized cells may be as recent as 1,000 million years ago, but may go back to 2,000 million years ago. The common stromatolite fossils nevertheless show that colonial organisms, perhaps resembling the large aggregations of marine algae now found in the oceans, go back much further. It is not unreasonable that aggregations such as these may have developed the knack of enabling different cells to acquire complementary and mutually beneficial functions. That is the mechanism by which, in modern organisms, different cells acquire the properties of say, skin or bone cells. The principles are now understood, and the details will be gathered rapidly in the next few years. But this biochemical innovation could not have been a simple step 1,000 or 2,000 million years ago. Not only does it require that some of the genes in each cell should be rendered ineffectual for the lifetime of the organism concerned, but there should be a well-defined process of development in which the different cells acquire their differentiated properties and also a mechanism for recovering single cells in which all the genes are active, to be used for the regeneration of a whole organism.[36] When more is known of differentiation mechanisms in existing animals and plants, it will be important to compare the details of the mechanisms in different plants and animals so as to pin down the time and manner in which the first multicellular creatures evolved. There is probably as much to be learned from a comparison of the way in which the cells of multicellular organisms are now held together (by complex and specialized protein molecules) and the methods by which the cells of colonial organisms hang together.

The first emergence of eukaryotic organisms, which include fungi and protists as well as multicellular organisms, is another significant landmark, probably about 2,500 million years ago.[37] Eukaryotic cells differ from those of bacteria most markedly in that their DNA is arranged as linear strands rather than in circular form, but they also have a markedly more efficient method of energy conversion (to be described) and distinctive and usually more elaborate molecular means for extracting information from DNA. Curiously, the selective advantages of this major upheaval in the organization of cells are not yet clearly understood, but may have much to do with the flexibility with which genes can be used, even in single-celled organisms such as the yeasts. Again, there is a need for comparative studies of the crucial genes from separate organisms so that the timing and manner of their emergence can be pinned down.

The tale of how eukaryotic organisms acquired their more efficient means of energy conversion is a detective story still only half complete. The cells of all modern organisms but bacteria convert the energy of simple chemicals into energetic chemicals that are usable within cells by means of specialized structures called *mitochondria,* which have the external appearance of bacteria and which are also equipped with DNA of their own (meaning that they do not have to wait on the duplication of the cell in which they are contained for their replication). In 1982, Lynn Margulis[38] put forward the idea that mitochondria are derived from bacteria specialized in energy conversion that may have been parasitic on more conventional bacteria, and which were in due course absorbed into the general constitution of the bacterial cells. The incorporation into plant cells, of the structures called *chloroplasts,* which are specialized in converting sunlight and carbon dioxide into chemical energy, is now supposed to have come about in a similar fashion.

These structures, which make some of the proteins they need internally, have a genetic code that differs slightly, but significantly, from that of the organisms in which they are found. That points directly to some stage in the early evolution of life in which there were at least two genetic codes in simultaneous coexistence. There are two further implications. First, the coexistence of two genetic codes indicates that at some stage before the emergence of eukaryotes, there was a time when there were organisms dependent on two different (if similar) genetic codes. Second, the manner in which mitochondria and chloroplasts have been acquired points to the potential importance of mutually beneficial parasitism (called *symbiosis*) as a means by which evolving organisms can acquire new functions.

Evidently it will be of great importance to learn more of these processes. It is tempting to suppose that the acquisition of mitochondria was the decisive step in the emergence of eukaryotes, but that is not necessarily the case. It could well be that the first organisms to incorporate mitochondria were bacteria which, in spite of their enhanced efficiency, were superseded in the competition for resources by presumably still more efficient eukaryotes. Whatever happened, it is important for an understanding of how life began to find out; the coexistence of two genetic codes is likely to be a telling pointer to the nature of the earliest life forms.

There are other clues to the nature of early life whose significance is not yet understood. In modern plants, the collection of sunlight and its conversion into energetic molecules is an exceedingly complex process, involving light-collecting molecules (chlorophyll) and a whole cascade of molecular enzymes with a complex structure closely linked with the membranes of the chloroplasts containing them. It is unthinkable that such an arrangement could have evolved in a single step. Before the origin of life is understood, the history of this complicated process will have to be disentangled by the comparison of the variations in the photosynthetic processes in various plants. One would expect such studies to throw up an explanation of the appearance of free oxygen in Earth's atmosphere 2,700 million years ago. It is also probable that the first light-collecting molecules were radically simpler than chlorophyll; the class of molecules called *opsins*, used as light-detecting elements in animal eyes, are a good deal simpler (despite structural afinities with modern light-collecting molecules).[39]

The ubiquity with which the ions of certain inorganic elements are used in routine life processes is probably a pointer to still earlier stages in the evolution of life. As part of the process of regulating the water content of a cell, there is a device (called a molecular pump) for expelling hydrogen ions from all cells. The hazards of excess water are great:

cells die by overdilution. The selective advantages of such devices would have been great once replicating entities with membranes had been formed. It will be surprising if the detailed construction of the molecular pumps in modern organisms does not throw some light on how the very first pumps were constructed.

Still other uses of inorganic ions in living things are likely to be even more pointed clues to the distant past. Some materials now rare in the Earth's crust (such as vanadium and molybdenum) are used in the construction of particular enzymes, cobalt ions are essential for the functioning of vitamin B_{12}, while there is a long-standing puzzle about phosphorus, which has several distinctive roles in modern cells.

Phosphorus is not now common in the accessible environment. Most phosphorus in the Earth's crust is either locked up in primeval rock or in ancient sedimentary deposits of calcium phosphate in, for example, Chile and North Africa. The extensive mining of those deposits for use as fertilizers in modern agriculture (as well as the recovery of bird droppings, called guano, from oceanic islands) is a testament to the importance of phosphorus in living things. Invariably, phosphorus turns up, linked with oxygen atoms, in groups called *phosphate* groups. In vertebrates, most phosphorus is locked up in the structural material of the bones, but the same chemical units provide the backbone for molecules of RNA and DNA.[40] Yet the more immediate role of phosphorus is in the molecules that serve as the universal currency of energy in cells, called adenosine triphosphate, or ATP. Weight for weight, ATP has about the same energy as the explosive TNT (but it does not explode). Being a compact molecule soluble in water and capable of taking part in chemical reactions with other molecules, it is a convenient way of driving biochemical reactions (and even of making muscle fibers contract) that otherwise would not happen.

The almost universal use of ATP for effecting most of the energetic chemical reactions in cells argues for its antiquity in the evolution of life. Indeed, the molecule differs from the nucleotides that form components of RNA and DNA only in the presence of two extra phosphate groups. It is therefore a likely candidate for a component of one of the prebiotic replicators.

The invariable role of phosphorus in the backbones of DNA and RNA molecules was illuminated in 1990 by the recognition that simple amino acids will serve as well as phosphate groups to hold together successive bases in the structure of DNA.[41] The binding between complementary molecules constructed in that way is even stronger than between complementary molecules of DNA. If, by an historical accident, molecular evolution had settled on such a molecule for storing genetic information, the result might well have been the more accurate perpetuation of genetic information between successive generations at the expense of restricted access to it during the lifetime of an individual organism. That is another way of saying that the effect of the molecular evolution that occupied the 2,000 million years or so of life on Earth was a series of compromises between perfection and the day-to-day needs of individuals to survive.

Another class of clues to the nature of early life is necessarily harder to define, consisting as it does of the unknown answers to the question, "Why does nature use in the construction of its biochemical machines some chemicals, but not others very like them?" The case of the 20 amino acids used in making proteins cries out for attention. All the naturally occurring amino acids belong to the class known technically as α-amino acids,[42] but in principle there is an infinite number of these molecules. Some of those not occurring in real proteins are so similar to others that are used that their neglect requires explanation.

There is another puzzle to do with proteins that says something— but nobody knows what—about the evolution of early life. The genes of eukaryotes are not simple undivided pieces of DNA, but are usually split into several pieces, which are separated from each other by irrelevant and apparently meaningless stretches of DNA called *introns*. In bacteria, on the other hand, introns are rare or nonexistent; on the face of things, introns arrived when eukaryotic cells made their appearance. But nobody knows, because the function of the introns is entirely ob-

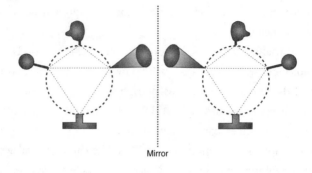

Mirror

Figure 4.2 Nature's Asymmetry Schematic architecture of an amino acid molecule in which the central sphere represents a carbon atom, the so-called α-carbon. The foot is the chemical group that gives the molecule the properties of an acid; the cone, the group that makes the molecule an amine. The irregular group at the top of (and behind) the sphere is the rest of the molecule. The vertical line represents a mirror; each figure is the refection of the other in the mirror. No amount of rotation will make the two figures coincide.

scure. That they have a function is, however, clear from the way in which the pattern of introns is preserved in similar genes from related but different species. Introns are also telling relics of early life.

So too, but in a still obscure way, is the asymmetry of the naturally occurring amino acids, which are all geometrically asymmetric in a sense first recognized by Louis Pasteur well over a century ago. Carbon atoms can simultaneously form four links with other atoms of chemical groups, which are usually arranged around the carbon atom as the corners of a triangular pyramid are arranged about its center.[43] If the carbon atom is linked in this way to four atoms or groups which are all different, there are two distinct ways in which the four groups can be arranged relative to each other; the two arrangements are related as one's left hand with its reflection in a mirror.[44] All the amino acid molecules used in living things have just that property, called *chirality* (see Figure 4.2).

So how do living things deal with this mixture of what are, in effect, essentially similar materials? Simply by ignoring one of them. Although strictly chemical processes for manufacturing amino acids produce mixtures of the two forms, cells synthesize only one form. If animals are fed amino acids manufactured in a chemical plant, they can use one form (called the L form) but not the other (called D).[45] The same rule applies to all the other molecules in the modern cell with which amino acids interact, and which for that reason need a geometrical shape matching that of their partners. The choice of L-chirality was probably arbitrary and not (as some have suggested) linked with the left-handedness of neutrinos. But an understanding of how chirality arose would be an important clue to the complexity of the self-replicating molecules in prebiotic times.

Extracting information about the earliest forms of life from clues such as these will not be straightforward, of course. Nor is the ambition novel. Since early in this century, generations of distinguished chemists and biochemists have paid close attention to the opportunities outlined above and many others. The novelty is that only now does data (about the structure of the DNA of genes) offer the chance of reconstructing at least the sequence in which known biochemical innovations appeared, perhaps even the dates of past landmarks in the evolution of life.

This program is essentially a collective enterprise. Individuals with an interest in the origin of life make the best they can of the information at their disposal. But the collection of this data is sporadic, while the direct support for research in the origin of life is meager to say the least. That is a shame, given the importance of an understanding of how life began on Earth. Will it not be galling if the mountains of data now accessible are found to conceal within them the solution of one of the most interesting riddles still unsolved?

So How Life Began . . .

The ultimate goal is an account of what prebiotic evolution was like together with laboratory demonstrations of its plausibility. There are two parts of the problem: what were the chemical components of the prebiotic replicators, and what were the physical circumstances in which they were first assembled?

Although Oparin's sea may have been amply supplied with chemical units capable of being linked together into materials such as poly-(X,Y) (see above), many have argued that the first processes of polymerization may have been more probable if carried out on or even within solid materials. In the late 1940s, the British crystallographer J. D. Bernal argued that the atomic structure of naturally occurring clays (which can incorporate small chemicals between layers of atoms) would have been suitable for this purpose. The idea has been carried further by Dr. A. G. Cairns-Smith of the University of Glasgow, who has argued that layers of atoms in clay could have guided the assembly of chemical units in an ordered fashion.[46] Separately, Dr. G. Wächtershauser, long since retired as a professor of biochemistry at the University of Munich, has argued that the surface of the common mineral iron pyrites "fool's gold" could have assisted early polymerization both of replicating molecules and cell membranes.[47] If the first laboratory demonstration of the spontaneous assembly of a prebiotic replicator involves the use of some such material as a catalyst, nobody will be surprised.

The identity of the chemicals involved in the first replicators is a more open question. The chemical building blocks of modern life include amino acids (which make proteins when polymerized), sugar molecules such as glucose (which makes starch), fatty acids (of which acetic acid is the simplest) which make cell membranes and the purine and pyrimidine bases which are the essential ingredients of DNA and RNA as well as of materials such as ATP.[48] Fashions come and go. Miller's demonstration in 1955 that electrical discharges in gases simulating the primitive atmosphere will yield amino acids provoked interest in proteins as the first replicators.[49] Ways in which the nucleotide bases and the fatty acids might have been formed on primitive Earth have since been worked out.[50]

The attention of the research community has nevertheless now switched to RNA molecules, largely because of the dramatic discovery in 1981 (by Thomas Cech of the University of Colorado at Boulder) that particular RNA molecules naturally function as catalysts, or enzymes.[51] Indeed, interest had swung in that direction before 1981, largely because RNA molecules have the potential to store genetic information much as DNA molecules do, by the particular arrangement of four chemical units along their length. In all modern cells, RNA molecules still serve as intermediates in the transfer of genetic information between the genetic DNA and the chemical apparatus—the ribosomes, made predominantly from RNA—at which protein molecules are assembled from amino acids.[52] Similarly, the small molecules that scavenge the cytoplasm of cells for amino acids, called "transfer" molecules or tRNAs, are made from the same material.[53] So there has emerged the view that before DNA-based life came into its own, there was a period when living things were sustained by genetic material consisting solely of RNA. That does not require that RNA molecules were the first prebiotic replicators —although there are many who believe that to be the likely outcome.

One question bearing on this issue is whether small RNA molecules can catalyze their own concatenation into longer molecules carrying significant amounts of genetic information. There has been some progress. For example, in 1996, Leslie Orgel at the Salk Institute in La Jolla, California, showed how to induce the formation of RNA molecules 50 units (nucleotides) long from RNA molecules that are only 10 units long.[54] Others have been able to show that analogs of RNA molecules just two nucleotides long are able to act as templates for their own replication.[55] But, for the time being, it is more important to reach a clear picture of what the organisms of the "RNA world" would have been like.

There is plentiful circumstantial evidence in the detailed biochemistry of modern cells. Many of the small chemicals[56] necessary as adjuncts to the working of protein enzymes are chemically related to the constituents of RNA units, while modern cells synthesize the nucleotides that are the chemical units of DNA molecules from those of RNA molecules. One curious feature of the cells of higher organisms is that the ends of chromosomes are maintained from one generation to the next only by the intervention of genetic information carried by RNA molecules. But the predominant use of RNA in modern cells is in the structure of the ribosomes universally used for the assembly of protein molecules, and whose mechanism is still not clearly understood. Is it possible that structures like these functioned as the main storehouses of genetic information in the RNA world of earlier times?

Telling what the organisms were like in the RNA world will be a gigantic problem of inference. Until that has been done, it will not be possible to tell whether the organisms could have sprung directly from prebiotic replicators. Indeed, at this stage it is not even clear what were the advantages for early organisms of the replacement of RNA by DNA as the genetic basis for life.[57] Mixing chemicals together in a laboratory vessel in the hope that they will show signs of life is an understandable response to frustration born of the difficulties that lie ahead. But there is no alternative to the rounded understanding of the processes that occupied the first 3,000 million years of the early biosphere. Still, we shall all be wiser when the mechanism of the ribosome is understood, which should not be long delayed.

Cooperation and Autonomy 5

Cells are the quasi-autonomous ingredients of life. The simplest cell, a single bacterial cell from the human gut perhaps, or a cell torn off the skin by rubbing it, is more complex than any machine yet built by people. Like all machines, cells are sustained by various parts—chemical parts consisting of different kinds of molecules. Part of the complexity arises from the numbers of the different parts. In bacteria, there are at least as many different kinds of molecular parts as there are genes in the genome of the species, which amounts to roughly 3,000 for the human and animal gut bacterium *Escherichia coli (E. coli)*.

What machines in our daily lives have as many components? And which of our machines can spontaneously adjust to changes in their environment, as cells do, by varying the numbers of these components? Who yet owns a car that sprouts an extra wheel when one tire goes flat? Yet the simplest cells continually adjust in that way to changing circumstances—a novel food supply, perhaps, or a threat to survival.

Moreover, there is no central control of the complex operations of a cell in the active sense in which, say, the mechanical systems of an aircraft are controlled from the cockpit. In a cell, the nearest thing to a controlling center is the set of genes physically embodied in the DNA, which taken together are called the *genome*.[1] The genome is both the place where the specification of all the chemical components of a cell is stored and the structure which is replicated when cells divide to yield either cells essentially like the original or germ cells that produce new organisms when they fertilize each other.

In organisms that reproduce sexually in that fashion, the DNA in the single fertilized cell not only specifies the myriad types of cells in an adult, but also the recipe that guides it through its development from embryo to adult.[2] In earlier decades, the idea that a single fertilized cell can have not only the potential to become, say, an adult butterfly, but also the caterpillar that hatches from the egg and the chrysalis in which the organism becomes a butterfly was regarded as almost magical. Generations of young children standing around the nature tables of the world's elementary schools have been invited to marvel at wonders of ontogeny, as development from embryo to adult is called. Only now, as the role of the DNA has become clear, has it come to seem a part of the machinery of life.

By comparison with the pilot in the cockpit of an aircraft, however, the genome plays a passive role, responding either to chemical signals indicating the current needs of its subsystems or directly to information about the external environment. For example, the dramatic act of cell division (called *mitosis),* in which cells divide into two virtually identical cells,[3] may be orchestrated by the genome, but its occurrence is not determined there. Mitosis happens only when signals from different parts of the cell concur; some of these signals provoke division, others are merely permissive.[4] A cell is a self-regulating biochemical democracy in which the several parts are continually casting votes in the form of the chemical signals they transmit. The genome is to the cell as a supreme court is to a national judiciary.

In the past few years, there has grown up a sense that all rational questions about the functioning of cells are answerable. This belief stems from the success with which the structure of DNA pointed directly to explanations of the properties of DNA molecules, notably their capacity to guide their replication in their own image. But the idea that the working of the cell has become an open book has been reinforced, in the past four decades, by the remarkable way in which the structure of DNA has been exploited to carry through in the laboratory chemical transformations that usually occur only in the laboratory. In recent times, only the hectic period in which the foundations of quantum mechanics were laid between 1925 and 1930 compares in élan with present research in molecular biology. In the search for an understanding of life, science now is pushing at an open door. We probably still grossly underestimate the benefits these developments will bring not only for understanding, but for the health and wealth of nations.

Nevertheless, and without disparaging the achievements of the recent past, there are difficulties ahead for now-exuberant cell biology, many of them the result of the sheer complexity of the simplest cells. The present state of this research has much in common with conventional chemistry in the early decades of the nineteenth century, when the atomic theory swept all before it. Then, those who manipulated simple chemicals to form novel compounds were able, for the first time in human history, to carry in their heads a picture of what happened when, for example, they treated benzene with nitric acid to form nitrobenzene and then used that to make the first synthetic dyes. Yet throughout the nineteenth century, chemists depended on empirical rules of thumb to predict what compounds would emerge from this or that chemical reaction, and what their properties would be. Only much later, in the 1940s and 1950s, have their successors won a deeper understanding of chemical reactions and their products.

The analogy is not exact, but the application of molecular biology to the understanding of how cells function is still at the rudimentary stage of nineteenth-century chemistry. It is as if an engineer were confronted with a machine of unknown design and asked to explain its function. What does he or she do? Why, first take the machine apart and make a list of its components, then do what can be done to describe the linkages between the components. (In the case of a cell, the linkages are chemical, sometimes electrochemical and even mechanical.)

That is where cell biology now stands: the *naming of the parts* (at least 2,000 in bacterial cells and some 100,000 in mammalian cells[5]) and the description of their linkages. To be sure, information is accumulating rapidly about the chemical characteristics of the parts. The challenge yet to be taken up is to resynthesize that great volume of information into an intelligible account of the dynamic working of any kind of cell.

The excitement in the field has even engendered an untrustworthy triumphalism together with a tendency to mistake description for understanding. These days, it is fashionable to scorn most nineteenth-century natural history done by gifted amateurs on the grounds that their efforts were mere "botanizing." (Darwin is usually exempted from the complaint on the grounds that he made something of the facts he gathered.) But much of contemporary cell biology is but high-level botanizing. It is potentially of huge importance, as will be seen in what follows. Yet it does not amount to comprehension.

Discovery Brings Technique

Even the naming of the parts has been possible only because molecular biologists have developed some of the neatest laboratory techniques that science has yet seen. From the outset, in the mid-1950s, people have deliberately used cells, or parts of them, to perform tasks well beyond the capacity of old-fashioned chemistry.

From the early days of molecular biology, it has been standard laboratory practice to use the contents of disrupted cells to replicate vital biochemical tasks. The idea is that the pooled contents of many cells will contain an ample supply of the enzymes needed for vital operations. By these means, the *replication* of double-stranded DNA molecules has been studied as well as the process of the *transcription* of the genetic information in DNA into the equivalent information in RNA molecules. The role of ribosomes in the *translation* of RNA molecules into equivalent protein molecules was first proved in the same way.[6] The crucial experiments in working out the genetic code in the early 1960s entailed feeding synthetic RNA molecules to cell-free systems of ribosomes and analyzing the protein molecules produced.

Elaborations of these techniques have been steady and rapid ever since, but there are two outstanding landmarks. One was the discovery in 1968 of the enzyme *reverse transcriptase,* which has the property of being able to convert RNA molecules into the corresponding DNA. In ordinary cells, DNA makes RNA, but RNA does not make DNA. So how do the viruses whose genomes are made from RNA, not DNA, make their way in the world?[7] Only by subverting the transcription machinery of the cells they infect, which means that their RNA must first be turned into a piece of DNA if the viruses are to be multiplied. With the discovery soon afterward (in bacteria) of enzymes that cut DNA at predefined places and of others that can join together separated pieces of DNA, the biotechnology industry was founded.[8]

A more recent technique has had a more remarkable influence on the practice of molecular biology. In 1983, Dr. Kary Mullis, then working for the Chiron Corporation in the United States, developed a technique for copying indefinitely any stretch of DNA (say a gene) in a longer molecule (say, the whole genome) provided that the arrangement of nucleotides at the two ends is known.[9] The technique hangs on the use of the enzymes from temperature-resistant bacteria normally used in synthesizing DNA; in the film *Jurassic Park,* DNA recovered from the stomach of a long-dead insect is amplified and used to construct extinct dinosaurs. More soberly, the technique was crucial in the analysis of DNA from the type-specimen of the Neanderthals, reported during 1987, from which it was concluded that the Neanderthals had no part in the ancestry of *Homo sapiens.*[10]

By these and other means, molecular biology has spawned a new kind of chemistry. For the first 150 years of traditional chemistry, everything hung on the development of ever more sensitive weighing machines to measure small quantities of materials; that was the only sure way of telling how many atoms of different kinds there are in particular molecules. But the weighing machines are now largely banished. Instead, the sizes of molecules are estimated directly and with reasonable accuracy from the speed of their migration through specially prepared gel-like films under the influence of electrical fields.[11] The amounts of the materials can often be estimated from the same data. The sensitivity of these techniques far outstrips what had previously been possible; people routinely manipulate mere billionths of a gram of complex biological molecules, extracting from such tiny samples details of how they are constructed from simpler chemical units.

The goals of the new industry spawned by these techniques, biotechnology, were at the outset to manipulate cells (often simple bacteria) to manufacture proteins that could be used as medicines. Imagination and ambition have since expanded: speculatively capitalized companies in many countries are now seeking to engineer changes in the properties of cells as such that will then be substituted in living things—people included—to correct inherited deficiencies of performance ("gene therapy" as it is known) or to improve on natural performance, notably in the engineering of plants with novel and advantageous genetic structure. The use of animals grown from embryos genetically engineered to secrete human biochemicals in their milk is also much spoken of, and will probably yield marketable medicines early in the next century. The only certainty about biotechnology is that it is still in its infancy.

Uncertainties remain hidden by the aura of success surrounding cell biology. Sheer complexity creates novel difficulties. Sheer success is also a distraction from understanding. The engineers taking cells apart, mechanism by mechanism, are understandably so excited by the marvelous delicacy of what they find that they have little inclination for the systems analysis that engineers in other fields insist upon. The big picture is in danger of being hidden by the detail.

CREATURES OF A SINGLE CELL

Diversity is the hallmark of the simplest creatures now alive: the bacteria, which consist of a single and almost autonomous cell. They live on and beneath the Earth's surface, but luxuriantly also near the surface of the oceans and sometimes at depths of thousands of meters below their surfaces; most species of bacteria have not yet been cataloged. Even by appearance (with the help of microscopes) they are a mixed bunch. Their dimensions are best referred to in micrometers, which are millionths of a meter or thousandths of a millimeter and written as μm. A single bacterium of *E. coli* is shaped like a cylinder about 1 μm across and three times as long; the dimensions are far beyond the acuity of the human eye. Cyanobacteria, which are capable of photosynthesis, may be similarly shaped but ten times bigger (so that their volume and therefore weight is 1,000 times as much). The cells of the nervous system are often bigger still, at least in length.

Most bacteria reproduce prolifically. *E. coli* cells, for example, divide every 20 minutes when the food supply is plentiful. That means

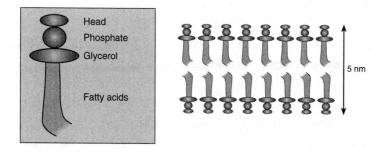

Figure 5.1 Keeping Cells Intact Arrangement of fatty-acid molecules (called lipid molecules) in the double-layered membrane of a cell. The three components of the hydrophilic head are (1) a small molecule such as serine (an amino acid), (2) a phosphate group and (3) the alcohol glycerol. Of the two long tail-like molecules, one is usually kinked.

8 ($=2\times2\times2$) cells after an hour, 64 ($=8\times2\times2\times2$) cells after two hours and 512 after three hours. If there is enough food to sustain this rate of growth, the total weight of the bacteria would amount to a kilogram after 13 hours. After 76 hours from the beginning, or in just over three days, the total weight would be comparable with the total mass of the observable universe. That is manifestly impossible; the bacteria would be on starvation rations long before that state of affairs could be reached.[12]

Bacteria, almost certainly the first true living things on Earth, have enough in common with each other to be recognizable for what they are and have enough in common with other kinds of cells to be archetypes for cells in general. First, their contents are enclosed in a double-layered membrane of characteristic structure in which parallel sheets of detergentlike molecules are arranged tail to tail; cells of all kinds have such a *plasma membrane*. (*E. coli*, like many other bacteria, also has an outer membrane constructed on similar lines.) Second, their genetic material consists primarily of a single loop of helical DNA—two complementary molecules twisted on each other. Third, bacteria derive energy from chemicals in their environment by the process of glycolysis, which is simpler but less efficient than the ways in which more evolved cells win energy. Like most other types of cells, bacteria use the material called ATP as the general currency of energy.

All other types of cells differ from bacteria in the way in which the genetic DNA is packaged: it is not arranged as a single circular molecule of helical DNA, but as several linear pieces called *chromosomes,* first recognized in cells towards the end of the nineteenth century. Second, the chromosomes themselves are packaged within a *nucleus,* visible in the microscope as a roughly spherical structure enclosed by a membrane essentially like that of the exterior of the cell. These cells also have a complicated internal membrane[13] that divides the interior of a cell into compartments and also provides anchorage for pieces of cellular machinery such as the ribosomes (which manufacture proteins) and what is called the Golgi apparatus (which sorts protein molecules according to the regions of the cell in which they are required).

Several other distinctive features of more evolved cells have come to light recently. For example, many types of cells keep a characteristic shape because they have an internal scaffolding framework made of aggregations of protein molecules and which are called *microtubules*.[14] Engineers embarking on the dissection of such a cell would be thoroughly approving of their discovery that these *microtubules* are structurally efficient hollow tubelike structures.[15] Remarkably, the scaffolding is not permanently in place, but is continually being formed and dismantled. The same skeleton is also probably a means of transporting chemicals, particularly proteins, from one place to another within a cell.

Organisms whose cells are organized in this way are called *eukaryotes*. Some of them, such as yeast and the parasites that cause malaria, consist of single cells, like bacteria. Others, which consist of myriads of cells, include all the animals and plants now conspicuous on the planet Earth. Single-celled creatures are by far the most plentiful organisms. They include the fungi which, when growing in large colonies, make edible mushrooms, but there are also single-celled fungi that cause human disease, notably in the lungs and genitals. Other single-celled organisms, collectively called protists or protozoa, are sometimes innocuous denizens of the environment, and sometimes the agents of disease, malaria for example. The diversity of these creatures is remarkable. They appear to have succeeded, by adaptation to particular environments, in occupying every conceivable niche on the surface of the Earth.

The diversity of single-celled creatures is matched by the diversity of the cell types cooperating in the bodies of multicellular animals and plants. In mammals, for example, at one extreme there are the red cells and the platelets circulating in the blood, which have the important functions of transporting oxygen and repairing damaged tissues respectively; they lack nuclei and thus the genetic DNA in the chromosomes of other types of cells.[16] That means that they cannot replicate themselves, that they have a relatively short lifetime and must be continually renewed in specialized organs such as the liver, spleen and bone marrow.

There are several opposite extremes, one of them represented by the cells in the major limb muscles of the human body. These cells, called *myocytes*, may be several centimeters in length and are also relatively thick (50 μm or more); they are elongated cells, perhaps 1,000 or more times longer than they are wide. Remarkably, they contain large numbers of identical nuclei, each with its own set of chromosomes, which comes about because they are formed by the end-to-end fusion of progenitor cells called *myoblasts*. In mammals, the major muscles are formed before birth, so that the fetus is born with all the muscle cells it will need for adult life, whether that of a weight lifter or a fashion model.

That versatility arises in an interesting way. The several nuclei in each muscle cell are able to rework and even expand the numbers of embedded contractile fibers in tune with the use made of them, as when a person is training to be a fashion model or a weight lifter, for example. Yet myocytes are not capable of cell division. So how can muscles regenerate, as after injury or a surgical operation? It is an astonishing business; a person's adult muscle tissues contain not only mature muscle cells that contribute directly to their function, but also naïve myoblasts left over from his or her birthday. After damage by a scalpel, for example, the damaged muscle cells degenerate but the myoblasts dutifully go through the processes of cell division and fusion by which the original muscle cells were formed *in utero* and, at least to some degree, the lost or damaged muscle is regenerated.[17]

Common Explanations

Despite the great diversity of single-celled organisms, the past few decades have shown that very different organisms use essentially the same biochemical machinery for carrying out similar functions. Thus the genetic code is essentially universal; the few exceptions, in mitochondria and chloroplasts, prove the rule: if the exceptions were originally parasitic bacteria, that merely shows that there were once at least two genetic codes in coexistence. It is also the case that the seemingly different energy conversions carried out by mitochondria and chloroplasts respectively are accomplished in essentially similar ways, using similar molecules for similar biochemical steps.[18] The molecules effecting particular functions in different organisms may differ slightly from each other, but the differences are usually insignificant. That the principles underlying life are in this sense universal is one of the intellectual prizes of the past few decades.

The universality of the biochemical machinery in cells cannot be complete, of course. Sometimes the common thread is a common chemical principle, as in the architecture of the external membranes of all cells or the way in which all animals with eyes rely on the same kinds of pigmented chemicals to collect light from objects in their sight. (Complete eyes have evolved differently half a dozen times in half a dozen different groups of animals.[19]) The common dependence of cells on the phosphorylation of proteins as a way of making them chemically active is another generally used mechanism.

Yet nature's ways of enabling cells to survive are not necessarily the most efficient there could be. Being a product of evolution by natural selection, once some mechanism has been honed to a passable degree of efficiency, it is improbable that a radically different way of doing the same job could evolve independently within the same lineage of organisms. An alternative mechanism, at least at the beginning, would probably be much less efficient than that already established. But lineages of cells are not in the business of incurring extra metabolic costs in the here and now for the sake of unspecified benefits many generations later. The lineage depends only on the success with which each generation divides into daughter cells to populate the next generation. The fate of the generation after that is immaterial.

There are now synthetic materials more efficient than hemoglobin at carrying oxygen around the veins of vertebrates, for example, but natural selection has never had a chance to work on them.[20] Variations on the hemoglobin principle are the basis of blood of all kinds. Similarly, all movement in or by animals depends on the action of fibers made of two muscle proteins called actin and myosin. There are obvious difficulties, but nature has not so far benefited from the advantages of wheeled locomotion. In this sense, the seemingly universal chemical machinery of the cell is not necessarily the most efficient that could be devised. This point is likely to be established by the genetic engineers; already there is great interest in improving the efficiency of industrially important bacteria by engineering the structure of the enzymes they produce.

There are further restraints on the inner chemistry of single-celled organisms: however specialized an organism may seem, all its cells must be capable of all the functions required of any one of them at any time. When starved of food, most bacteria and fungi are transformed so as to use less energy. These forms, called *spores*, look different and behave differently from the usual forms. The spores of bacteria can survive in normally inhospitable conditions for months or years on end, the spores of fungi are those dispersed by wind from the umbrella-shaped caps of mushrooms. Each cell of such a species must be capable of existing in the normal state or as a spore; it must be *totipotent*.[21] Cells may differ from each other genetically, meaning that they carry out particular functions in slightly different ways, but single-celled organisms would not survive if crucial functions were entirely lacking.

The enthusiastic naming of cell parts has revealed one other striking feature of life as it has evolved on Earth: its quite astonishing intricacy. Despite large strides made by cell biologists in their search for deeper understanding, they have hardly begun to grapple with making this intricacy intelligible—of welding it into a description of what life is that makes a coherent story to tell to themselves and the rest of us.

The Great Division Story

In the operation of an aircraft, different mechanical functions are carried out by different subsystems, often recognizably distinct from one another. Even without a cockpit or a pilot, the same is also true in cells; the difference is that the subsystems are interconnected in ways that are poorly understood at present. But already it is clear that most cellular subsystems are exceedingly intricate. And there is a very large number of them. DNA has to be replicated (at cell division), monitored and repaired when necessary, stretches of DNA must be transcribed into their equivalents of RNA, food or sunlight must be converted into energy, the internal skeleton must be maintained (and then modified if the cell is capable of movement).

The machinery of cell division, still not fully worked out, is a good illustration of the problems created by intricacy that can only be multiplied in the years ahead. One cell will from time to time divide into two almost identical cells, inscrutably called "daughters." In a general sense, the process is cyclical: in due course, each of the daughter cells can divide again, and so on. All single-celled creatures must be capable of autonomous cell division or they will not be perpetuated. The cells of multicellular organisms, on the other hand, can lead a more leisurely existence: some, such as the blood cells, must divide to schedule if the organism to which they belong is to survive, but others may never divide at all. What drives this process?

Most studies have been carried out with fertilized eggs of species of the toad *Xenopus* and with the single cells of various yeast species, especially the bakers' yeast used for making bread,[22] but there is enough information from other organisms to suggest that yeast cells typify the general process. Successful cell division requires two steps in sequence. First, the genome must be replicated to provide the daughter cell with a genome of its own, and then the whole cell must be divided physically into two (which requires the prior production of enough of the proteins and other materials to ensure that both daughters will survive).

In the nineteenth century, the doctrine of vitalism was partly sustained by reference to phenomena such as cell division, which seems to epitomize life itself; how can a cell be capable of autonomous division if it is simply a collection of molecules whose structure can be factually described? The remarkable discovery of the past few years is that each stage of cell division is indeed driven by specific protein molecules whose structure is encoded within the genome, and which can be factually described. The key discovery in the late 1980s is that there is a class of proteins called *cyclins* whose concentration in a cell changes markedly as cells go through their division cycle[23]; since then, more than 50 proteins involved in the regulation of the process have been identified even in the relatively simple cells of yeast. The pace at which information is accumulating is quite extraordinary; since about 1990, it has been rare that one of the weekly or biweekly issues of five reputable scientific journals in this field[24] has not contained an important contribution to the description of some molecule or molecules involved in the regulation of cell division, yet the listing is far from complete.

What provokes a cell to divide? The impetus may come from outside or from within. Free-living single-celled organisms may be driven to divide merely by their size, which is probably a good indicator that the nutrient supply in the environment is sufficient to sustain offspring cells. Single cells belonging to colonies (such as the fungal cells in mushrooms) can be provoked into cell division by chemical signals put out from elsewhere in the colony; the signals are again encoded in the genome. The same is true of some of the cells of multicellular organisms such as people; liver cells, for example, will be provoked into division when the liver as a whole is under pressure. But the provoking event may also be internal to a cell: that is the case with fertilized eggs where cycles of cell division follow quickly on each others' heels, which is possible only because the maternal egg is supplied with the proteins and other cellular components necessary to sustain several cycles of cell division.

Cell division does not always produce a pair of identical daughter cells. Even in single-celled animals and plants, cell division can lead to daughter cells that look and behave differently from each other—the spore cells produced by fungi and many bacteria are examples. But multicellular organisms make greater use of asymmetrical cell division: as will be seen, cells acquire their specialized functions in a hierarchical process in which cells called stem cells divide asymmetrically, each time producing another stem cell and one with more specialized functions of the kind that contribute to the functioning of the organ to which it belongs.

The hectic research of the past decade has shown that there is nothing in a cell whose function is to regulate cell division; remember, the dividing cell has no pilot. The closest analog of a controlling device is an intricate biochemical switch involving cyclin molecules.[25] Curiously, cells also have a mechanism for getting rid of cyclin molecules, one that turns them back into amino acids, but they can be rescued from that fate by another protein molecule with which they are capable of forming a loosely bound complex.[26] The internal trigger for each stage of cell division is that the amount of this complex should reach some threshold; then, it seems, there is an internal transformation of the structure into a form that can activate the replication of a cell's DNA in the first stage of cell division and the physical division of the cell in the second.[27]

The process is in a sense a democratic adaptation by cells. The continual production of cyclins (and their destruction) marks the passage of time, which may be relevant. The sequestering of cyclins into complexes with another protein, which can then be activated to set in train the next phase of cell division, allows signals from elsewhere in a cell to influence the course of events. External signals that indicate the need for new cells probably exert their influence at this stage. So may internal signals—indicating, perhaps, that there are already cellular components enough to see the cell through the next phase.[28] A remarkable feature of this collective control is that the pause between the replication of a cell's DNA and its physical division is occupied by a systematic investigation of the integrity of the newly synthesized genome and, when necessary, by its repair.[29] There is a similar pause during the physical division of the cell into its daughters, when biochemical steps are taken to ensure that there are two pairs of each chromosome.[30]

Many decades will pass before all the details of cell division are fully understood for cells in general and for the exceptions that will no doubt come to light. For cells that belong to organs as in the human body, there are probably signals specific to that organ reflecting the need for new recruits and there are probably organ-specific receptors on the surfaces of the cells concerned that contribute to the outcome— cell-division or otherwise; certainly the delicacy with which an organ such as the liver retains its size, often tuned to the habits of the person concerned, is a proof of the effectiveness of cell-by-cell regulation along these lines. But the most striking of all the stages of cell division (in eukaryotic cells) is that preceding the physical division into daughters, when all four like chromosomes are held in physical proximity. Cell division does not proceed beyond this point unless there are equal numbers of similar chromosomes. How is that arranged? By exchanging corresponding pieces of supposedly similar chromosomes, in a process called *recombination*.[31] Many of the proteins and RNA molecules regulating these steps (and the genes responsible) have already been identified; there are many others to find. But the obvious unanswered question is the route by which this particular subsystem of the typical cell evolved. There are already snippets of information, but we shall be lucky if there is a clear account of what happened before the end of the next century.

Meanwhile, there is the issue of comprehension to consider. The intricacies of these subsystems and their linkages are already so great that even people working in the field are unclear about what they have learned. The central problem of cell biology now is not so much the gathering of information, but the comprehension of it. What remedy can there be for that state of affairs?

CATALOGS OF SYSTEMS

The intricacies of cell division are mirrored by those of other cellular processes now dissected. What follows is a list of some of parts of the standard cell whose working has been uncovered in recent years. In earlier decades, each item in the list would have been a signal triumph for biology; each is a collective triumph, the work of small armies of people working in laboratories across the world.

DNA Processing

In cells, DNA is manipulated in two ways: it is replicated in preparation for cell division and transcribed into molecules of RNA in ordinary life. Each process requires a complex of several enzymes to be assembled at an initiation site (just one on the circular chromosome of bacteria, several on eukaryotic chromosomes). Replication is the more intricate, if only because there are two strands of DNA with opposite directionality to be copied; little is known of what sets the process in train unless it is the complete assembly of the replication complex (in which case, how that serves the purpose is unclear). Transcription, a simpler process, is better understood, but the question of how the needs of a cell for the products of particular genes are translated into the replication of those genes remains unclear. People are just now in the thick of identifying transcription factors that help to activate particular genes—and there appear as many molecules that prevent expression of the same genes. The catalog entry is incomplete.

RNA Editing

RNA molecules transcribed from genes within the nucleus of a cell are usually modified before they leave the nucleus, not least so that the RNA equivalents of the introns can be removed. The job is done by complexes of protein and RNA molecules, but little is known of the working of this machinery, of the influences that control it or of its evolutionary origins. Nor is it clear what function is served by the routine addition of a long string of adenine ribonucleotides to the RNA molecules put out from the nucleus.

The strangest feature of newly transcribed RNA molecules in the eukaryotic nucleus is that, in some circumstances, the RNA molecules transcribed from what are supposedly authentic genes are deliberately modified, sometimes by the insertion of extra nucleotides at specific places, sometimes by the replacement of some nucleotides by others. So far, this unexpected happening has been found chiefly in the mitochondria of plant cells and in those of the organisms called trypanosomes, the infectious agent of sleeping sickness. The result is that a gene eventually produces protein molecules whose structure is not exactly that specified by the DNA. The puzzle is to know why these changes, which are presumably advantageous to the organism, have not been incorporated in the genes themselves, thus avoiding the need for editing by way of afterthought—not to mention the need for a separate biochemical mechanism for carrying it out. Much the same question is provoked by the recognition that even in mammalian cells, the RNA molecules produced by some genes may be processed in the nucleus in more than one way, meaning that a single gene may yield more than one protein product.

Protein Assembly

It has been known since the early 1960s that RNA molecules are converted into the equivalent molecules of protein in the cytoplasm of the cell by the structures called ribosomes. Progress is slowly being made in unraveling the roles of the two RNA molecules and the several protein molecules known to be involved. A decade or so from now, the functioning of the ribosomes may be so well described that it will be thought to have been understood. Not the least taxing of the questions then asked will be how the combination of natural selection and adaptation hit on this particular way of assembling protein molecules. Both ribosomes and the small RNA molecules (tRNA) used in all cells to scavenge amino acid molecules from the cytoplasm, physically to carry them to the ribosomes, are obvious relics of an RNA world preceding the evolution of early life.

External Connections

Cells may be autonomous in that they are adapted to carry out the functions required of them without external direction, but that does not remove the need for external communication. At the very least, they must take in food or some other source of energy, but being part of an organ in, say, the body of a mammal also requires adjustment to the needs of a whole aggregate of cells. The function of a cell's membrane is to separate its contents from the chemically hostile exterior; the membranes evolved by natural selection serve that purpose excellently, with the result that only very simple molecules such as water or carbon dioxide can pass freely into a cell from the exterior. Even protons cannot.

Communication between a cell and the outside world therefore depends on protein molecules embedded in the external membrane. The past few years have uncovered a huge variety of such molecules. Some are physically shaped to form channels in the membrane, allowing specific chemical molecules (glucose, for example) to enter a cell. Others span the cell membrane in such a way that particular molecules can react with the part of the molecule lying outside the cell, and with the consequence that the part of the molecule lying inside the cell brings about a significant change of function. Still other transmembrane structures are adapted for the excretion of particular chemical components (sodium ions, for example). Collectively, these structures are known as *channels* or *receptors,* as the case may be. Cells from different tissues or organs have different complements of these molecular components on the external surface, and are distinguishable from each other by that means. There is some way to go before the catalog of receptors and channels is completed for any single organism.

Differentiation

Even within a single multicellular organism, cells differ from each other, often remarkably. Who would think that a human nerve cell and a human skin cell belong to the same species? And who would believe that they, and the muscle cells, the bone cells and those of the blood, are all derived from a single cell—the fertilized egg which became the embryo from which a whole adult has grown?

The relationships between these different types of cells were studied and to some extent established even before the structure of DNA was put forward in 1953. The process is called, for obvious reasons, *differentiation*. Recent years have shown that the process is a continuation into adult life of the ordered manner in which embryos develop, otherwise known as *ontogeny*. To a large degree, molecular biology has united these two previously separate fields of study.

The essence of what happens was plain at the end of the nineteenth century. At a very early stage in the development of an embryo, the cells destined to become the reproductive organs of the adult are physically separated from the others, while the embryo acquires regions destined to be the top and bottom, or head and tail, of the adult. The classical experiments showed that moving a few apparently identical cells from one place to another would lead to the development of an abnormally shaped organism, proving that the transplanted cells had already required a sense of where they belonged. But it was then unclear how distinct properties were embedded in the cells concerned.

The triumph of recent years has been the discovery of molecules that interact with immature cells and thereby endow them with the properties necessary to mature into the full repertoire of the cells required to sustain an adult. The process is strictly hierarchical, which is apparent in the formation of the 20 or so different types of cells circulating in mammalian blood. All of these are derived from cells in the bone marrow which have the potential, on division, to form cells specialized in some function or another. Red cells, for example, are packed with hemoglobin, but lack a nucleus and so cannot renew themselves. Other blood cells, for example the white cells called macrophages, are adapted for the removal from the blood of foreign objects, dead or even live bacteria perhaps, while the cells called lymphocytes play a crucial part in the immune system's defense against infection.

At the outset, the molecules responsible for this hierarchy of specialized transformations were only vaguely characterized, and were usually called "factors" of one kind and another. Now the molecular constitution of most of the materials is fully defined. At any level in the hierarchy, cells must of course carry on their surfaces receptors that will allow them to respond to the regulatory chemicals that signal the need for further specialization. In mammalian blood, the regulation of the populations of different cell types must be immediate. During infection, for example, lymphocytes are copiously produced by the secretion of the appropriate "colony stimulating factor."

But how does a specialized cell, say a lymphocyte or liver cell, maintain through several cell divisions the differentiated character it has acquired? There are several possible answers. The hallmark of a differentiated cell is that only some of the whole complement of genes can be active. Both in bacteria and yeast, particular protein molecules can suppress the activity of some genes. If enough of this regulatory protein is passed onto daughter cells at cell division, they will then function as did the differentiated cell. The more probable mechanism of differentiation is that one set of genes is permanently switched off by a chemical modification of the DNA[32] which has the property of being faithfully replicated in the daughter cells formed at cell division.

The unraveling of the mechanisms of cell differentiation in the past few years is one of the reasons why molecular biologists sense that their science has become an open book. The genes whose protein products in a developing embryo give a sense of place to the cells of a mature animal or plant—the mechanism by which some segments of a developing insect produce wings and others legs, for example—have been mostly identified. In principle, development from embryo to adult is now understood. It remains to understand what exactly happens.

And there lies the rub. The details will not quickly fall into the catalogs. There are several different types of cells in the human body, each has acquired its differentiated character by a distinctive route, involving perhaps as many as a dozen different steps of hierarchical subspecialization. Inevitably, many of the proteins regulating these specializations will have complex interactions with each other. The task of coupling them together in an account of the development of, say, a human being will outdo the intricacy of even the cell-division cycle.

The practical benefits of a biochemical map of the process of differentiation will nevertheless be immense. Although the process seems to be irreversible, there is every likelihood that some of the mechanisms that make cells specialized will be reversible once the details are understood. The outcome could well be techniques for regenerating tissues of one kind from organs of another, avoiding in the process the difficulties arising in transplantation surgery over the compatibility of organs from other individuals.

Once again, it seems, molecular biology has created a huge amount of data, all potentially interlinked. Intricacy abounds, but what is to be done to make the data intelligible?

MODELING LIFE

In other fields of science, it is well understood that such complexities can be accurately understood only by constructing quantitative, often mathematical, models. When, for example, will a cloud in the sky turn to rain? Despite appearances, clouds are not static entities, but are in a state of dynamic balance in which water droplets or particles of ice move vertically, both up and down, under the influence of the temperature at the base of the cloud and at its top. Meteorologists have worked out rules of thumb to predict when water droplets will begin to fall as rain from the base of a cloud; the cause is that the dynamic balance has become unstable—and only careful modeling can reveal all the influences that matter.

Much the same is true of current understanding of epidemic outbreaks of infections such as measles in urban populations, which arise because of instabilities in the dynamic balance between those who are vulnerable to infection and those who are immune. The complex balance between the influences that propel and, alternatively, inhibit the passage of cells through the successive stages of cell division are essentially similar.

Despite the energetic effort now under way to identify the chemical components of cells, little has been done to construct models that would help to explain what the data mean. The cell-division cycle and differentiation are but two subsystems in the typical cell. There are also vital processes such as the replication of DNA, its transcription, the role of ribosomes in producing protein molecules, the segregration of proteins into different parts of the cell, the maintenance of the external membrane and so on. And there is the largely untouched question of how all the subsystems hang together.

One of the most urgent needs is to make comprehensible the complexities of the response of cells to external influences, called *signal transduction* in the trade. That, again, is all done by molecules. The general principle is that signaling molecules from other cells, hormones such as insulin in the human body for example, interact with the specialized molecules called *receptors* embedded in the membrane of a cell and spanning the space between the outside and the inside. An interaction occurring outside the cell induces a physical change in the part of the receptor molecule lying within the cell. There follows a cascade of chemical tranformations of molecules within the cytoplasm, resulting in a particular gene or genes being transcribed in the nucleus. What is entirely puzzling, and will remain so until somebody has built a realistic model of the process, is how the specificity of the cell's response matches that of the external signal it receives.

Does this mean that the exploration of the working of the cell is about to become a branch of mathematics? Not quite yet. One obstacle is that the techniques that have so admirably proved suitable to the naming of the parts of cells in the past quarter of a century are less good at collecting quantitative information about the working of the molecules involved in vital processes.

Another obstacle is psychological: those working in the field are disinclined to follow that path; there are still so many remarkable things to be discovered about the working of protein molecules in cells. In 1995 and 1996, for example, a number of proteins were found that assemble as ring-shaped structures around other protein molecules or even around strands of DNA; their function seems to be to hold their target molecules in place while the target is manipulated by another protein enzyme.[33] They are nothing but molecular vises. While there are still enthralling tales like that to tell of the inventiveness provoked by natural selection, molecular biologists will not lightly abandon the enthusiastic naming of the parts. But in the end, there will be no choice.

Fortunately, a few brave spirits have taken up the challenge. To the extent that cell division is a rhythmic, or at least a repetitive, process, it has something in common with the 24-hour clock by which daily life is ordered by living creatures as different as bacteria, plants and people (who complain of jet lag when they have to reset their biological clocks). So there is a long-standing interest among biologists in rhythmic processes—familiar in the physical sciences as *periodic motions* (such as the revolution of Earth about the Sun or even vice versa).

One of the pioneers in the transfer to biological systems of what had been learned of oscillating mechanical systems is the British biologist J. A. Winfree, whose first models of oscillating events in biology go back a quarter of a century. In 1996, Albert Goldbeter from the Free University of Brussels, published a book[34] that summarizes what has so far been done (much of it by himself) to model repetitive phenomena such as cell division. The book is a landmark in two senses: it demonstrates how much can be learned from modeling about real problems in biology —and it shows how little has been done in a quarter of a century.

Goldbeter's study of the strange behavior of a eukaryotic organism is a marvelous example to us all. The slime mold *Dictyostelium discoideum* is made up of single cells (amoebae) that depend on aquatic bacteria for food. When food is scarce, however, up to 100,000 of them aggregate into colonies looking a little like small slugs, which may then organize themselves into upright structures supporting a spherical ball from which spores are released into the environment—much as from a mushroom.

What attracts normally free-living amoebae together, and keeps them as a colony? A molecular signal, inevitably. In 1967, the molecule was identified as *cyclic AMP* (written cAMP), which is nothing but a slightly rearranged version of the familiar molecule of ATP robbed of two of its three phosphate groups.[35] Single cells are adapted to move toward the source of this simple chemical, and after a time begin secreting the same material into their environment. Laboratory studies have shown that the process of colony formation has an unexpected rhythmic quality; successive waves of single cells move towards the aggregating center every few minutes or so.

And the explanation? Single cells have receptors on their surfaces for recognizing cAMP; how else could they be attracted to the source? But having recognized the call to help form a sluglike colony, they are stimulated to produce more cAMP themselves and to secrete that into the environment; that has the benefit of amplifying the original signal for the benefit of more distant cells. But the nearest receptors will be those on the surface of the cell itself, which will thus be further excited to produce still more cAMP.

In the language of physics and engineering, this is a feedback system —its response to an external influence is to modify that influence in some way. In this case, in response to external cAMP, cells increase the amount of cAMP in their environment. It is analogous to the way in which, with a freely swinging pendulum, the velocity of the swinging bob is *increased* (by gravity) whenever it is already *increasing* (and *decreased* when it is already *decreasing*). It is therefore not surprising that the single cells of *Dictyostelium* during aggregation into colonies should show rhythmic behavior.

Why then bother with tedious mathematical equations to model the oscillations of such a system? For several reasons. First, the equations lead to predictions of the behavior of the slime mold under exceptional conditions—chaotic aggregation. Second, predictions of the aggregation behavior when large amounts of cAMP were added to the environment were not confirmed, leading to the idea that the receptors on the cell surface became desensitized to cAMP under these conditions, which in turn led to speculations (eventually borne out by experiments) that similar receptors in human somatic cells that signal the presence of hormones such as insulin would also become less sensitive with prolonged use. (Human responses to narcotic drugs take a similar form.) This seemingly recondite work also proved to be a clue to the discovery that cAMP is one of the small group of small molecules which are active *within* cells of all kinds as agents of signal transduction.

Goldbeter also has a model of the cell cycle as, for that matter, do other people, notably John M. Tyson at the Virginia Polytechnic Institute. At this stage, neither model has much predictive power, probably because the oscillating systems they describe are essentially autonomous, and are not coupled to the rest of the essential machinery of the cell— the genome on the one hand and the general state of metabolism on the other.

The research community holds two views of these models. One is that it is a considerable achievement to have simulated, however crudely, the functioning of a phenomenon as complicated as the cell-division cycle and that the next step must be to make the models more realistic by extending them to embrace the other proteins known to be involved. On this view, the crude models now devised are potentially the centerpieces of much more elaborate schemes that will eventually encompass the whole working of the cell. Biologists habitually shudder at the prospect.

They are protesting at the inevitability of the unwelcome. Such models are necessary not merely to make sense of what has been learned in 25 exciting years, but to make use of it as well. Building such a model would be a gigantic undertaking. The difficulty is not simply the large number of the quantities and equations that would be linked together; some of the models already built for simulating the chemistry of Earth's upper atmosphere already involve some thousands of equations. The more serious difficulty would be the need for detailed knowledge, which would have to be collected by experiment, of the interactions of real-life molecules with each other. Several decades will pass before such a project will be feasible, but the outcome would be a much more stringent way of testing ideas about the working of the cell. It would be well worthwhile making a small beginning even now.

The research community's other and predominant view is that modeling is an arid exercise, merely a way of putting what is at present known about the workings of the cell into a needlessly complicated and unfamiliar language. That opinion overlooks the way in which models help make a coherent story from disparate sets of data that benefits not only practitioners, but bystanders as well. And models are also pointers to the design of experiments that test current understanding incisively. Cell biologists' collective neglect of modeling has become a grievous handicap.

One result is that the record of the exciting study of the working of the cell seems to be an aimless anthology of anecdotes: "Imagine, *this* molecule does *this* to *that* molecule!" One obvious defect of this rapidly growing collection of particular laboratory observations is that it is almost certainly incomplete; almost always, further experiment will make it necessary to add, *"this* molecule also does *something else* to several *other molecules."* That is yet another reason why model building, mathematical or otherwise, has become a necessary means of making sense of what has already been learned. That it is potentially a way of finding out what further elements of learning are required will in due course also sink in.

THE PHYSICS OF BIOLOGY

There is a separate and more teasing obstacle facing those with the ambition to understand why cells behave as they do: accounts of how some molecules interact with others are simply fairy stories. What happens can be described; why it happens remains a mystery. Ask a yeast geneticist exactly why the molecular targets of cyclin molecules (known as "cdk" molecules) are more effective at activating others when they are coupled. "Something to do with the structure," he or she will say, before inviting the next question.

The plain truth is that there is as yet no single protein molecule of any complexity whose mechanical action is as clear to ordinary chemists as how equal numbers of molecules of hydrochloric acid and sodium hydroxide will "neutralize" each other. (H^+ions from the acid and OH^-ions from the alkali are supposed to form water molecules, leaving a dilute solution of Na^+ and Cl^-ions.) There are no such tales to tell for protein enzymes, which are commonly supposed to change their shape when they catalyze a chemical reaction in a cell, perhaps splitting into two the molecules to which they are bound. Of course, cell biologists have been busy on other things—identifying the genes whose products regulate the cell cycle, for example. But the longer ignorance of how protein molecules function persists, the more like a scandal it will seem.

People who skate on thin ice dare not pause to think how thin the ice may be. That is how it is with those who are busily naming the parts of cells and identifying the linkages between them. They are forever proclaiming the discovery of a new part, or describing a new linkage, expecting that the announcement will in itself be regarded as an explanation. They are not often disappointed, because their most severe potential critics are in the same dilemma as themselves—eager to pass off the identification of another molecule that plays a crucial part in the working of a cell as a shaft of light that illuminates the mystery of life itself.

The physics of the problems faced by cell biologists are nevertheless horrendous, chiefly because real molecules in cells are immersed in the most peculiar material: water. Although water molecules are among the simplest there are, consisting of just a single oxygen atom and two hydrogen atoms represented by the formula H_2O, a drop of liquid water must not be regarded as a crowd of molecules thrown together like a collection of golf balls in a bag. Instead, the identity of the individual molecules is endlessly compromised by the difficulty of telling whether a hydrogen atom belongs to one oxygen atom or to a neighbor.

That is a direct consequence of quantum mechanics; telling whether a hydrogen atom belongs to *this* molecule or *that* is exactly like the difficulty of telling whether an electron is *here* or *there*. The result is that water molecules are held together more strongly than would be expected,[36] by what are called "hydrogen bonds." Exactly the same mechanism comes into play when other simple molecules are immersed in water, which explains why substances such as alcohol and sugar are without restraint mixable with water. (In each case, hydrogen atoms attached to oxygen atoms mimic half a water molecule.)

All proteins respond a little like that in water; indeed, the characteristic link between amino acids generates at least two hydrogen atoms (attached to a nitrogen atom and an oxygen atom respectively) that can be likened to a one-sided water molecule. But protein molecules have complexities of their own; some amino acid units contain further peripheral sites for making links with the surrounding water network, while all the peripheral hydrogen and oxygen atoms can make links between one amino acid and another, possibly a long way away from each other.[37]

No wonder, therefore, that the problem of protein structure in the real world remains unsolved, even after 25 years of effort by dedicated research groups across the world. In principle, the shape adopted by a protein molecule in water or in a cell should be a consequence of the order in which amino acids are strung together along its length; would not such a molecule simply curl up into the least energetic shape? The immediate trouble is that the energy of the most favorable arrangement is usually not very different from that of many different arrangements; by accident, molecular biologists have rediscovered one of the most difficult problems in mathematics, as will be seen later. Nature (and evolution) appear to have anticipated the difficulty. Why else would there be molecules in cells whose function is to ensure that other protein molecules fold correctly?[38]

Explaining the interaction between different protein molecules in the working cell is even more difficult than predicting shape from a sequence of amino acids. It is not yet possible to tell how such a complex problem will eventually be brought within the compass of the analytical techniques used successfully in other fields of science. Nor would any-body pretend that the full rigors of the techniques used elsewhere to calculate the interactions between molecules could (or should) be applied as a matter of routine to the complex interactions between molecules in cells; all the world's computers are not yet powerful enough for that. But even a few demonstrations that the special properties of proteins are determined by the way their atoms are linked together, possibly with the intervention of water molecules, would help to bridge the present gap of understanding. In their absence, there is a danger that empirically founded statements such as "cyclin molecules bind preferentially to cdk," will continue to be mistaken for an answer to the question of what it is about the structure of cyclin and cdk molecules that ensures that outcome.

Some help may come from a technique called *molecular dynamics* (one of whose chief exponents is Martin Karplus of Harvard University), which makes it possible to predict the conformation of molecules and even the strength of their interaction with each other, at least if the forces between all pairs of atoms can be represented by a formula of some kind. There are also techniques by which, in principle, the interaction between two molecules can be described even after allowing that the presence of one will affect the movement of electrons in the other, and vice versa. Both these techniques are great thieves of computer power. Sheer cost explains why so little use has been made of them. But there is more than that to say. While the naming of the parts of the working cell yields such rich dividends, computation of any kind is lumped together with model building as the parasitism of the noble creature that is biology.

The point is well made by Walter Düchting from the University of Siegen in Germany, who with two colleagues has built a computer model of a cancerous tumor and its response to radiation therapy and surgery.[39] One purpose is the better design of these therapeutic techniques by making predictions that can be tested, initially in experiments with animals (although the authors look forward to the time when computer simulation will be both cheaper and quicker than the use of laboratory animals for testing). They conclude, however, that:

> An important precondition *(conditio sine qua non)* of all modelling activities is a stepwise reduction of antipathy against the systematic modelling approach which is created by scientists predominantly working empirically. The authors sincerely hope that this review will help to inspire more confidence in modelling.

How long will it be before that prejudice is banished in the face of the imperatives of understanding?

THE WAY AHEAD

The case for a more quantitative approach to the working of the standard cell goes deeper than the modeling of the cell-division cycle and the other subsystems in a working cell. All living things are aberrations in the sense that they do not conform to the second law of thermodynamics as it applies to isolated systems. They are not in equilibrium with their surroundings in any ordinary sense, for then they would be dead and decomposed. They are sustained in their exceptional condition only because they are intermediates in the conversion of flows of energy from one form into another. Plant cells absorb sunlight, producing low-grade heat and atmospheric gases in return. Animal cells take in food that is ultimately derived from plants and excrete chemicals of lesser complexity. Only the flux of energy from the Sun makes life possible.

For practical reasons as well as philosophically, we need a better understanding of the relationship between the output of energy from the Sun and the complexity of the biosphere on Earth—not simply of the intricacy of the construction of particular species but of their mutual interdependence. Over the years, Ilya Prigogine at the Free University of Brussels has been seeking a philosophical framework within which these questions could be accommodated. Others believe that the study of complex systems will eventually yield an understanding of how a flux of energy through the biosphere leads to the evolution of organized life-forms of greater or lesser complexity. Yet an understanding of this relationship between the Sun's radiation and the diversity of life on Earth remains beyond reach.

No discipline in science has ever ridden as high as cell biology does now. It is also remarkable that so much has been learned about cells and their self-regulation in the past decade or two. But the general scorn of model building is both an error and a serious waste of laboratory time. The desultory character of the search for how proteins really function is similar in kind, but may actually impede understanding of how cells function in real life. If cell biology becomes a part of physics, it will have only itself to blame.

The Genome and Its Faults 6

The practical and intellectual consequences of the structure of DNA are without precedent in the whole of science. When Copernicus put the Sun and not Earth at the center of the solar system, he set in train a program of inquiry that has still to yield an understanding of how the universe is constructed. But the structure of DNA provides not just an understanding of the mechanism of inheritance and of the origins of Darwinian variation, but seems to have made it possible to answer any question about the mechanism of life. The genes in living things assure both that organisms survive from minute to minute and are perpetuated from generation to generation. That idea, that the genes must have both functions, was a long time distilling out of classical genetics in the first half of this century. It first became crystal clear in 1944, in a book by the physicist Erwin Schrödinger *What Is Life?*[1] Now, with the understanding of the role of DNA and its chemical cousin RNA, there seems no other plausible way of designing the machinery of life.

In 1953, a plausible model of the atoms in molecules of DNA was published by James D. Watson and Francis H. C. Crick, then two young men at the Cavendish Laboratory at Cambridge. The goal was to identify the seat of inheritance. Everything that has since been learned has confirmed the accuracy of the model. That it should also have revealed the mechanism by which living things function seemed at the time to be a kind of extra.

What are the implications of this view of life for the mechanism of inheritance? To a good approximation—a very good approximation—we are the products of our genes. They are replicated when our cells divide and replicated, sorted and rearranged when we beget offspring of our own. That is why molecular geneticists are now hunting through the genomes of several organisms, seeking not merely a deeper understanding of human physiology, but also a better understanding of the evolutionary relationships between species.

For the time being, the old argument about the relative importance of *nature* and *nurture* in the development of the human beings seems to have been settled in favor of *nature*. But that is almost certainly an illusion. It may not be long before the external influences on the genes are well cataloged enough for the importance of *nurture* to become apparent again.

How the Genome Works

The model of DNA has been a stunning success. Each strand of a duplex molecular helix of DNA is built from chemical units called nucleotides, which are chemically linked into chains. Each nucleotide is a composite molecule in which the distinctive component is a chemical entity called a base. Famously, there are just four bases in natural DNA, most often known by their initial letters A, T, G and C. Genes, and indeed whole chromosomes, consist of two DNA molecules placed side by side in register and held together in a right-handed helix by the propensity of the bases to form hydrogen bonds with each other.[2] In a duplex molecule, A on one strand and T on the other invariably pair together, as do G and C, which (because A and G are bigger than T and C) ensures that the resulting helix is geometrically regular.[3]

Each strand of DNA has *directionality,* which comes about because the end of one nucleotide makes chemical links with an intermediate part of the next.[4] In a duplex molecule of DNA, the two strands have

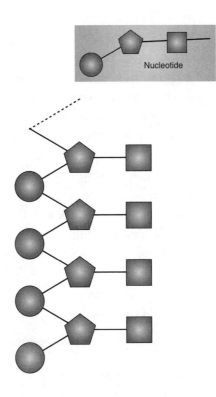

Figure 6.1 Directionality in DNA A schematic drawing of the coupling between nucleotides in a single strand of DNA. Each nucleotide (see inset) consists of phosphate (circle), deoxyribose (pentagon) and a purine or pyrimidine base (square). Directionality arises because successive phosphate groups link to specific and distinct atoms of the deoxyribose. The drawing is schematic because the nucleotide atoms do not really lie in a plane; if the phosphate groups are imagined to lie beneath the plane of the deoxyribose molecule to which they are attached, the tendency of the strand to form a right-handed helix can be visualized.

opposite directionality. RNA molecules are also constructed from four nucleotides, but the base called T in DNA is replaced by another called U in RNA.[5] At least in modern cells, RNA molecules are not produced in duplex form, but often acquire a characteristic shape by forming internal hydrogen bonds and often short stretches of double helix where the bases are complementary to each other (in that A pairs with U and G with C).

Why is directionality important? For one thing, it defines the direction in which genes are processed, as when part of the DNA strand is transcribed into an RNA molecule carrying the same genetic information. But the head-to-tail arrangement of the two DNA molecules means that each strand can carry genetic information that can be processed by the same enzyme molecules working in opposite directions.

Watson and Crick, in a celebrated sentence,[6] recognized in 1953 that their structure was a model for inheritance. Organisms must be able to replicate duplex molecules of DNA when cells divide, so equipping each of the two offspring cells with essentially the same DNA. When organisms reproduce asexually (as with bacteria and many plants) that is sufficient. Sexual reproduction is more complicated, involving in the act of fertilization the union of two cells (called *gametes)* containing only half as many chromosomes as ordinary body cells *(somatic* cells). So much had been established in the 1920s, long before the structure of DNA appeared, but the duplex character of DNA is in itself suggestive of the pairwise complement of chromosomes in somatic cells.

The bare bones of how the genome functions are now made plain. The genes are identifiable stretches of DNA embedded in the circular chromosome of a bacterium or the chromosomes of eukaryotic cells.[7] During the life of an ordinary, somatic, cell they can be transcribed into molecules of RNA carrying the same genetic information (corrupted, perhaps, by occasional accidents). Either the RNA molecules are functional in their own right, or the information they embody is used to direct the synthesis of protein molecules in their own image. The outcome is that each cell in the body of an organism behaves in a manner that reflects the arrangement of the bases A, T, G and C in the nuclear DNA.

Inheritance is similarly straightforward. In single-celled organisms, the genome is replicated early in cell division and the offspring cells go their separate ways with almost faithful replicas of the original genome. The same is true of multicelled organisms that reproduce vegetatively, but sexual reproduction is more complicated.[8] For one thing, the tissues of the germ line are separated from the somatic tissues from an early stage in an embryo. Moreover, the cells of the gonads are prone to a distinctive process of cell division called *meiosis* in which the number of chromosomes is reduced by a half. In animals, the outcomes are sperms (in males) and ova (in females).[9] Consequently, an embryo springing from sexual fertilization has a full complement of chromosomes, each in a pair derived from one or the other of the parents.

Pairs of chromosomes look similar to each other under a microscope, but there is an important exception. People, for example, with a total of 46 chromosomes, have 22 outwardly similar and functionally identical pairs and two dissimilar chromosomes, called the *sex chromosomes* and labeled X and Y. (The X is larger than the Y.) In mammals, the somatic cells of females contain two X chromosomes; the somatic cells of males, on the other hand, contain both an X and a Y. (In birds,[10] the rule works backward: male birds have two similar and females two different sex chromosomes.) It emerged in the 1980s that, in males, the Y chromosome carries a gene that causes the embryonic germ-line cells to form male gonads; if that gene is inactivated, even XY individuals develop as females.[11] In computer language, to be female is the default condition.

By the 1970s, after two decades of accumulating knowledge of the properties of DNA, it seemed only a matter of a short time before there would be an understanding of how the genome of a species orchestrates the fine details of inheritance. But that was not to be.

THE WHOLE OF INHERITANCE?

Several surprises have come to light. One is that even in bacteria, inheritance is not determined solely by the single circular chromosome. In bacteria there are also much shorter duplex molecules of DNA which are replicated separately from the main chromosome; they are passed on to the daughter cells at cell division. *Plasmids,* as they are called, use the cell machinery of their hosts for their own reproduction, but do not have to wait on the cell cycle to do so. They are parasites which have no existence outside the cells that harbor them. And they enormously complicate bacterial inheritance. Their genetic material may include genes that make bacteria resistant to particular antibiotics and genes responsible for the exceptional virulence of otherwise innocuous bacteria. How did they and their genes get there in the first place?

Too little is known of the natural history of these parasitic elements for our comfort—and for our safety from infection. Because plasmids consist only of DNA, there is no obvious way in which they can migrate from one cell to another. But they do not stay in the cell in which they are lodged and in whatever progeny it may produce. In the real world, the genetic material of the plasmids is repeatedly transferred between bacterial cells—even between cells of different species.

Viruses are also agents for the promiscuous transfer of genetic material between cells. Bacteria (like people) are plagued by viruses consisting of a package of genetic material surrounded by a coat of protein molecules whose structure is specified by the genes they carry—but which must be manufactured by the biochemical apparatus of the infected cell. That can be done in two ways: either the virus replicates as if it were a plasmid, or it insinuates itself into the chromosome of the infected cell and is replicated with it whenever it divides until it breaks out again as a virus. When that happens, the genome of the virus may be carrying part of the genetic equipment of the infected cell which can then be transferred to some other cell, even one of a different species.

Understanding the "horizontal" transfer [12] of genetic material in bacteria is (or should be) an important goal of public health. For bacterial viruses to be effective in this role, they must be able to infect more than one species. Many do.[13] The result is that inheritance in bacteria does not depend exclusively on the structure of the genome. Genetic information can also be transferred horizontally among bacteria during the phenomenon called *conjugation*. This happens when distinct bacterial cells briefly merge, exchanging some genetic material in the process. Far too little is known of these shadowy happenings, which may have been crucial in the early evolution of life.

Bacteria are not the only organisms to have acquired genetic information horizontally, from other species. The mammalian genome is full of evidence of the import of genetic material from elsewhere, possibly by viruses. It is not yet known whether this transfer of information has been of central or marginal importance in the course of recent evolution, nor is it known whether these processes are still active. Although the genome of each species is a good approximation to the mechanism of inheritance, it is certainly not the whole of it.

The supremacy of the genome of eukaryotic organisms is further attenuated by the mitochondria and chloroplasts in their cells (the latter only in plants). Both types of organelles have a rudimentary genome of their own together with a stripped-down version of a cell's biochemical equipment for making proteins. It is significant of something, although nobody knows of what, that neither kind of organelle has remained self-contained: some of the genes coding for proteins that are essential for the functioning of the chloroplasts and mitochondria are found in the genome, presumably because they have migrated there for the greater good of the economy of the cell.

The genetic endowment of any lineage of eukaryotic cells must include the mitochondrial genes, some of which have been linked with human disease. The mitochondrion comes into its own in sexual reproduction, because sperm, adapted for efficient fertilization, do not carry mitochondria. Ova, by contrast, are adapted to get the eventual embryo off to a good start. During their often long maturation in the ovaries, they accumulate enough of the general contents of a cell for the first few or even several cell divisions which have to wait only on the replication of DNA at every cycle. Among the general contents are the mitochondria, perhaps enough of them to satisfy the needs of 1,000 embryonic cells.[14] Mitochondrial inheritance is thus unambiguously *maternal inheritance*. That may not be of great practical importance in human medicine, whatever its symbolic significance. But we may be surprised on that opinion.

Even genes are not as simple as they seem. That is aptly illustrated by the discovery, in 1978, of an entirely unexpected feature of the organization of the genes in eukaryotes. In the single circular chromosome of bacteria, the genes are arranged, anthropomorphically speaking, sensibly and economically. They are laid out in groups with a common function (in which case they are arranged head to tail), with gaps between them not significantly larger than are required to provide a foothold for the protein molecules that activate and then transcribe them into RNA. The beginning of each gene is marked by one of two specific triplets of nucleotides.[15] The end of the gene is similarly marked by a *stop* triplet; transcription stops there. Who would arrange things differently?[16]

Eukaryotic cells seem to have taken an alternative path. In 1978 it was found that in their chromosomes, individual genes are arranged in pieces separated from each other by stretches of DNA apparently irrelevant to the protein molecules eventually made from them. The whole length of the DNA is transcribed into a molecule of RNA, the apparently irrelevant pieces (called *introns*) are removed, and the pieces used in the eventual translation of the protein (called *exons*) are then joined together correctly. One bizarre case of such a gene is *BRCA1*,[17] which is involved in most cases of inherited propensity to breast and ovarian cancer (and in perhaps 5 percent of all breast cancer cases). The whole gene consists of 100,000 nucleotide bases, but there are no fewer than 22 exons comprising exactly 5,592 nucleotides in aggregate. In other words, only one twentieth of the total gene is functional.[18,19]

Then in 1981, Thomas Cech at the University of Colorado at Boulder demonstrated that, at least in the protozoa called *Tetrahymena,* the unwanted introns are excised by a previously unsuspected enzymatic activity of the RNA molecules of which they are initially a part. (That partly sustains the now-common conviction that catalytic function of RNA must have been relevant to the origin of life on Earth.)

By any standards, this is a strange business. The general strategy for making proteins in eukaryotic cells seems to follow that evolved in bacteria: transcribe or copy the genetic information in a molecule of DNA into RNA and then use that molecule for running protein synthesis. The complication in eukaryotic organisms is that the unwanted segments have to be removed—in some genes in some organisms by the RNA itself. However the job is done, the enzymes responsible must be able to recognize where to take the appropriate chemical scissors to the length of the RNA molecule.[20] At the very least, this complication is an extra metabolic cost for eukaryotic cells. It is also potentially a source of error. What countervailing selective advantage can there possibly be in this arrangement?

As yet, there is no convincing explanation why most eukaryotic genes are split genes. Instead, there are only speculations. One (offered by W. Gilbert of Harvard University) is that the exons represent primitive stand-alone genes from earlier organisms that were recruited into eukaryotic genomes early in the course of evolution, and whose proximity to others made it possible for organisms to make sophisticated protein molecules by running two or more such genes together. There is some evidence, not entirely compelling, that such elements can be seen to have survived more or less intact in the structure of modern protein molecules. A more plausible line of argument is that the persistence of introns in modern genes has something to do with the way that molecules of DNA are packaged in linear chromosomes. The DNA molecules are twisted into a double helix, but the helixes are then wound onto roughly spherical packages of protein molecules called *histones,* with two windings of the helix for each protein package (called a *nucleo-*

some).[21] On the molecular scale, a chromosome is a little like a necklace of nucleosomes held together by a thread of DNA wound on the outside of the structure.

The functional significance of this complicated structure remains a riddle. One obvious awkwardness of the arrangement is that individual genes must be partially unwound if they are to be transcribed into RNA, but that at least suggests a possible role for the introns: their function could be to provide spacers between the exons of a gene so as to simplify the unwinding. However, in the present state of ignorance of this vital feature of the organization of chromosomes, that is mere speculation.[22]

The true selective advantage of organisms with split genes may lie in the errors arising in the excision of the unwanted parts of RNA molecules and the joining of their functional parts together again. Some error is inevitable, but it is also a way of generating protein molecules not otherwise produced. Here, too, the influence of the genome on what really happens in the cell is, to some degree, a matter of chance. And it is far from clear why some RNA molecules transcribed from genes are edited within the nucleus before being translated into proteins.

The ends of linear chromosomes are potentially of immediate medical importance. The circular duplex chromosomes of bacteria have no loose ends, so that duplex DNA molecules are not at risk of unraveling. The linear chromosomes of eukaryotes, by contrast, do have loose ends which, in the ordinary course of cellular life, would be expected to become untwisted and then degraded. That could result in the loss of essential genes. Eukaryotic cells have evolved a way of preventing that by means of an enzyme called *telomerase* whose effect is to add a length of genetically meaningless DNA to the end of each of the strands of a duplex molecule of DNA. But the extension of the strands is not symmetrical. One strand may end up 20 or 30 nucleotides longer than the other. What happens to the piece left over? It doubles back on the end of the chromosome, forming a short stretch of triplex DNA with extra stability. Sailors safeguard the ends of a piece of rope in the same way.

In 1995, a research group from the Cold Spring Harbor Laboratory on Long Island (New York) claimed that the telomeric ends of chromosomes become progressively shorter as the organisms to which they belong become older.[23] Yet cell biology is still looking for an explanation of senescence—the phenomenon of aging—which affects not only individual animals but cells maintained in laboratory culture. If these claims should be confirmed, it is easy to imagine the zeal and enthusiasm with which the international pharmaceutical industry will prosecute the search for an enhancer of telomerase production in everybody's cells.

A ROLE FOR JUNK

The genes in the genomes of eukaryotic organisms specify the individuals of the species concerned, but the genes constitute only a small fraction of the DNA in the whole genome. Simple arithmetic shows that. The 3,000 million nucleotides of the human half-genome embody between 80,000 and 100,000 genes,[24] which works out at 30,000 nucleotides for every gene. That is an overestimate of the length of the average gene, which is more like 1,000 nucleotides. In other words, it may be that only 3 percent of the human complement of DNA is functionally significant. Some of the remaining 97 percent or so of the human genome is given over to the sequences of DNA required for regulating individual genes; understanding of how this is done is far from complete, although it seems clear that two, three, four or perhaps more protein molecules must stick to distinct and particular regions near and sometimes within a gene before it can be transcribed. Other stretches of DNA are given over to telomeres, to centromeres (by which sister chromosomes are pulled apart at cell division) and to places in the chromosome that signal where replication of parts of the whole structure must begin. But that organizational overhead in the human genome cannot amount to more than, say, a further 2 percent of all the DNA, leaving the function of 95 percent or so unexplained. As distinct from bacteria, in eukaryotes the genes are not only divided into segments, but are separated from each other by such long and apparently meaningless stretches of DNA that it has been called "junk." [25]

The junk is not all without meaning. For example, some of the nonfunctional DNA is structurally similar to active genes, but lacks the places to which regulatory proteins normally stick, and is therefore inactive. The pieces of DNA may be evolutionary relics of once active genes. The striking feature of the remaining junk DNA is its repetitiveness. In the human genome, there are approximately half a million (500,000) copies of a substantially invariable sequence of nucleotides roughly 300 nucleotide bases long and called *Alu.* It occupies roughly 5 percent of the whole genome.

Do these repetitive elements have a function? And how, in any case, did they get there? The best guess is that they were at some stage in evolution genetic elements capable of moving from one place in the genome to another, duplicating themselves in the process. In other organisms, as different as the fruit fly *Drosophila* and the mouse, there is direct evidence that repetitive elements like these are capable of moving within the genome, sometimes disrupting functional genes as they do so. Called *transposons,* some of them even include in their DNA short stretches that correspond to enzymes (called *integrases*) whose function is to insert one stretch of duplex DNA into another. There is some reason to believe that the short *Alu* sequence is really a nonfunctional copy of a gene that plays a part in the processing of RNA molecules in the cell, and that the 500,000 copies of it now in the human genome have all arisen in the past 60 million years or so.[26]

Some of the transposons now in the genomes of modern organisms clearly derive from past infections by viruses. The genome of the infectious agent may become part of the genome of the host cell, either temporarily or permanently. The herpes virus that causes cold sores in the human mouth lies dormant in this way, as does the hepatitis virus that causes liver cancer in some of those infected by it. One of the many reasons why HIV cannot easily be detected in infected blood cells is that a DNA copy of its RNA genome has been incorporated into the host genome. It is striking that some of the nucleotide sequences of genes that now play crucial parts in human physiology have an uncanny similarity to the genomes of infectious viruses in quite different species.

The implication is that, while the role played by the genome remains central, it is neither the sole influence on inheritance nor the exclusive source of the biochemical recipes that direct the functioning of ordinary cells. The role of split genes and of transposons in the evolution of genomes as they now are remains unclear. Have bursts of transposition been responsible for past bursts of rapid evolution?[27] And may the very large number of repetitive elements in the human genome betoken some kind of long-term instability? To this legitimate and somehwat alarming question—there is no answer yet.

IMPERFECTIONS OF THE GENES

Long before the structure of DNA was known, genetics had been widely used both in the commercial breeding of plants and animals and in the understanding of human disease. Indeed, the search for the determinants of genetic disease was taken up by physicians in the 1900s. Towards the end of that decade, Archibald Garrod at the University of Oxford had drawn attention to the importance of what he called "inborn errors of metabolism," caused by inherited deficiencies of some vital enzyme.[28] That grew, by the 1930s, into the idea that each "normal" enzyme is the product of a "normal" gene—and that metabolic deficiencies are the products of aberrant genes. By the 1930s, that had become the doctrine of "One gene, one enzyme": the function of genes is to make enzymes. We now know that genes make other things as well —ribosomes, for example.

The use of the word "normal" is ill-advised. Genes are particular stretches of DNA at particular places in a chromosome, characterized by particular sequences of the bases, or the letters A, T, G and C. What happens when one of those bases is replaced by another (called a "point-mutation" in the trade) depends on which gene is involved and which base within it. A single substitution can lead to a change in one of the amino acids in the protein that results, but that does not always happen because of the inherent redundancy of the genetic code.[29] More commonly, a point mutation will lead to a change of the amino acid in the resulting protein without a perceptible change in its function as an enzyme. So variations of the precise nucleotide sequence of a gene are commonplace, run in families and have no physiological or medical consequences whatsoever.[30]

What, in that light, is the "normal" gene? The implications of the question are important for the language of modern genetics. Different nucleotide sequences are called *alleles* (of the particular gene). Ordinarily,[31] all individuals in a species have the same genes, but differ from each other in the particular alleles of the genes carried in their version of the genome. It is therefore incorrect to say that an individual, say a person, "is the product of his or her genes"; the alleles are what matter. Moreover, the characteristics of a particular population can be described only statistically, as the frequency with which the various alleles occur.

So far as is known, populations of African blacks and Caucasian whites all have the same genes, but differ in the frequency with which the alleles of the genes that determine skin pigmentation occur. To describe these differences as "genetic" is commonplace, but "allelic" would be more accurate. It is surprising that modern geneticists, who mostly and genuinely deny that their work has racial implications, and who bemoan public misunderstanding of what they are about, do not themselves use the more precise terminology they have invented.

All alleles are not equal. Some point mutations in a gene may have profound implications for physiology. The classic case is the disease called sickle-cell anemia, first recognized among West African blacks and the descendants of West African slaves in the Americas. The disease is a *recessive* allelic disease, striking only those who inherit the sickle-cell allele from both parents. Hindsight has shown that those who have inherited one sickle-cell allele (from one parent) are better able than those who have inherited none to resist infection by malaria parasites. Why then is it so common in populations with their roots in West Africa? Simply put, its prevalence is a kind of living fossil of the past effects of natural selection in a region where malaria was endemic.

Most inheritable genetic diseases, however, are unmitigated disasters for those who inherit them and for their families. That explains the excitement in genetics in the past two decades: techniques for identifying the alleles of genes associated with inherited disease are rapidly improving. Prenatal diagnosis is now possible, allowing mothers who so choose to avoid the birth of children handicapped by inherited disease.[32] By 1996, more than 100 inherited diseases had been linked with known alleles in such a way that prenatal diagnosis would be practicable.

So is it not just a matter of time before every inherited disease can be diagnosed in advance and then avoided? Unfortunately not. The essential difficulty is that there are several inherited genetic conditions in which not just a single gene, but several, are involved. For example, people may inherit a disease by inheriting aberrant alleles of one or another of two or more distinct genes. Familial breast and ovarian cancer is such a case. There are several known aberrant alleles of *BRCA1* (see above) that predispose a woman to the development of breast or ovarian cancer, but there is also another gene, called *BRCA2*,[33] whose several aberrant alleles have similar but distinct effects. Moreover, it is still not known what proportion of familial breast cancer is accounted for by these two genes; there may well be others. In circumstances like that, physicians cannot confidently offer unambiguous diagnosis to anxious potential parents.

Many inherited conditions are even more complicated. Some but not all diabetes, for example, is familial—several members of the same family may develop the condition. So far, three separate genes have been linked with the condition. The difficulty is that aberrant alleles of each of them seem to be required for predisposition to the disease. Prenatal diagnosis of the condition will be complicated (and expensive), and that is likely to be the case for most inherited conditions. The notion that it may ever be possible to carry out prenatal diagnosis for every inherited condition is an illusion.

The contentious issue of the inheritance of intelligence in human beings falls under that prescription. Quite apart from the disputed usefulness of the standard measurement of intelligence quotient (IQ) as a proxy for intelligence, there are at present no worthwhile ideas of how IQ is correlated with the necessarily biochemical functions of genetic alleles, aberrant or otherwise. What little is known about the higher functioning of the brain, however, suggests that exceptionally high intelligence would require that many conditions would have to be satisfied independently, and that as many genes would be involved. Environmental influences, both prenatal and in early infancy, are also likely to be important determinants of adult intelligence. Prenatal screening of fetuses for intelligence is about as likely as the invention of an antigravity machine.

The general truth is that when the inheritance of some characteristic involves a group of genes scattered throughout the human genome, the classical techniques of genetics may not be powerful enough to tell which are the genes involved. None of that diminishes the importance of the discovery in the past few years of genetic alleles linked with cruel diseases, nor the value of their prenatal diagnosis. The implication is that our capacity to shape our genetic constitution, at present limited to the avoidance of a small number of inherited diseases, will never amount to the complete design of offspring.

There is a further oddity about the character of human genes whose significance for the long-term future of *Homo sapiens* may be profound. The disease called Huntington's disease (HD) runs in families; its symtoms include dementia in middle life and death a few years afterwards. By means of a decade-long collaborative research program organized at the Harvard Medical School, the gene whose aberrant alleles are responsible has now been found and the aberrant alleles identified. The result was entirely unexpected. The gene concerned is called *huntingtin*. Little is known of its normal function except that it seems to be especially active in brain cells. One distinctive feature of the nucleotide sequence of the gene is that it begins with a series of repeated triplets CAG.CAG. CAG . . . and so on. If there are fewer than 40 such triplets (each of which codes for the amino acid called glutamine), the carrier of that allele will be healthy. But if there are more than 40 repetitive triplets, the outcome will be HD. And the greater the number of repetitive nucleotide triplets, the earlier in life the onset of dementia and the more rapid the course of the disease. Those who carry 80 or so repetitive triplets—twice as many as what seems to be the threshold between normalcy and disease—develop HD in their early teens.

In people free from the disease, the alleles of the *huntingtin gene* usually have either 15 or 17 of the repeating triplets. Where do the extra-long triplet sequences come from? It appears that they are added during the meiotic cell division that leads to the formation of the half-set of human chromosomes in sperm. The biochemical machinery for replicating strands of DNA, faced with a length of repetitive DNA, slips out of register and makes more of them than are on the allele it is copying.

The evidence so far is that the copying error may increase, but seldom decrease, the number of triplets. That explains what physicians have long known: when HD runs in families, those stricken in successive generations tend to succumb at an earlier age, and with more severe symptoms. Other inheritable diseases have the same property (called "anticipation") of seeming to become more severe as the generations pass. What seems to be amiss in HD is that the machinery for replicating DNA can cope with the standard complement of 15 to 17 repeating units reasonably well, but that occasionally there are accidents, the repetitive stretch of DNA is lengthened and then, at each generation, there is an inexorable increase of the amount of junk DNA at the beginning of the *huntingtin* gene until the threshold of about 40 repetitive triplets is reached, when symptoms of disease appear in the individuals concerned. That accounts both for the sporadic occurrence of HD and the phenomenon of anticipation.[34]

Whether the tendency of the repetitive triplets to increase in number, well documented in respect to HD, applies to the other diseases is not yet clear. Does the underlying error of the DNA-processing machinery betoken an inherent instability of the human genome, leading to an increasing burden of inherited neurological disease as the generations pass? In HD, even if the genes can only lengthen, but cannot shrink, the phenomenon of anticipation implies that within a particular family, HD will eventually die out; ever-longer alleles will cause those who carry them to die before they reach childbearing age (which shows that natural selection is still at work even in well medicated societies). Not enough is known of the other repetitive triplet diseases to know whether similar built-in safeguards apply to them, or, for that matter, what are the influences that determine the spontaneous mutation rate that leads to HD in the first place. Jean Weissenbach, the distinguished French geneticist, wrote in 1996, that "it is too soon to be alarmed about the stability of the human genome."[35] But the danger is one well worth watching.

As with the possible effect of transposons, there is at present no evidence, one way or the other, to suggest that evolutionary disaster is on the cards. But there is no reason why the genome of any species now alive should be indefinitely capable of faithful replication. The survival of a species and of its genome is entirely conditioned by its capacity to produce viable and fertile offspring here and now. There is no way in which the continuing adaptation of species to present exigencies can be informed by the future—either by foreknowledge of significant environmental change or by the recognition that the genome that sustains the species spells genetic catastrophe in the future. It would be bad luck, not least for our view of our place in nature, if the human genome had the potential for such an outcome. The prevalence of transposons in the genome and the tendency for repetitive triplet nucleotides to lengthen does not imply that the human genome is more vulnerable than that of other mammals. It is, however, a question to keep in mind.

NAMING 100,000 PARTS

The Human Genome Project, which is a loosely collaborative international effort to work out the nucleotide sequence of the whole human genome, may or may not provide more pointed clues to the long-term viability of *Homo sapiens,* although its chief purposes are more immediate. By the early 1980s, when it had become feasible to discover the sequence of nucleotides in whole genes, brave spirits across the world, but notably in the United States, began to nurse the ambition that the nucleotide sequence of the whole human genome might be quickly discovered. Walter ("Wally") Gilbert, the Harvard biology professor who had earlier shared a Nobel Prize for a scheme for telling the sequence of nucleotides in DNA, argued in 1985 that the research community could carry through the project commercially, collecting fees from pharmaceutical companies in return for information about the structure of the human genome, and using the surplus cash to sequence the genomic "junk" as well.

In the event, the project became a more deliberate enterprise. The international research community did what all beleaguered groups do when they cannot tell the best way forward: it created a committee, called the Human Genome Organisation or HUGO. The first chairman was James D. Watson, one of the co-discoverers of the structure of DNA. In the mid-1980s, the committee advocated beginning with simpler organisms—the fruit-fly *Drosophila,* yeast, the groundworm or nematode called *Caenorhabditis elegans* and *Arabidopsis,* or kitchen cress. The idea was to "prove the technology" for the large-scale sequencing of the human genome. The truth is that, in the mid-1980s, nobody could have been sure that the task would be technically feasible. Now the doubts have been exorcised; there will be a complete sequence of the half-genome by about 2005.

The project is gigantic. There are 3,000 million pairs of nucleotides in half a human genome[36] (the DNA content of a single sperm or ovum). A stack of that many single U.S. dollar bills would stretch roughly a quarter of the way around the globe; that is also a crude estimate of the cost of the project.[37] But the image of a huge stack of identical bills does not represent the complexity of the information that will be gathered. For that, imagine that the dollar bills come in four different colors, say yellow, red and blue as well as green (corresponding to the four nucleotides, A, T, G and C), and that the precise order in which the differently colored bills are stacked will be the essence of the project's data. Few personal computers sitting on people's desks have enough storage capacity to keep that amount of data, let alone to process it.

Then there were two surprises. The first was due to Craig J. Venter of the U.S. National Institutes of Health (NIH), who described a technique for identifying all the human genes by fishing RNA molecules from cells of different tissues (brain, liver, spleen and so on), turning those molecules into molecules of DNA by the use of reverse transcriptase, and then analyzing the sequence of nucleotides in the DNA. That, said Venter, would more quickly yield a list of all the human genes than the Human Genome Organization had believed possible.

For good reasons and bad, the research community went into shock. Its indignation stemmed from the suspicion that Venter had thought of something that their committee meetings had overlooked, but indignation was reinforced by the decision of the NIH, guided by its rather impulsive director (Dr. Bernadine Healy) to seek patent protection for Venter's nucleotide sequences even though nobody at the time had the faintest idea of their usefulness, or "utility" in the language of the patents trade. The international row that followed led to a humiliating *volte face* by NIH and Venter's resignation to found a privately funded not-for-profit research center, the Institute for Genomic Research.

The other surprise was also Venter's doing. Using techniques for working out the sequences of genes automatically and for manipulating the data that results by computer, between 1995 and 1997, the Institute for Genomic Research published complete nucleotide sequences for four separate, if simple, organisms—two bacteria, an Archaeabacterium and a mycoplasma.[38] For good measure, 1996 also saw the publication of the complete nucleotide sequence of bakers' yeast, accomplished by an international collaboration of more than 40 laboratories in Europe and the United States and organized as a collaboration with funds provided by the European Commission, the executive branch of the European Union. Then, early in 1997, the sequence of the *E. coli* genome was completed.

The most striking features of the complete genomes so far available is that they contain large numbers of genes whose function cannot be guessed at. The standard procedure for telling the function of an unknown gene is to look in one of the rapidly growing databases for a gene whose sequence of nucleotides is similar. If the look-alike gene has, say, 80 percent of its nucleotides in common with the unknown gene, that will be counted a spectacularly good match—so much so that the unknown gene will amost certainly have a function similar to that of the gene already found. Otherwise, it may be necessary to compare the amino acid sequence predicted from the nucleotide sequence of the unknown gene with the contents of other databases, making due allowance for the substitution of one amino acid by another that is chemically very similar.

At this very early stage, it is no scandal that people are at a loss to know what exactly is the function of about a third of the genes so far identified in complete genome sequences. Presumably the proportion will shrink as the databases become more comprehensive, but the high proportion is a reminder of how much remains unknown.

Of what value, then, will the Human Genome Project be? First, it will invert the present technique of searching for the genes whose aberrant alleles are linked with inherited disease. When there is a list— admittedly a long list—of all the human genes, it should be possible to identify which genes are responsible for known genetic diseases, and what the aberrant alleles are, by the thorough examination of samples from a few patients rather than by studies involving thousands of people (patients and their relatives) from different countries and lasting several years.

Yet, the importance of the search for disease alleles can easily be exaggerated; those responsible for the common inheritable diseases have been or are being tracked down. To be sure, there remain problems such as those with diabetes and, more seriously, with the major psychiatric diseases, schizophrenia and manic-depressive or "bipolar" illness. Although there is strong evidence of a genetic component in the causation of schizophrenia, almost a dozen attempts to locate the gene by classical methods have identified as many potential sites, all of them different, scattered through the human genome. Something odd is happening, but nobody yet knows what.

The more durable value of the project in medicine will be in helping to understand the mechanisms that cause organic disease, whether inherited or not. There is already one striking case of the successful treatment of an inherited disease, *familial hypercholesterolemia,* which leads to death by heart attack at a relatively early age (in middle life) and whose cause appears to be the synthesis of too much cholesterol. Cholesterol is an indispensable component of all cell membranes rather than the poison it is often held to be, but too much of it is damaging. So why not devise a drug that inhibits (or "down-regulates," as the cell biologists say) the synthesis of cholesterol? That thought, in the minds of Joseph Goldstein and Michael Brown, professors at the Medical Center in Dallas, Texas, led to the development of a simple drug, the survival of thousands of people who would otherwise have died and the award of a Nobel Prize to Goldstein and Brown in 1985.

In due course, there will be many other stories of that kind to boast about. Even sickle-cell anemia may have a happy outcome. The disease comes to light in early infancy, coinciding with a significant change in the character of the blood; human beings have one kind of hemoglobin during fetal life and another after they are born. Evidently the switch from one kind of blood to another is an adaptation to the different

physiological conditions of the human uterus and the big wide world—
an evolutionary luxury, perhaps. What happens in the genome is that
the gene for a protein called γ-globin is switched off and those for
making β-globin are switched on; the former is a component of fetal, the
latter of adult hemoglobin. Only β-globin is affected by the sickle-cell
mutation. So even a person who has inherited the sickle-cell gene from
each of his or her parents will have a perfectly good γ-globin gene[39] that
would make satisfactory hemoglobin if it were not inactivated. Why,
then, not think of treating people with sickle-cell anemia by reactivating
the dormant γ-globin genes? The search for a drug to do that is underway.

If a research group at the University of Oxford has its way (and the
luck it will also need), the most common form of inherited muscular
dystrophy may be dealt with similarly. The genetic cause of muscular
dystrophy, whose victims have little muscular strength and who die in
early adulthood from heart failure or some other sequel of muscle
failure, is one of several aberrant alleles of the gene for the protein called
dystrophin, which is crucial in the interaction between the ends of
nerves and the muscle cells they activate. In people with Duchenne
muscular dystrophy, the mutant form of dystrophin is inadequate for
normal function.

So why not treat the disease by injecting patients with artificially
made dystrophin? Unfortunately, that would not work; not previously
having experience of the protein, the body's immune system would
promptly neutralize it. So what else to do? The first stroke of good luck,
for Professor Kay Davies and her team at Oxford, is that nature has
overprovided once again; not only does the body make dystrophin, but
also a very similar material apparently involved in the regeneration of
muscle and called utrophin. So Davies and her colleagues have carried
out a striking experiment: take a mouse (or several), disrupt its equivalents of the *dystrophin* gene (so that they will inherit the mouse's equivalent of muscular dystrophy) and then also give them several extra copies
of the *utrophin* gene to see whether that makes up for the loss. It does.[40]

So the search is now underway for a simple chemical that will activate the *utrophin* gene in those who suffer this terrible disease. And even if that leads nowhere, there may well be some other way in which what has been learned of muscular dystrophy can be put to therapeutic effect.

From the first recognition of the *dystrophin* gene, it has taken the best part of a decade to get to the point at which drug palliation is on the cards. A further decade will pass before there could possibly be an actual remedy. This is the sense in which the research community has not yet come to terms with the scale of the effort that will be needed to understand the functions of the 100,000 or so human genes that will eventually be listed in the databases. If $3,000 million is a reasonable estimate of the cost of carrying through the Human Genome Project, the cost of understanding what all the genes do is certain to be very much greater. But as things are, skill rather than money is the impediment.

Better medicine will not, however, be the whole harvest of the Human Genome Project. The most important benefit will be the understanding this project will provide of how the human genome came to be what it is. Even the junk that occupies most of the genome will throw new light on the relationship between *Homo sapiens* and our evolutionary cousins, the great apes in particular (as discussed in the following chapter). From the prevalence of potentially movable genetic elements of different kinds, we shall learn something of the degree to which viruses and other infectious agents have contributed to the evolution of organisms. When the human genome is a computer printout, we shall also know more confidently than at present about the functioning of the eukaryotic cell. At that point, the project may well seem one of the best investments in understanding ever made.

There is only one fly in the ointment—the manipulation, inevitably by computer, of the vast amounts of data the project will generate. The simplest tasks can be hugely time consuming. Suppose a certain sequence of 12 nucleotide bases is particularly significant, perhaps because it plays a part in the regulation of a gene. Finding every occurrence of this sequence in the human genome with a Pentium microprocessor running at a clock-speed of 100 Mhz (and a hard disk capable of storing the 3,000 million items of information in the whole genome), would take 10 minutes of high-speed computation.[41] A more realistic problem that will be attempted when the human genome is a computer listing will be to work out the evolutionary relationships between all the Alu sequences in the human genome caused by mutation (which can happen freely in nonfunctional stretches of DNA), so as to tell whether there has been a pattern to their spread, and perhaps even to identify the element from which they are all derived. That would be a major computational exercise (for each element is roughly 300 nucleotides long). Answering intelligent questions about working genes will entail still greater difficulties unless clever people are able to devise some form of artificial intelligence that will meet needs now only guessed at.

FROM EMBRYO TO ADULT

Fortunately, it has not been necessary to wait on the completion of the grand sequencing schemes to understand how some of the most remarkable processes in biology are orchestrated by genes. At least in outline, there is now a molecular description of how an embryo becomes an adult—ontogeny as it is called. Much of this has been accomplished in the short spell since 1981, although the geneticists responsible are the first to acknowledge their debt to previous generations of embryologists.

In animals, ova are larger than sperm because they contain in premaufactured form many of the chemicals required to sustain the embryo through several cell divisions. In the fruit-fly Drosophila, for example, the fertilized egg divides exactly 13 times, yielding an embryo with some 8,000 cells in the form of a hollow cylinder, before it has to start manufacturing its own protein molecules.[42] It now emerges that eggs also contain substantial amounts of RNA molecules, transcribed from maternal genes, whose presence accelerates cell division at this early stage by directing the synthesis of particular protein molecules.

Why the haste? For many animals, successful reproduction depends on the survival of free-living embryos, often scattered in hostile environments. Speeding through the early stages is a huge selective advantage. Even so, the adaptations represented by the biochemical construction of the eggs of real animals to secure this advantage are remarkable in their variety and subtlety, perhaps best illustrated by the observation that when a bird or reptile hatches from its egg, there is almost nothing left inside. Virtually the whole of the egg together with such water as may have been taken in through the shell has been transformed into infant animal.

The early stages in the development of the embryo are similarly well calculated. Classical embryologists had discovered that an egg has a recognizable top and bottom which become the head and tail of the developing embryo and eventually of the adult; this is mirrored by the polarization of RNA molecules in the single-celled embryo. That polarity persists in the second stage, by which time the single embryonic cell has become two, and into the third stage, when there are four cells. It is important (and wondrous) that at this stage any single cell from a developing mammal embryo is capable of developing into an intact adult; that is how identical twins are formed.

During this process of repeated cell division, genes specify the protein molecules that the multiplying cells of the embryo secrete and which literally hold the cells together. It is also necessary that different parts of the growing embryo should acquire specialized functions and that those functions should be appropriate to their physical location. That is brought about by the secretion of the protein products of other genes, many of which have now been identified in a variety of organisms.[43] In the fruit-fly, there seems to be a highly organized system in which the successive segments of the larval stage (a grub or small caterpillar) embody in dormant form the tissues that become, after metamorphosis, the components of the intact adult—the head, the thorax and the structures that support the wings and the legs. These arrangements are controlled by a set of genes called *homeobox* genes (discovered in 1981) whose protein products appear to give the different segments of the larva their distinctive properties.[44]

The essentials of this magical process have now been understood: embryos develop into adults by the action of genetically determined protein molecules, many of which have been identified. When the Human Genome Project is completed, many of the 100,000 human genes will be found essential for the development of a person from embryo to adult. A process once regarded as mysterious has been reduced to one of genes and their products. In retrospect, it is also clear that, for the best part of a century, embryologists were on the right track; they searched for chemicals in, say, the growing limbs of chicks that would ensure that legs and arms (or wings) developed joints at the appropriate places, with toes or fingers or their equivalents at the extremities. We now know that there are chemicals, proteins made by genes whose concentration varies from one end of a growing limb to the other, giving cells at all points along the limb a sign of where they are (or are supposed to be).

In the 15 years since the discovery by Walther Gehring and his colleagues, at the University of Basel,[45] of a gene involved in deciding the character of each of the 13 segments of the larva of the fruit fly *Drosophila,* people have grown confident that the principles of ontogeny have been uncovered. Even for the fruit fly, that is not a simple business; each segment of the larva contains a small package of tissue (called an *imaginal disk*) that will, on maturity, develop into a specific part of the adult fly. Gehring's gene was the first of a now-large family of genes that are known to guide development. (Each insect stage carried with it a kind of embryo from which its successor will spring.)

Although the catalog of the developmental genes is unlikely to be complete even for the fruit fly, a couple of general principles have been established. First, the genes are organized in a hierarchy; the activation of one activates a whole group of others. Second, genes that regulate development in *Drosophila* have analogs in other organisms. The gene regulating the production of the 13 segments of *Drosophila* larvae has an analog in people that controls the structure of the hind-brain—between the brain itself and the spinal cord which, like insect larvae, has a distinctly segmented appearance. Similarly, the fruit fly gene called "sonic hedgehog" has an analog in people (and vertebrates generally) involved in the left-right asymmetry of the body plan—the heart on the left, the liver on the right, and so on.[46]

That seems unambiguously to show that the recipe for the development of an embryo to an adult is encoded in the genes. Yet two important caveats are necessary. First, the recipe specifies the ingredients, but says nothing of how they are to be mixed together. Second, virtually nothing is known of the manner in which one gene will switch on, or activate, a whole suite of genes beneath it in the developmental hierarchy —nor of the manner in which the gene products give individual cells their specific character. A grander question, hardly asked as yet, is how this complex system evolved, and when. Does it date from the Cambrian Explosion more than 500 million years ago or is it even more ancient?

GENETICS AND CANCER

Cancer, one of the most common causes of death in rich countries, is also a genetic phenomenon. The cells that rapidly proliferate to form a tumor differ genetically from normal cells, and several of the changes have now been cataloged. As well as throwing light on the natural history of the disease, this has engendered confidence that genetically identifying the aberrations of tumor cells will enable cancer to be treated more effectively.

Tumors arise from a kind of Darwinian natural selection within the body of a single animal or plant. In a healthy individual, the sizes of organs and tissues are maintained at an appropriate level by the regulation of the cell-division cycle as well as by the regulated death of surplus cells.[47] Cells that are exempt from the regulatory processes, or that acquire an exemption, may proliferate more quickly than others and will thus, in due course, overwhelm the complement of normal cells. What the past few years have shown is that there is a great variety of molecular means by which cells can be liberated from restraint, all of which involve genetic changes of some kind.

Several of the chemicals known to cause cancer (called *carcinogens*) damage somatic cells by interacting directly with the DNA, perhaps attaching themselves to one of the nucleotide bases and preventing the use of that gene. Sometimes, the cause of a cancer may be the physical disruption of a chromosome: there is a classic case of a form of human leukemia caused by the transfer of a small part of human chromosome 22 to the end of chromosome 9 (which can be detected with a microscope).[48] Given that confident and early diagnosis is often the best route to treatment, one can understand the decision of the NIH, at the beginning of 1997, to spend several million dollars to establish a database of the chromosome anomalies linked with cancer.

But not all cancers are visible in the chromosomes of a cell. Most are caused by genetic changes or mutations in a gene whose function is essential to the orderly (and restrained) division of somatic cells. The pioneering study of the rare cancer called *retinoblastoma*, the cause of which was worked out in the late 1970s by Dr. Robert Weinberg at the Whitehead Institute of the Massachusetts Institute of Technology, illustrates how mutations can lead to unrestrained growth of groups of somatic cells. Retinoblastoma is a tumor of the retina of the eye, and can arise at any time during a person's lifetime; the unraveling of Weinberg's complicated tale was possible because there is an inherited form of the disease in which tumors appear (often in both eyes) during infancy.

Using modern gene technology, Weinberg found that all the somatic cells of those with hereditary retinoblastoma carry a mutation of a gene whose function was then unknown and which he called the *retinoblastoma* gene, otherwise *Rb;* aberrant alleles of this gene seemed to predispose the cells of the retina to uncontrolled growth, presumably because the effect of its normal protein product (called "Rb") is somehow involved in the regulation of cell division. But why should even those who had inherited the aberrant allele from only one parent be susceptible to binocular tumors? In sickle-cell anemia, after all, those who inherit the abnormal allele from just one parent are free from anemia and have the advantage of immunity from malaria. Why should *Rb* be different?

The explanation is that Rb does indeed have an important function in the regulation of the cell-division cycle; the less of the normal protein in a cell, the more quickly it will divide. But mutations of all genes crop up more or less at random in all dividing cells. Cells carrying a damaging mutation in one of the two *Rb* genes will divide more often than others because they have only half the normal amount of the product of *Rb*, so that it will be only a matter of time before there is also a mutation in the other *Rb* gene. Pure chance will see to that. And then the regulatory function of the *Rb* gene will be entirely lost to the cell and its progeny, allowing them to multiply *ad lib*. Collectively, they will found a malignant tumor.

Now, after a decade's study of naturally occurring cancers, it turns out that the function of the protein product of the *Rb* gene is not confined to cancers of the retina. Aberrant alleles of *Rb* are found in the cells of a large proportion of all kinds of tumors, suggesting that the Rb protein plays a part in the regulation of the cell cycle in all cells in the body.[49] Accordingly, *Rb* was named, by reference to the function of its naturally occurring form, a *tumor suppressor* gene; several are now known.

The best known is a gene called *p53*, which is again found in mutated and ineffective form in a large proportion of naturally occurring tumors;[50] the normal product of the gene comes into its own at the end of the first phase of the cell-division cycle, when the total DNA content of the chromosomes has been duplicated. Present understanding is that *p53* is activated only if the DNA has been wrongly replicated, when its protein product has the effect of activating another gene, called *p21*, whose protein product prevents the activation of the trigger of the cell-division cycle, but in a decade or so, there will be more intelligible accounts of the array of genes, gene products and extraneous chemicals that keep the cell cycle orderly and whose mutation can cause tumors.

Since Weinberg's study, several similar features of the formation of tumors have come to light. In colorectal cancer, for example, one of the most common genetic mutations involves a gene whose protein is involved in the DNA repair machinery. In the familial (inherited) form of the disease, again, there is a mutation in one or the other of the two inherited alleles in all the cells of the body, followed, after what may be an appreciable interval of time (and many cycles of cell division), by the appearance in colon cells of a chance mutation of the second allele. Regulation of the cell cycle is then thoroughly undermined. Malignant growth, including the invasion of neighboring tissues by the aberrant cells, follows.

The natural history of much tumor formation is thus neatly explained: although a potentially damaging mutation may be present in the group of cells from which the retina is formed, an overt tumor appears only after a second mutation, which gives the lineage of cells concerned a decisive advantage over others in their capacity to proliferate. But neuroblastoma is almost the simplest kind of tumor; in the origin of many others, half a dozen successive mutations may be required before a clone of cells is equipped to subvert the whole life of an organism. The long induction period required for most cancers is thus made intelligible.

Mutations are not the only roots of tumor formation. Viruses may also be carcinogenic agents. The landmark case is that of the Rous sarcoma virus (which infects chickens), first described in 1907 by the late Peyton Rous more than half a century before he was awarded a Nobel Prize for his discovery. It is now known that the human genome contains many genes derived from past infection by cancer-causing viruses. The viruses concerned are the *retroviruses,* whose genetic information is embodied in RNA rather than DNA. They reproduce themselves only by integrating their genetic information (as DNA) within the genome of the infected host. One such, called *adult T-cell leukemia virus (ATLV)*, actually causes blood cancer in people and is prevalent in southern Japan. Soon after its character was recognized in the late 1970s, it was briefly regarded as a possible cause of the human AIDS.[51]

The human genome contains more than 50 genes that closely resemble genes carried by retroviruses which are responsible for causing cancer in at least some mammals or vertebrates. There is, for example, the gene called *ras-A,* which is part of the normal human genome; its protein product plays an important part in signaling between the outside and the inside of a cell. But the nucleotide sequence of the gene is very similar to that of a viral gene (called *v-ras*) responsible for causing cancers in rats, and which on that account is called an *oncogene*. The useful version of the gene in the mammalian genome, on the other hand, is known as a *proto-oncogene* because only a mutation of the normal sequence can yield a protein product involved in causing cancer.[52]

As yet, too little is known about the ways in which proto-oncogenes can be activated to cause tumors. Certainly the movement of a section of human chromosome 22 to the end of chromosome 9 disrupts an oncogene known as *myc,* the result of which is a form of leukemia. The migration of transposable elements into positions that interfere with the normal function of a proto-oncogene is also a potential cause of trouble. But the natural history of many cancers of this kind remains to be revealed, as does the origin of the proto-oncogenes.

The practical issue is how, and when, new knowledge will eventually be used to treat cancer more effectively. Genetic analysis can provide more certain diagnosis which will be the most immediate benefit. Knowing that a person has inherited a susceptibility to a particular kind of cancer will often suggest prophylactic treatments, such as the removal of the ovaries from women with familial breast/ovarian cancer. Soon, it will be routine for physicians treating patients with new tumors to order a genetic analysis of the quickly growing cells. The bugbear of cancer treatment, whether by drugs or immunotherapy, is that supposedly curative agents are less than specific to cancer cells, raising the question of whether the side effects are worse than the disease. Now, with the recognition that all tumor cells have their own genetic identity, there is a realistic hope that they may be attacked for what they are.

Although the best strategy for the avoidance of cancer will remain the avoidance of known environmental causes, cigarette smoking for example, the speed with which the mechanism of many types of cancer is now being understood raises the prospect that the coming decades will see ways of snuffing out many kinds of tumors that cannot now be treated. The knowledge that failure of the regulation of the cell cycle plays a central part in the growth of tumors will stimulate the search for drugs to make up for the loss of the activity of the tumor suppressor genes. People then will look back on the period since the early 1980s as the time in human history when the causation of those same cancers was first understood. For some years, there will then be the usual fuss about the affordability of new procedures, but costs will decline as physicians become familiar with new techniques. Then, we shall have no choice but to worry about other conditions of which we are likely to die.[53]

MISUNDERSTANDING OF GENETICS

One sequel to recent deveoopments in genetics has been widespread misunderstanding, both generally and within the research community (which should know better). Merely the availability of prenatal diagnosis suggests to some that communities in which the techniques are already available are about to practice eugenics of the kind ignorantly advocated in Germany in the 1930s. Others fear the imminence of the "designer baby"—an individual equipped with superior faculties, both intellectual and physical. Both fears are groundless. There are important ethical questions to ask, but these are not them.

An issue strictly unrelated to genetics now obtrudes—the manipulation of human embryos to produce children. The technique of *in vitro* fertilization has been used successfully for at least two decades; IVF allows infertile couples to have children naturally after the artificial fertilization of an ovum by the male partner's sperm. Many religious people avoid the practice for themselves, on the grounds that it is not natural,[54] but that is not an ethical objection of general character. Even when the sperms come from often anonymous third-party donors, there is no difference in principle from preexisting practices; similarly, the use of ova from donors is essentially the preexisting practice of surrogacy.

Ethical issues nevertheless arise in three ways: children are not necessarily the genetic offspring of their supposed parents while, as IVF becomes more common, the proportions of such people will increase. And because several embryos are created in IVF, but only a few are implanted in the uterus, the procedure permits genetic screening (in particular, sex determination) as well the storage of fertilized embryos for implantation at a later time. Because the genetic origins of children may be important in their health care as adults, there is a strong case for allowing individuals on maturity access to information about their genetic parents; that is now accepted under British law, but is not commonplace in the United States. Telling the sex of embryos, or telling in advance of implantation what other genes they carry, is technically feasible but would require manipulation of the embryos at the four-cell stage.[55] The storage of embryos (technically straightforward and apparently free from risk) raises the possibility that genetic siblings with very different ages might coexist, but the first occasion when, under British law, IVF clinics were required to dispose of stored but unwanted embryos also raised a furor on the grounds that embryos (even though stored at liquid nitrogen temperatures) are living things. Public concern centered on issues such as these will persist for as long as IVF is practiced.

In reality, genetic diagnosis of artificially fertilized embryos should be preferable to the now standard technique of prenatal diagnosis followed by abortion (except that the cost would probably be higher). It is relevant that IVF permits only the passive genetic manipulation of people; investigations based on one cell from a four-cell embryo would allow putative parents to choose, from among all the genomes provided by the lottery of genetic recombination, those that happen to be free from particular disease alleles.

There is, however, one exception. In principle, the manipulation of early embryos could be used for producing identical copies of particular adults, known as "cloning." Single cells taken from embryos at the four-cell stage could be stored until the remaining three-celled embryo had matured into an adult, when it could be decided whether or not to grow another (or several) with the same genetic constitution. While so little is known of uterine influences on fetal development, and because the genetically identical embryos would presumably be implanted in different uteri, adults grown from different cells of a single embryo would not be identical in all respects.[56] Cloning people by this route will always be a hazard.

Since early in 1997, it has also been possible to produce identical animals by starting with cells of ordinary tissue. In February of that year, a sheep famously called "Dolly" was produced from the nucleus of a cell taken from a sheep's udder at an agricultural research station in Scotland. So far as is known, Dolly is genetically identical with the sheep from which the tissue was taken.[57] Procedures like that could probably be used to clone people; because every somatic cell contains a complete copy of the human genome. In countries with a legislative framework for regulating embryology, even research directed at cloning people is already prohibited, which is desirable. Whether that is the case elsewhere, and whether the policing of research can be fully effective, is another matter.

Meanwhile, genetics proper is also riven with controversy, chiefly on two counts—the discriminatory use of genetic information about individuals seeking employment and by the insurance industry and the application of techniques now available to the deliberate improvement of the genetic characteristics of a population, called eugenics. Much fuss reflects the novelty of recent developments.

The employment issue turns on the suspicion that genetic information could be used unfairly to deny employment to particular individuals, but should not people who have inherited a susceptibility to a cancer of some kind be discouraged from working in conditions in which carcinogens abound? The remedy is to use law to prevent the use of screening as a cloak for discrimination on other grounds. Similarly, despite the efforts of many legislatures to prevent life and health insurance companies from requiring genetic information as a precondition of insurance, consumers as a whole are likely to resent the higher charges they will have to pay for insuring those who know of a discouraging genetic diagnosis and for whom nondiscriminatory insurance will seem a good bargain. At the same time, insurance companies themselves presumably cannot be prevented from offering policies at advantageous rates to those who volunteer favorable genetic information about themselves.

That set of dilemmas points to one way in which the new genetics will, in the decades ahead, change our regard for our place in the world. Knowledge (or the mere fact that genetic knowledge is obtainable) will force us to recognize that some disadvantageous conditions are unavoidable. Genetic bad luck has come to stay.

On the other side of that coin, knowledge of genetic constitution is potentially a starting point for better health care. Take, for example, the human gene *APOE*, whose product is involved in transporting cholesterol in the blood. Of the known alleles of the gene, one—*APOE4*—is associated with heart disease in middle life and with Alzheimer's disease in a person's seventies. But in neither case is the correlation exact. For people known to carry this allele, there are obvious prophylactic measures to avoid heart disease. Whether other steps might avoid late-onset Alzheimer's disease is not known, but measures like that would be of great value in human medicine. As things are, there are grounds for suspecting that physicians are not pursuing valuable opportunities out of diffidence at revealing unwelcome genetic information to their patients. That diffidence is proper when an allele predisposes unambiguously to a serious disease for which there is no effective treatment, Huntington's disease, for example. But diffidence can be taken too far.

The other recurrent worry is that new techniques in genetics can and will be used to reinstitute eugenics programs of the kind that gave Nazi Germany a bad name. Eugenics is the word coined by the Briton Francis Galton in 1883 to refer to the supposed improvement of the genetic endowment of a population by taking deliberate decisions about the patterns of reproduction within a group of people. Galton, a cousin of Charles Darwin, urged that there would be great benefits if, say, people of high intelligence produced more children than others; he set in train the "eugenics movement" long before the concept of a gene had been defined.

In the event, the movement led to legislation preventing intermarriage between people with psychiatric illnesses (called "mental defect" in the argot of the times) in many U.S. states and Canadian provinces, as well as legal powers compulsorily to sterilize people (usually women) with similar loosely defined conditions. Among European states, Germany and Austria and some of the Scandinavian countries followed the same pattern in the 1920s. Notoriously, in the 1930s, the government of the Third Reich extended policies of this kind until they were subsumed in the bizarre goal of eliminating the Jewish people from Europe.

Ironically, Britain, where eugenics as a concept was invented, never followed suit: by the 1930s, geneticists such as J. B. S. Haldane had marshaled powerful arguments to show that Galton's dream could not be realized. The seemingly desirable outward characteristics of particular individuals would always be compromised by the recessive alleles in their genetic makeup, capable of reappearing in succeeding generations, while statistical arguments (founded ironically by Galton) undermined the concept of a pure and ideal lineage of people by the doctrine of "regression to the mean." In sexually reproducing animals, the sets of genes contributed by the two parents are so thoroughly reassorted by genetic recombination that individuals who are exceptional by some arbitrary criteria will usually have offspring that are less so.

The clinching argument against positive eugenics of this kind is the obvious analogy with animal breeding, where new breeds of cattle or dogs are indeed repeatedly produced. The goals are usually clearly defined—for cattle, more milk production or better beef, for example. The cost in animal life is huge. A cow may produce 15 or so calves in a lifetime of breeding, of whom one or two will be recruited to the nucleus of a breeding herd. Several generations later, there may be a viable herd of cattle uniform by the criteria chosen at the outset, but in the process perhaps 90 percent of the animals born will have been discarded—sold to farmers with no direct interest in the breeding program or even sent to market as calves. And that is the cost of breeding for such goals as milk yield or the quality of an animal's muscle, which are measurable and which can be identified early. Breeding for the less tangible characteristics that make desirable human beings would present the committees in charge of such a hypothetical program with the need for decisions they would not be able to make objectively, and therefore which they could not consistently communicate to their successors. The matchmakers who arrange marriages in many parts of the world, or the elite social groups who make a practice of intermarriage, are probably at least as good at practical eugenics as the present state of understanding of the science of genetics would allow. And that is not saying much, for the record of these marriage patterns is hardly creditable.

If positive eugenics is impracticable, negative or passive eugenics has already arrived and is here to stay. Genetic counseling has been practiced since the 1930s with the objective of helping would-be parents to avoid genetically or congenitally handicapped births. Since the recognition in the 1960s that the congenital handicap of Down's syndrome is caused by the presence in the embryo of three rather than two copies of human chromosome 21, in jurisdictions that allow abortion, mothers have been allowed to terminate pregnancies destined to lead to such a birth. Now a much wider range of strictly genetic diseases can be avoided in that way.

This practice is much disputed, even in rich countries where it appears nevertheless to be growing quickly. Against it are religious objections, objections to abortion and, more generally, the argument that caring for disadvantaged children is a civilizing influence in society. There is also the view that, while genetic knowledge is incomplete, there may be dangers to society at large in ridding people of disadvantageous genes if, in some circumstances, a single allele inherited from one parent only gives a person an advantage[58]—as with the allele for sickle-cell anemia.

The arguments for prenatal diagnosis and prophylactic avoidance are as strong or stronger. In most rich countries, it is a criminal offense not to send a child to school, but it is taken to be a matter of parental choice whether he or she should be born incapable of benefiting from schooling. More than that, it is a matter of common experience and solemn social research that a genetically handicapped child can potentially damage not only the lives of its parents but its siblings. Certainly, when given an informed choice, most putative parents of genetically handicapped children prefer to opt out. That, indeed, is why the practice has come to stay; it will not materially change the proportions of disease alleles in rich societies.

WHAT NEXT FOR GENETICS?

The harvest of understanding in genetics won in the past 20 years has outdone even the expectations of the early enthusiasts; the likely benefits of these discoveries for human well-being are yet to be realized. Perhaps forgivably, there is on that account a temptation to suppose that genetics is "the book of life," and that all questions about the nature of life (human behaviors included) will eventually be answerable in the language of genetics. But that is an illusion. Recent discoveries, such as that most eukaryotic genes are split genes, were a surprise in the early 1980s; it will not be the last of its kind. Meanwhile, there are several technical problems crying out for solution. How are groups of related genes regulated in concert? What exactly are the influences that prevent the activity of some genes, both in differentiated cells and those in which some genes are permanently switched off, as in the inactivation of one of the two X chromosomes in mammalian females? To list problems like these is not to suggest that they are barriers to understanding, but merely that they will take time to solve.

The more serious difficulty will be to know what to make of the complete genome sequences that will become available in the next few years. The computational problem has been discussed above. Understanding how a particular sequence of DNA determines the character of the organism to which it belongs will be much more difficult. Already it seems probable that the protein products of most genes will turn out to have more than one function—that for which they are named and several others. Like the cell biologists, geneticists will be driven to make computer models of the genome before they will be able to make sense of all this complexity.

Naturally enough, the enthusiastic identification of genes in recent years has concentrated on those known to play crucial roles in the functioning of cells. Less attention has been paid to the genes that determine a cell's reaction to its environment by receiving and processing chemical signals from elsewhere, from other cells and from the environment more generally. Yet that is how we will establish the balance between nature and nurture in the development of an organism. Geneticists should not be surprised if nurture turns out to be more important than they now allow.

Perhaps the biggest prize in genetics will come from the comparison of the genomes of more or less distantly related organisms which will make it possible to reconstruct the history of evolution. But there is an important lesson that molecular geneticists should learn: in a field that engenders great public sensitivity and triumphalism, the idea that everything is determined by genes is their Achilles' heel. Most people, including those whose health may be powerfully improved with the help of new genetic knowledge, know that there is more to life than genomes.

Nature's Family Tree

Charles Darwin's great book *On the Origin of Species by Means of Natural Selection* was the most influential document of the nineteenth century. It permanently changed our regard for our place in nature by demonstrating that all the life-forms now on the surface of the Earth are products of the same processes—chance variation and natural selection. But the *Origin of Species* did very little to deepen understanding of what it advertised as its central theme—the origin of species. And that remains a problem unresolved.

Darwin is not to blame: in the 1850s, the general understanding of living organisms was only meager. It is true that Gregor Mendel had already noticed the regularities in the color of pea plant flowers, but his accounts of his experiments was published obscurely and few knew of them until they were rediscovered at the turn of the century. The *Origin of Species* left the mechanism of inheritance an open question. Even the question whether species have an objective existence was disputed in Darwin's time.

That life-forms on Earth had changed, however, was not in dispute. Geologists described a fossil record clearly showing that the pattern of life had been different in the past. Darwin himself had traveled the world, noting the similarities and the differences between related groups of animals and plants. The finches of the various islands of the Galapagos Archipelago (in the eastern Pacific), evidently of the same ancestral stock, differed so much from island to island by the time Darwin arrived on the H.M.S. Beagle in 1835 that they appeared to be different species. The question, for Darwin and others, was how things had become like that.

Now, almost a century and a half after Darwin's book appeared, it is plain that the identity of an individual and the character of the species to which it belongs are largely determined by the genomes of the organism. The question of how life evolved has thus been converted into the historical question of how the genomes now extant acquired their distinctive characteristics. Luckily, much can now be learned by comparing modern genomes with each other; even the redundant DNA, the "junk," in the eukaryotic genomes will provide telling clues to evolutionary history. The fossil record of evolution is no longer just the collection of scraps of fossils accumulated by the world's museums, but the DNA in the cells of all the creatures still alive (not to mention the DNA that can sometimes be recovered from dead animals and plants).

DARWIN MISREPRESENTED

Darwin offended his readers in many different ways, but not least for having attributed what he called the "Descent of Man" to people's supposed common ancestor with the great apes. More substantially, many of his contemporaries took offense at the idea that the characteristics distinguishing novel species stem from variations that arise by chance, which flies directly in the face of the idea that the diversity of life is the product of some Grand Design. To the extent that Darwinism is literally a godless theory of evolution, it is inevitable that it should have provoked controversy among religious people. After more than a century of argument on this point, it may now be a source of some civility that Pope John Paul II declared in 1996 that Darwinism should be allowed to stand or fall by its value as a scientific theory.

Darwin's argument is easily (and is often mischievously) misrepresented. The most overlooked point in his theory of evolution is that it is not a theory about individuals, but about groups of individuals belonging to the same species—a population.

Consider a population of animals or plants in which sexual reproduction is the norm and consider also the environment in which these creatures are placed. The environment includes the other species living there, which will certainly compete with all members of the population for common resources such as food, and may even be predators of them. The progeny of their matings will generally resemble their parents, but will also by chance embody small variations from the norm. It is crucial to Darwin's argument that these variations are heritable. That is the first keystone of Darwinism: chance but heritable variation.

Many variations will be disadvantageous for the individuals concerned (which may then die or fail to reproduce), but some will be advantageous in a particular environment at a particular time. The advantageous variations give the progeny carrying them a greater chance than others of producing offspring of their own; these characteristics will therefore become more common in the succeeding generation. That is natural selection. Although the process hangs on the survival and fecundity of individuals, what matters for evolution is the contribution of individuals' progeny to the next generation. As the generations go by, some characteristics may become more and more common, perhaps conspicuously so. Natural selection, which moulds this process of adaptation, is the second cornerstone of Darwinism.

The power of natural selection has been consistently marveled at since 1858. Often it has seemed to be a purposive force, generating remarkable physiological innovations such as eyes, ears and the organs that allow animals to smell. Yet there is no purpose in the sense of a long-term goal of any kind. The variations embedded in the makeup of organisms in an evolving population are those that make them better suited to the environment in which they happen to be placed. Even the declaration that the "purpose" of natural selection is to bring about "greater harmony between living things and their environment" is not a purpose in the ordinary sense, for the environment keeps changing.

The tyranny of natural selection takes a further twist. Important adaptations, such as the emergence of some kinds of Late Devonian fish 365 million years ago with the capacity to spend time on dry land, entail many successive changes of body plan and function, each one of which must, in itself, run the gauntlet of natural selection. At no stage in the evolutionary process can the members of an evolving species suffer heritable variations that entail a decrease of fitness relative to that of their fellows. That means that the very first fish to live amphibiously (perhaps only briefly on dry land) must have derived an immediate benefit from their newfound faculty—access to a new food supply, perhaps, or escape from aquatic predators. That benefit is measured in the only currency recognized by natural selection—enhanced fitness, usually itself measured by the number of offspring that survive to reproduce.

Of course, the replacement of fins by primitive legs would not have been a sufficient adaptation in the long run. An elaboration of the nervous system of the aquatic animals making the first adaptations to life on dry land would also have been required.[1] The transition is evidently one of the most important in evolutionary history; all the vertebrates now living, *Homo sapiens* included, owe their existence to it.

The question of whether an eye or even a whole brain (of which the eye is but a part) can arise by chance is misstated. It is based on a frequently deliberate and even perverse misunderstanding of the role of natural selection—and the meaning of "chance." The question should not be whether chance can produce an eye or a brain, but whether a particular organism already manufactures for other purposes protein or other molecules from which, with variation, eyes and brains could be fashioned. Many forms of bacteria can respond to light, thanks to the absorption of light by molecules called opsins; different ways of packaging these materials in some of the cells of multicellular organisms account for the at-least half-a-dozen occasions on which, in the course of evolution, vision has independently evolved. And that is not a sign that animals have an innate propensity for sight, but is a testament to the selective advantage of seeing—in gathering food, catching prey or hiding from predators, not to mention reading books.[2] It is central to Darwin's argument that there is no way of predicting the ultimate physiological results of the accumulation of advantageous variations in an evolving lineage.

Darwinism differs from Newton's mechanics in that it is not a predictive theory.[3] Will the variations conspire to make an eye, or lead less conspicuously to enhanced efficiency in the use of chemicals as a source of energy? Or will the variations be advantageous only in the environment in which they arise, so that lineages that seem to have been lucky to acquire some adaptation are found, in a changed environment, to have the bad luck of dying out.

The flightless birds called dodos illustrate how adaptation can lead to extinction. These creatures, relatives of the pigeon (but larger), were found on many islands of the Indian Ocean until the eighteenth century, when they were exterminated by European sailors hunting them for food. The ancestors of the dodo could fly: how else could they have populated so many widely spread islands? But once dodos were established on their islands, natural selection would have favored the individuals that gave up flying, saving the huge metabolic cost that expends. Flight muscles would atrophy, and in due course there would be genetic variations more permanently disabling the apparatus of flight. At no point could natural selection have anticipated that a time would come when flight might again be an advantage, as when European sailors came hunting.

Natural selection has and will repeatedly lead lineages of organisms into evolutionary blind alleys. Like the dodo, species may become so exquisitely adapted to a particular environment that their members cannot survive even modest changes. It is even possible that the evolution of a species may be successful over the short run—say 1,000 or 10,000 generations—but unsuccessful in the longer run, say after 100,000 generations. If we learn that the human genome is unstable in the long run, it would be a somber warning that the accumulation of even the most remarkable adaptations does not guarantee indefinite survival.

The collection of living things on the surface of the Earth at each stage in the past 4,000 million years must therefore have been the product of all previous historical accidents in earlier evolution. On the authority of the Book of Genesis, which gives human beings a special place, that view of life is inconsistent with Judeo-Christian religion. But Buddhists are less likely to be offended that people have the same status as other animals and plants. In reality, the accommodation between modern Christianity and the fossil record turns on what may be called the "God of the Plan": the emergence of the first living things, along the lines outlined earlier, and even the circumstances that made possible that sequence of events, were designed in such a way that *Homo sapiens* would eventually emerge. That is no doubt how the Vatican will eventually rationalize Pope John Paul II's acknowledgment that Darwinism should stand or fall on its merits; is it not in the interest of religion as well as science the better to understand the working of the Plan? The result would be a welcome liberation of evolutionary studies from theological constraint—and a transfer of the conflict of science and religion to the field of mathematics.

To be fair, it is not only religious people who resist the notion that the presence of *Homo sapiens* on Earth is an historical accident. It is natural that we should regard the surface of the Earth as it now is, with ourselves a conspicuous part of it, as being a natural state of affairs. It is a test of the human imagination to guess at what would happen if *Homo* sapiens were suddenly to disappear. In due course, it will be a stringent test of our understanding of the dynamics of the "web of life" (ecology) if accurate prediction can be substituted for imagination.[4] But the truth is that the whole span of human evolution, since the presumed separation of the human lineage from that of some species of the great apes (probably the chimpanzee) has occupied a mere 4.5 million years—a tiny fraction of the time during which there have been living things on Earth. Is it not therefore a great conceit to suppose that Homo sapiens is here for the rest of time? It will be safer for people to talk of having "inherited the Earth" when they have a longer and more robust tale to tell.

The case for regarding the human race as an historical accident has been consistently and vociferously made by Stephen Jay Gould, the Harvard paleontologist.[5] Gould's mission seems to be to demonstrate the marvels of evolution and to emphasize that heritable variation and natural selection together constitute a natural process without purpose in any ordinary sense. In his book Life's Grandeur, Gould writes:

> If a small and tenuous population of protohumans had not survived a hundred slings and arrows of outrageous fortune (and potential extinction) on the savannas of Africa, then Homo sapiens would never have emerged to spread throughout the globe. We are the glorious accidents of an unpredictable process *with no drive to complexity,* not the expected results of evolutionary principles that yearn to produce a creature capable of understanding the mode of its own necessary construction. (Italics added.)

Few quarrel with Gould's conclusion that there is no "drive to complexity," but merely sets of circumstances in which different organisms successively evolve to occupy conspicuous positions. But there is one logical flaw. What if some adaptations contribute hugely to the fitness of an evolving lineage? What about brains, even those of a "creature capable of understanding the mode of its own necessary construction"—Gould's oblique reference to *Homo sapiens?* Precisely because people have acquired the capacity to influence not only their own Darwinian fitness but that of other creatures—remember the dodo?—they have also limited opportunity to opt out of the succession of historical accidents that makes the still sketchy record of the past 4,000 million years so rich.

That statement is emphatically not a Panglossian denial of current and future problems, which are dealt with more carefully later in this book. The nineteenth century failed to appreciate the complexity of the web of life (revealed in part by Darwinism) and also exaggerated its own capacity to change the world. But it remains an interesting, perhaps even an important question whether the capacity of human beings for tool-making as well as for considered thought and decision free them from the vagaries of the historical accidents that limit the lifespans of other species, Gould's bacteria possibly excepted.

FIRES OF CONTROVERSY

Among the controversies spawned by Darwin's book, little attention was paid to a seemingly technical protest by Thomas Henry Huxley, Darwin's great supporter in the years following the publication of *The Origin of Species.* Darwin had written of the spontaneously arising variations as small variations, implying that the evolution of a species would be gradual and, often, imperceptibly slow. One of his goals was to demonstrate the great antiquity of the evolutionary record. Huxley warned Darwin, on the eve of publication, that he had "loaded" himself with "an unnecessary difficulty in adopting *Natura non facit saltum* ["nature does not make jumps"] so unreservedly." Events have shown that Huxley was right.

By the end of the nineteenth century, gradualist Darwinism had to contend with the rediscovery of Mendel's earlier study of the genetics of pea plants in the Brno monastery garden.[6] The conflict was between the smallness of the variations Darwin supposed to be the raw material of natural selection and the discrete nature of the genetic variations that

Mendel had observed—the flowers of his pea-plants could be either one color or another, but not some color in between. The early geneticists believed that their goal, the full understanding of the mechanism of inheritance, would allow them also to improve on Darwin's theory of evolution.

The pioneers were Hugo de Vries in the Netherlands, who replicated Mendel's work by crossing distinct strains of the evening primrose, and Thomas Hunt Morgan, Alfred Henry Sturtevant and Hermann Joseph Muller[7] who made Columbia University in New York into the citadel of the new study of inheritance. Morgan was the one who seized on the usefulness of the fruit fly *Drosophila* as an object for studies in genetics; it breeds quickly in captivity, and remains one of the workhorses of genetics.

Inheritable characteristics (say hair color) are determined by genes arranged along the length of the various chromosomes. But in the absence of an understanding of what genes are made of, they were essentially abstract entities. Even so, Sturtevant, while still a graduate student, made the crucial observation that the arrangement of the genes is linear; they are arranged in the chromosomes as if on a continuous line, not in some kind of patchwork. Particular inherited characteristics (say, red hair) are determined by particular variants (or alleles) of the gene concerned with hair color. For different genes, there are different rules for telling how the two parental alleles interact. Sometimes a particular allele will dominate all possible partners, sometimes an allele will exert its effect only if the other allele is identical.[8]

The achievement of the Columbia school is undisputed,[9] but the original suspicion that the new genetics would amend Darwin's theory of evolution proved unfounded. The tussle turned on the point on which Huxley had warned Darwin, that gradualism might prove an encumbrance. Continuity and gradualism are built into Darwin's view of the living world. The Columbia school, by contrast, was looking at inheritable change through the other end of the telescope, concentrating on the often dramatic consequences of genetic change for individual members of a population, such as fruit flies without wings, which in the real world, could not survive and so would form no part of the living population.

What followed, in the 1920s and 1930s, was a dramatic intellectual rapprochement between two potentially warring factions. Earlier this century, R. A. Fisher (whose academic interest was the mathematics of probability) had shown how evolutionary processes could be discussed numerically; his prodigious student J. B. S. Haldane carried the argument further, showing how discrete genetic changes can be incorporated in mathematical language without offending against Darwinism. In the United States, Sewell Wright (who spent a year laying tracks for the Canadian Pacific Railway before being employed by the U.S. Department of Agriculture to study the genetics of guinea pigs) similarly argued that Darwinism and Columbia's version of discrete genetics are not inconsistent.

The resolution of the conflict was in the end straightforward. Certainly some inheritable variations are discrete and recognizable as such. But inheritance then affects the frequency with which alternative variations appear in an evolving population. Whether or not the naturally arising variations are discrete and substantial, on the average, the character of the population is changed to some degree from one generation to the next.[10]

How recent these events were! Just over half a century ago, people were still bitterly at odds over the precise significance of the link between genetics and evolution. Only a little later, T. D. Lysenko was able to capture and corrupt most of Soviet genetics with disastrous results for successive Soviet grain harvests and ultimately for Nikita Khrushchev, secretary-general of the Soviet Communist Party from 1958 to 1965.

Darwin's gradualism has recently been more directly questioned. In 1972, Gould and Niles Eldredge mounted an attack on Darwin's notion that the evolution of a species represents the accumulation of small variations—precisely Huxley's anxiety on Darwin's behalf. They marshaled evidence from the fossil record showing that it is the rule rather than the exception that life-forms remain unchanged for long periods of time, often millions of years. Gould and Eldredge called that state of affairs *stasis*. They say there are few signs in the fossil record of the gradual changes that would betoken, on Darwin's view, the successive adaptations of species to make them ever better adjusted even to a stable environment.

Then how and when does change occur? Suddenly and episodically, according to the "punctuationalists." The apparent equilibrium of life-forms during the long periods of stasis is punctuated by times when the pace of change is rapid, often dramatic. That is the doctrine of punctuated equilibrium. Since 1972, it has seemed that evolutionists have been divided by the question into two bitterly warring camps.

Despite the heat engendered by this argument, it is no more than an extended comment on the nature of the fossil record. The early Darwinists were repeatedly challenged to produce fossil evidence of the supposedly once-existent transitional forms between mammals and, say, the early primates. The absence from the fossil record of a "missing link" between human beings and their supposed common ancestor with the Great Apes became a nineteenth-century scandal. Properly, but with unconvincing regularity, the Darwinists had no choice but to reply that the fossil record is manifestly incomplete.

One of the bugbears of paleontology is that it is a matter of chance whether a particular animal or plant will be fossilized. Everything depends on what it is made of, how and where it dies and on how dry or wet the ground is. The Burgess Shale fauna is a rich deposit of middle Cambrian animals named after a now famous quarry in the Canadian Rockies that was once the site of repeated mud-slides at the edge of a shallow sea which buried living animals in conditions that prevented access by oxygen and thus their putrefaction. The inhabitants of the city of Pompeii in Italy, whose corpses now draw flocks of tourists to the excavated ruins, are a proof that a thick blanket of volcanic ash (from the eruption of Vesuvius in 69 A.D.) is another route to fossilization.

Although punctuated equilibrium has its roots in paleontology, support for it has also come from more recent evolutionary studies, stimulated in the past half-century by successful attempts to link paleontology with biology more generally, with ecology for example. There is the case of the antelopes of southern Africa, which Eldredge claims in his support, citing a study by Elisabeth Vrba (now at Yale University).[11]

The African antelopes called the gnu or wildebeest, like the hartebeest and five other species now living in southern Africa, have all evolved from an ancestral impala stock, which is still widely distributed throughout Africa. Fossil evidence of the evolution of these creatures is conveniently available in the form of the durable horns they cast (heaps of which are stored at the Transvaal Museum at Pretoria, where Vrba formerly worked). The fossil record of the impala stretches back for more than 5 million years and (so far as horns can tell) the creatures have been substantially unchanged during that time. Yet the fossil record shows evidence of no fewer than 25 species of animals derived from the stock, only seven of which have survived. Why so many species of derived animals, but only one apparently unchanging impala? Vrba's explanation is that all of them, both alive and now extinct, evolved from impala by specialization, the use of particular vegetation as food; they each occupied an ecological niche. The result is that they were more vulnerable to historical accidents (such as changes of climate) than are impala, with their less specialized dependence on the environment.

As with much else in the fossil record, this evidence can be read in two ways. The apparently unchanged form of the cast-off impala horns may be a sign of stasis, and thus support for the idea of punctuated equilibrium. But there may have been progressive changes in other attributes of the impala during the past 5 million years that are not reflected in discarded horns. And while the repeated appearance and extinction of other antelopes may be a mark of repeated punctuation of the equilibrium, it is also consistent with gradualist Darwinism. Species adapted to an ecological niche would be expected to disappear from the record when the niche itself vanishes.

The argument about punctuated equilibrium has occupied a great deal of time and energy. And the outcome? Stalemate born of exhaustion. Gould and Eldredge did a public service by drawing attention to the great differences between the speed of evolution of species simultaneously occupying the same places on Earth; some may be in equilibrium with the environment, others may change relatively quickly as accumulating variations allow their members to change their habits so as to occupy emergent niches. But none of that is in conflict with Darwinism. Prolonged stasis is familiar to conventional evolutionists: John Maynard Smith, for example, cites the case of the lungfish adapted to survival in temporary bodies of water such as intertidal coastlines by burying themselves in mud: they have a fossil record 300 million years long.[12] The weakness of the position taken by Gould and Eldredge is that it has all the sound and fury of an iconoclast attack, yet it does nothing to undermine the essential principles of Darwinism.

Several caveats put this controversy in perspective. The first is technical: the phrase "the speed of evolution . . ." has no absolute meaning. On the Darwinian view, even in a species in equilibrium with a stable environment (or in stasis), new variations will continually arise at random, but will not usually change the general character of the species. That does not mean that surveillance by the evolutionary mechanism of the match between the species and its environment has been suspended. Even stasis is evolution.

The second caveat concerns the adaptations arising in the course of evolution, which may be impelled by several influences of which environmental change is probably the most common. Suppose that a population of some species has reached equilibrium in some stable environment, and that the ambient temperature then begins to increase. The members of the population carrying the alleles best suited to higher temperatures will prosper in comparison with their fellows, so that those alleles will be more common in the next and later generations. Mutations that might previously have sharply reduced the fitness of the organism may also appear and be perpetuated. The consequence will be a changed population, with different frequencies of the old alleles and perhaps novel alleles as well.

This can never be a painless process. The organisms carrying alleles disfavored at increased temperature will produce fewer offspring than previously, but even individuals carrying the favored alleles may suffer reduced fitness until novel alleles appear, of necessity by chance. Almost inevitably, the result is that the population will decrease in size as the generations pass, and will decrease more rapidly the faster the temperature increases. Adaptation is a hazardous business. It is therefore a matter of chance whether the adapted species eventually survives (in which case the population may grow again). The chance that the species will be snuffed out will be greater if the environmental change is rapid. Genetic diversity, on the other hand, will assist survival. Interestingly, the character of the population emerging from such a crisis will be shaped not only by the alleles favored at higher temperatures, but also by the other genetic characteristics (which may be irrelevant to the adaptation) of the relatively small number of organisms when the population is a minimum. This is known as the founder effect.

The process of adaptation is essentially that described by Darwin, but since the 1920s it has also been described in mathematical language, thanks to the example of R. A. Fisher and J. B. S. Haldane. ("Population biology," as this art-form is called, is one of the few uses of mathematics in biology to which all biologists are reconciled, often with unaccustomed enthusiasm.) The conclusion is that successful adaptation to changing circumstances (or the opportunities offered by a novel niche) is not inevitable. The tyranny of natural selection is too powerful. The only acceptable adaptations are those that increase the fitness of the species in the here and now. And genetics also takes a hand: adaptations must be fashioned by mutation from the genes that lie to hand and from their arrangement within the genome. Pigs will not fly anytime soon; they lack the developmental genes that make wings rather than forelimbs.[13]

Darwinism as such offers no general rules for telling what course adaptation will take, nor could it. What matters is the interplay between the genetic potential of an evolving species, the character of the environment and the nature of the changes being brought about. The ending of the Ice Age between 18,000 and 12,000 years ago may have been the reason why mammoths disappeared from the Northern Hemisphere.[14] Why were they apparently incapable of throwing up genetic changes that might have made survival possible? Complex questions such as these cannot be tackled with the information now available. Similarly, Gould's implied question about the causes of the rapid appearance of novel species at the Cambrian Explosion must eventually be answered in the same way, as the interplay between environmental and genetic change—not magic.

That will be possible only when the process of speciation is fully understood in modern genetic language. Heritable variation and natural selection, the central tenets of Darwinism, will then be found to account for what is known of the patterns of evolution. The obvious place to look is in the evolution of *Homo sapiens*. As will be seen, the human lineage has changed more markedly in the 4.5 million years of its independent existence than has that of the great apes during the same period.[15]

That points to another caveat: continuing ignorance about the difference between a species and the ancestor species from which it has evolved. The operational distinction between living species is straightforward; members of distinct species are mutually infertile.[16] Darwin's observation of the finches of the Galapagos Islands led him to conclude that prolonged geographical isolation is one route to the emergence of new species. Two identical populations may drift so far apart that their members can no longer interbreed. So much is accepted still. Darwin wrote at great length about what he called sexual selection, dependent on supposed ingrained mating preferences, which he recognized to be another means by which a population might separate into at least two parts that are reproductively isolated from each other. With the passage of time, subgroups of the population would then be affected differently by natural selection. Darwin supposed that the passage of generations would yield mutual infertility and new species. What was missing, and is still missing, is a rounded understanding of the genetic causes of mutual infertility. A lack of understanding of the mechanism of inheritance is no longer an excuse: is this not the age of the gene?

THE ORIGINS OF SPECIES?

Despite the rapid accumulation of information about genes and the functions of their products in cells, there is still no clear picture of the genetic basis of the differences between related species. Comparison of the genetic constitutions of species evolved from a common ancestor is one way to tackle the question, but the genomes of related species as they are now may be a confusing guide to the reasons why they originally became mutually infertile—the process of speciation as it is called. The difficulty is that once species have become self-contained breeding entities, further evolution may bury the original genetic trigger of separation in a welter of later evolutionary change.

There are nevertheless a few well-documented cases in which point mutations in some gene, the simplest form of genetic change, has led to separate species. One of the most remarkable is the cause of the separation between the two fruit-fly species *D. melanogaster* and *D. simulans,* where the difference between the two species consists exclusively of four point mutations of a gene called *per,* whose function is to regulate the rhythm with which the wings beat.[17] Because the mating behavior of the fruit fly depends sensitively on the rhythm of wing beating, this leads to reproductive isolation of the flies with and without the mutated allele. Much the same accounts for the distinct species of snails whose shells are coiled in opposite directions; mutation of a single gene causes the coiling direction to change, creating mechanical difficulties in copulation and thus the reproductive isolation of species.[18] No doubt it will also be found that mutations within the regions controlling the activity of genes can also be causes of speciation.

Most speciation events are more complicated than that however, affecting more than a single gene. Jerry A. Coyne from the University of Chicago, writing in 1992, concluded that several genes are usually involved in establishing the reproductive isolation of two groups within what is originally a single population.[19] Coyne also concludes that the genetic changes accompanying the formation of a new species almost invariably include an important role for the sex chromosomes (X and Y).

It is possible to create fruit-fly hybrids between different species but they are not always fertile. Coyne refers to one striking study in which males obtained by mating hybrid female fruit flies with one of their parental species were invariably infertile when the sex chromosomes were from the two different parental species, but otherwise were capable of producing viable sperm.[20] Infertility due to defective sperm is a proxy for incipient reproductive isolation, but there seems to be no general rule for telling in what way the sex chromosomes will influence the reproductive isolation leading to the formation of new species.

It is not surprising that the sex chromosomes are involved. The Y chromosome (in fruit flies as well as mammals) is known to regulate the development of male gonads, for example. What better way of building reproductive isolation into the genome than by a change ensuring, say, that hybrids always produce inviable sperm or even that they cannot be produced at all? There are two difficulties with that question. One is that not much is yet known of the functions of the genes in the X and Y chromosomes, or of the influences upon them of genes located elsewhere, let alone of how the sex chromosomes influence the others. The second difficulty is that the question is improper, implying as it does that natural selection can shape the structure of the genome in the long-term interests of the species. In reality, natural selection looks forward only generation by generation. The question we must ask is

"What is the selective advantage of reproductive isolation that accompanies speciation?" or, even, "Why do species exist?" The answer is that reproductive isolation preserves intact the genetic changes that have given an evolving group of animals (or plants) enhanced[21] fitness, ensuring that they are not diluted by the genes from the other creatures in the lineage from which they have separated.

The overriding question is when (and then how) sexual reproduction itself evolved. Despite decades of speculation, we do not know. The difficulty is that sexual reproduction creates complexity of the genome and the need for a separate mechanism for producing gametes. The metabolic cost of maintaining this system is huge, as is that of providing the organs specialized for sexual reproduction (the uterus of mammalian females, for example). What are the offsetting benefits?

The advantages of sexual reproduction are not obvious. One view is that sexual reproduction makes it easier for an evolving organism to get rid of deleterious genetic changes. That should certainly be the case if there is more than one genetic change and if their combined effect on the fitness of the evolving organisms is greater than the sum of the individual changes acting separately.[22] But there is no direct evidence to show that this rule is generally applicable. Indeed, a recent experiment with the bacterium E. coli suggests otherwise.[23] The telling feature of that experiment is that it was reported only in 1997, and concerned the most familiar organism in laboratory genetics. That shows how little has yet been done to found even rudimentary evolutionary speculation on laboratory investigations.

Sexual reproduction is also a more powerful way of generating genetic diversity than by the simple replication of existing organisms—for example, by growing individual trees and shrubs from pieces (called "cuttings" by gardeners) taken from a growing plant. Apart from the few mutations that may have accumulated, the new plants are true genetic copies of the originals. Sexual reproduction, by contrast, combines together in each cell a set of genes from two different individuals; moreover, each of those is a random assortment of the genes derived from that parent's own parents. Genetic diversity is the hallmark of sexual reproduction.

How did this process (and the complexities it entails) evolve? In a uniform and unchanging environment, lack of diversity is not a handicap; variations from the norm will usually be disadvantageous, the individuals carrying them will be less likely to survive to reproductive age and the norm will remain the norm. The dilemma is that natural selection cannot anticipate changes in the environment, and so arrange for the development of specialized sexual organs as a safeguard against environmental change. But even when the environment does not change from year to year, there are many circumstances in which genetic diversity is a selective advantage for an organism, as when fungi and many plants rely on the wind to scatter their spores and seeds, establishing their own genetic image in distant and different sites from those in which they grow themselves.

Much more must be learned of the course of evolution before it is known how (rather than why) sexual reproduction evolved, but meanwhile the yeasts provide a clue. These single-celled eukaryotes can exist with either a diploid or a haploid set of chromosomes—either 16 pairs or just one each of all 16 chromosomes respectively. Cells of both types can survive and multiply by ordinary cell division. The haploid cells come into being when nutrients are in short supply and do so in a manner resembling that in which gametes are produced in the gonads of animals. Although there is nothing in the yeast genome corresponding to the X and Y chromosomes of sexually reproducing animals, the haploid cells acquire a "mating type" (the next best thing to sex for yeasts), which reflects the configuration of a particular gene. These same cells can then "mate" only with other haploid cells of the appropriate type to reconstitute a diploid cell. This simple version of sexual reproduction has the property that the genetic constitution of the diploid and haploid forms of a yeast remain largely unchanged (within the limitations of the fidelity of the DNA repair mechanism), but that the formation of the haploid cells allows the exchange of genes between the corresponding parts of the pairs of chromosomes. That may be entirely appropriate when nutrients are in short supply, so that a degree of genetic diversity may be advantageous. Many other single-celled animals, species of the pond organism paramecium, for example, have an essentially similar mechanism of sexual reproduction.

There is a huge gap between this rudimentary form of sexual reproduction and that in the simplest multicellular animals and plants, yet it is probable that the first led to the second. It will be possible to be sure only when there is a detailed knowledge of how the development of the sexual apparatus is controlled genetically in a variety of living organisms, when it may be possible to learn from the structure of the genes concerned something of the order in which they made their appearance in the course of evolution. That task will require intricate work by future generations of biologists.

Meanwhile, it is important that there are other routes to speciation than the appearance of mutated alleles in one or even several genes. The rearrangement of genes among the different chromosomes in an organism appears commonplace, to judge from comparison of the genomes of separated species. That is true of *Homo sapiens,* whose genome differs most obviously from those of the great apes in having 46 rather than 48 chromosomes. Whatever happened to the chromosome that disappeared from the great apes? Genes closely similar to those on the missing ape chromosome are now mostly to be found on the long arm of human chromosome 2. But that is not the only rearrangement; the human X chromosome contains a region analogous to a similar region of the ape X chromosome which has been inverted and which is also copied, at least in part, by the human Y chromosome. There is no way of telling whether these rearrangements were the original causes of the separation of humans from the apes, or whether they are consequences of later evolutionary change. That will also have to wait on the listing of the genes, an understanding of how they function and even a detailed comparison of their sequences. Given the complexities of the roles of the sex chromosomes, that will not be an easy task either. The understanding of speciation is not much deeper than in Darwin's time.

KIN SELECTION AND ITS CONSEQUENCES

There is a further refinement of the mechanism of sexual reproduction to be found in the so-called social insects—bees, wasps, ants and so on.[24] Children at elementary schools are well drilled with accounts of how the social organization is maintained. Reproduction within, say, a hive of bees is entirely the responsibility of a single female, called the queen. The other insects in the hive are females, or workers, who forage for food and dance attendance on the queen, and males, called drones, who do nothing except eat. The British biologist W. D. Hamilton[25] was the first (in 1964) to ask what selective advantage there could be in an arrangement whereby all except one of the females in a hive supply all the others with food, sacrificing their own opportunity to breed in the process. "Altruism" was Hamilton's word.

The explanation of this behavior lies in the peculiarities of sexual reproduction in bees. The queen mates usually only once, storing a lifetime's supply of sperm with which to fertilize eggs. Unfortunately, queens' eggs come thick and fast; many of them escape fertilization and develop into males, which are haploid creatures with only one set of chromosomes. When such a male produces sperms to be stored for the reproductive lifetime of a queen, they are all genetically identical with each other. The queen's eggs are also haploid cells, but the genes carry a random assortment of the alleles from the queen's own parents. It follows that the male drones in a hive, derived as they are from an unfertilized egg, will differ from each other genetically—there is a 50 percent chance that they will carry one or an other of the queen's alleles for a particular gene. But female workers, derived from fertilized eggs, will carry all the alleles of their genetic male parent and, like the drones, half the alleles of the queen. What that means is that female workers will be more closely related genetically to their own sisters produced by the queen than they would be to their own offspring if they were to mate with a male in the hive. So, the inference is, worker females are reproductively self-denying because that behavior ensures the production of further females genetically more akin to themselves than would be possible by independent mating.

Hamilton's explanation rested on the notion of *kin selection,* or on the assertion that organisms behave in such a way as to favor the survival of close genetic relatives. That is the spirit in which parents defend their offspring against danger, for example. In this view, the peculiarities of reproduction in bees lead directly to the self-denying reproductive behavior of the female workers (maintained by the queen's secretion of hormones, which are volatile chemicals called pheromones which affect the whole hive). An arrangement in which the female workers do not reproduce themselves, but assist the queen to do so, is a more efficient way of perpetuating their own alleles.

Three vibrant controversies have sprung from Hamilton's explanation of the behavior of bees, one of which centers on the doctrine of the "selfish gene" put forward by the Oxford biologist Richard Dawkins in 1976.[26] The essence of his argument is summed up by the aphorism that "the organism is the gene's device for propagating as many copies of itself as possible." Dawkins has accurately described[27] the transformation (in the decade following first publication) of the general perception of his book from that of a restatement of the strict Darwinian position to that of a radical and extremist tract. Many biologists are offended by the notion that the animals and plants on which their studies are lavished are no more than vehicles for the propagation of genes. Still others protest[28] that the selfish gene necessarily implies "genetic determinism," or that the properties of organisms are almost mathematical consequences of the construction of their genes. But Dawkins's "gene's-eye" view of evolution is a perfectly valid restatement of Darwinism which has the virtue of giving extra insight to the understanding of a variety of biological phenomena—the relationships between parasites and their hosts, for example. The only sustainable criticism of Dawkins is that his use of the word "selfish" as an attribute of a gene is dangerously anthropomorphic.

There has been more substantial controversy over the notion of "sociobiology," which is ostensibly the application to human behavior of Darwinian principles and necessarily incomplete genetic knowledge. The idea first crystallized in 1975, in a book called *Sociobiology*[29] by the Harvard professor E.O. Wilson, who (then and now) has a distinguished reputation as an entymologist and naturalist. What more natural, it seemed at the time, than that the first explanations of the behavior of social insects (such as bees) should be extended to cover human behavior? There is nothing improper in such an ambition. Indeed, it must be a long-term goal for evolutionary science that there should be a more rounded understanding of how the patterns of human behavior recorded by psychologists over the past century have been shaped by the interplay of natural selection acting on the human genetic endowment and of the particular environment that affects individuals. But that time has not yet arrived, and will not do so until there is a much fuller understanding of how the human brain works than anything now in prospect. In 1974, the program of sociobiology was, to say the least, premature.

Inevitably, the debate triggered by the publication of *Sociobiology* caused a great commotion, much of it based on misunderstanding of what Wilson had written. Hamilton had earlier explained the behavior of the social insects in terms of what he called "kin selection"—the solicitude of individual insects for their genetic relatives. The same concept was now applied to human social behavior such as nepotism (the temptation of people in positions of power to appoint their relatives to senior jobs) and even to the supposed indulgence of nephews and nieces by their homosexual uncles.[30] The truth is that the social situations accounted for are poorly described—the literature of sociobiology is heavy with anecdotal information—but also warrant much more complicated explanations than those applicable to bees in a hive. Sociobiology never graduated to be an academic discipline, as is much of the remainder of science.[31]

Regrettably, the place of sociobiology in the intellectual firmament has now been succeeded by "evolutionary psychology." Again the purpose is laudable enough: to study the behavioral patterns of animals of different kinds with such care that it will be possible to work out how behavior has been molded by natural selection. But again the temptation to interpret social phenomena, from drug abuse to the prevalence of divorce, appears to have proved irresistible for the practitioners of these stripling pursuits. Plausibility rather than proof seems to have become the touchstone of what constitutes an explanation. One must sympathize, of course, with professional psychologists, much of whose work now centers on gathering information that will throw light on the working of the human brain (and who tend to call themselves "cognitive scientists"), and where understanding will in due course make it possible to rebuild classical psychology on new foundations. But the speculations of evolutionary psychology are so manifestly lacking an empirical foundation that they trivialize the reputation of science.

Meanwhile, there is a more substantial (because recurrent) controversy about the concept of "group selection," first advanced in the 1960s by the British biologist V. C. Wynne-Edwards.[32] The assertion is that natural selection may act on whole groups of animals or plants, and not exclusively on individuals as strict Darwinism supposes. The distinctive feature of Wynne-Edwards's position is the remark that most groups of animals do not breed so freely that they exhaust the food potential of their environment, starving as a consequence. He argued, for example, that the occasional habit of many species of birds to form large flocks (or shoals) is a means by which the animals sense the size of the population to which they belong, and in the process learn to follow prudent reproductive behavior. The complex behavior of the social insects, which is evidently the product of evolution, was originally cited in favor of group selection only to be undermined by Hamilton's notion of kin selection.

The difficulty with group selection is that there is no obvious way in which the environment can influence the welfare of a whole group of organisms rather than that of its individual members, while many of the supposed illustrations of group selection turn out to be explicable in terms of Darwinian natural selection. For example, the size of the population of a species is often regulated by the territorial habit, which assures breeding pairs an adequate food supply while preventing reproduction by those who have failed to capture a territory. Nevertheless, group selection is likely for some time to be advocated as an explanation of the social behavior of animals and plants. From the point of view of the selfish gene, of course, it is a nonsense: if genes, or rather alleles, are selfish, why should they cooperate in the survival of other alleles, with which they are in competition?

That great controversies of this kind should repeatedly arise in no way detracts from Darwinism as an organizing principle in biology. The doctrine is nothing more than a statement that the emergence of all forms of life has been shaped by the interplay between natural selection and adaptation (itself restricted by the genetic complement of all organisms at any time). It is rather a measure of the usefulness of Darwinism that it appears to be a perennial source of stimulation in biology (not to mention sociobiology and affairs more generally). That is the spirit in which evolutionists have recently embraced the notion of "co-evolution," which amounts to little more than a recognition that, for living creatures, the most significant part of the environment in which they must evolve will consist of other creatures, either predators or prey. Thus it is probable that the characteristics of the creatures now alive have been shaped, at least to some degree, by the nature of the living as well as the physical environment. This concept has potentially profound implications. Searching for empirical confirmation of what are still largely theoretical expectations will enliven the study of evolution. But it will be noticed that the mere notion of co-evolution multiplies to a huge degree the historical accidents that have shaped the present composition of the ecosphere.

TELLING GENETIC TIME

The sequences of genes have already been used to ascribe dates to recognizable events in evolutionary history. The principle is straightforward. Mutations of DNA occur whenever cells divide, and are not always removed by the mechanisms for repairing errors. No biochemical mechanism can be absolutely free from error. So why not estimate the evolutionary relationship between two distinct but similar organisms by comparing the structure of genes whose products (either proteins or RNA molecules) carry out similar functions? The greater the differences in the structures of the genes, the more time is likely to have passed since their common ancestor was alive.

People have nursed this ambition since the 1950s, when the only data available were the order of amino acid units in the chains of a few protein enzymes. Now there are large amounts of data about the structure of DNA and RNA molecules, which makes the goal of constructing evolutionary clocks more realistic. Those who practice this art call it molecular chronometry. The practical difficulty is that there is no universal molecular clock, nor can there be. For the time being, the techniques are best used for constructing what are essentially family trees showing the relationships between related species. This research is known as molecular phylogeny.

There is, for example, the question of the relationship between the various globin genes (whose protein products are components of hemoglobin) in the human genome, which contains six globin genes. The protein molecules called α-globin and β-globin are found (in pairs) in adult hemoglobin. (The other versions of the gene contribute to fetal hemoglobin.) Comparison of the structure of these genes suggests that the α and β versions first arose by a duplication of a primordial gene some 500 million years ago, and that the other variants have arisen by mutation of the α-globin gene within the past 100 million years. Evidently, information of that kind is invaluable not only in tracking the course of evolution, but for the understanding of life more generally. How reliable is it?

The first problem is that the ancestral gene no longer exists; all that can be done is to compare the structure of the genes contributing to hemoglobin in the blood of living animals. Then it is necessary to make assumptions about the likelihood that a particular nucleotide will change into some other. Mutations that affect the function of one or another of the globin molecules for the worse will not be perpetuated and so will seem never to occur. Indeed, the likelihood that a particular nucleotide will be changed into some other will depend on where it is placed within a gene. The result is that the evolutionary clock cannot be calibrated absolutely, as if it were an ordinary clock, but only by reference to other information—the structure of comparable proteins and the relationships between the organisms in which they occur.[33] That is why molecular phylogeny is, for the time being, more reliable than molecular chronology.

Even so, the potential of this technique is already plain. The most ambitious exercise so far attempted is due to Russell F. Doolittle (at the University of California at San Diego). With a group of colleagues, he compared no fewer than 57 proteins with comparable functions in organisms so diverse that they span all the various kingdoms of living things.[34] The main result is that all creatures now alive are descended from a common ancestor living roughly 2,000 million years ago.[35] In passing, this monumental comparison assigns dates to recent evolutionary events—the emergence of land animals (tetrapods) 400 million years ago, for example—which agree with the fossil record.

So has that provided paleontology with a framework within which all evolutionary events can be dated? Sadly, not yet. Many of the obvious snags have been raised in discussion of Doolittle's ambitious comparison. The crucial difficulty is the need for assumptions about the rate at which mutations occur at different sites along the length of protein molecules (or in the corresponding triplet codons in the genetic DNA).

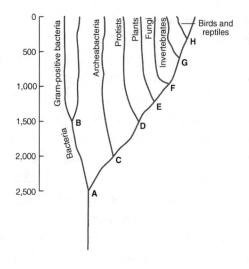

Figure 7.1 Doolittle's Family Tree. The family tree constructed by Russell F. Doolittle amended to allow for the recognition that eukaryotes made their first appearance at least 2,500 million years ago. (Freely adapted from the cladogram in the original article, with adjustment for the earlier ages for the emergence of eukaryotes allowed by Doolittle in later controversy, referred to in the text.)

Some sites are more accommodating of variation than others. In molecules of protein enzymes, for example, it is usually necessary that there should be specific amino acids at a number of particular places (called active sites), while the functions of other parts of the molecules are much less specific. There is a further twist: the particular sites in a protein where variations from the norm destroy the function of the protein may not have been so sensitive at earlier stages in evolution. Some of Doolittle's critics say that neglect of this complication has led him to an excessively late estimate for the first appearance of (single-celled) eukaryotes, claiming that a better date would have been as early as 3.5 billion years ago.[36] Another obvious complication is that Doolittle's argument makes no allowance for the transfer of genetic information from one species to another by means of infectious agents—

plasmids (for bacteria), viruses (for all kinds of organisms) and transposable elements. In response to his critics, Doolittle has accepted that a date of more than 2.5 billion years would be more appropriate for the separation of eukaryotes and bacteria.

None of that implies that molecular chronometry is too unreliable to be taken seriously. On the contrary, it is the means by which history of evolution will eventually be anchored in time as accurately as, say, the reigns of the successive kings of England. But the rapid accumulation of data about the sequences of genes and proteins cannot yet throw an accurate light on the timing of evolution over 4 billion years. The molecular clocks now being built may be little better than the timepieces in use 2,000 years ago, based on the dripping of water through a fixed orifice into a measuring vessel. The problems that now beset molecular chronometry are exactly comparable with those that have perplexed cosmologists in the past thirty years in the measurement of intergalactic distances. Like cosmologists, the molecular timekeepers will find it necessary to work backward in time one step after another, hoping in the process to learn which genes or proteins are the most reliable indicators of the passage of time, and what has been the recent influence of the horizontal transfer of genetic information.

The moral is that molecular chronometry can never be simple. Chemical analysis of a gene or protein will not in itself yield an estimated date for the separation of two species from each other. Fossils dug from the ground and often detailed knowledge of the biology of the organisms concerned will always be required to make good sense of what is learned from molecular chronometry.

Nevertheless, a particular application of molecular chronometry deserves attention. Molecular clocks based on stretches of DNA with no known function would be expected to keep ticking at a uniform rate. Beginning in the 1960s, the late Allan T. Wilson from the University of California at Berkeley seized on a stretch of nonfunctional DNA from human and other mammalian mitochondria for the purposes of chronometry. The difficulty of recognizing the corresponding parts of the

DNA molecules limits the length of past time for which reliable dates can be obtained. (Because mutations are unconstrained, the clock runs faster than Doolittle's but, for the same reason, its reliability is limited to less than a million years or so.) Such an analysis has led to the claim that the separation of the Caucasian and Oriental lineages from the African peoples occurred relatively recently, perhaps only 125,000 years ago. That estimate is surprisingly recent, which is why it is arresting.

The challenge for the years ahead is formidable. Nobody now disputes the value of sequence comparisons in reconstructing the relationships between species—the phylogeny—but even there it is important that the frequent inconsistencies arising from different comparisons should be understood. Making molecular clocks more reliable is not simply a matter of more data and more computer time, but of a better understanding of the constraints on allowable mutations. Moreover, the agents of the horizontal transfer of genetic information are interesting in themselves: were there, as some have suggested,[37] episodes in evolutionary history when this was one of the mainsprings of evolution? That remains to be discovered.

THE HUMAN CONFUSION

The problem of human evolution is simply stated. All species of the great apes have very similar genes to those in people: the globin genes are all essentially the same, for example. The DNA structures of corresponding genes differ slightly from each other, but by no more than can be accounted for by the mutations accumulating in 5 or 6 million years. To be sure, the catalogs of genes are still incomplete—there is no Great-Ape Genome Project yet. Yet the genomes of the species differ in a dramatic way: people have only 46 chromosomes while the great ape species all have 48 (23 and 24 distinct pairs respectively.)

Whatever happened to the "missing" chromosome of the great apes? The chromosome has not disappeared, but is substantially incorporated at the end of the long arm of the human chromosome called chromosome 2. It is a "translocation" resembling those responsible for many types of cancer, as described in the previous chapter. In this case, the translocation is the most conspicuous difference between *Homo sapiens* and the great apes, although it is not yet known why the effect of the translocation should be so profound. It may be possible to make educated guesses when the Human Genome Project is complete; the great-ape genes on human chromosome 2 may turn out to be regulated differently. It is also possible that some of them (or some other genes) have been disrupted or duplicated. Only detailed comparison will tell that—and only when there is a fuller catalog of the functions of human genes than there is now. Nevertheless, one of the mechanisms of species formation seems to consist predominantly of a chromosome translocation. It remains to be seen how that can be squared with Darwin's gradualism.

The timing of the divergence between *Homo sapiens* and the great apes agrees well with the evidence of the fossil record. In 1995, a group of paleontologists from the United States (from Berkeley, California) and Ethiopia described their finding of a hominid[38] skull and part of a femur (thigh-bone) in central Ethiopia that was dated with reasonable confidence to 4.3 million years ago. The fossilized remains are by far the most ancient hominid fossils yet found, while the creature concerned (called *Australopithecus afarensis*) seems to have resembled the chimpanzee more closely than any of the other great apes,[39] which again agrees with the equivocal evidence the geneticists marshal.

Paleontologists have always held the habit of walking upright on two feet to be the hallmark of *Homo sapiens;* great apes use their knuckles, but are better adapted for climbing trees. With the continuing discovery of fossils (particularly of skulls), it became plain in the 1930s that there was a progressive threefold increase in the size of the skull in the course of human evolution. Now the faculty of using language (understanding spoken speech, reading written words, making sense of them and articulating a reply) would be taken as an even more specifically human attribute. Can a simple translocation of a chromosome account for all that? Not by itself.

Confusion about the origin of *Homo sapiens* arises because the fossil record of the great apes is meager or almost nonexistent, for reasons that are not well understood. Fossils of *Australopithecus* from 4.3 million years ago appear to be much like modern chimpanzees: the great apes have not changed much, but *Homo sapiens* has had a much more checkered history. The direct lineage to *Homo sapiens* appears to run through the hominids called Australopithecenes up to *Homo erectus* (from about 2.3 million years ago), but along the way, there seem to have been many occasions where more than one potentially hominid species coexisted. Several different Australopithecene fossils are classified as belonging to distinct species, but it is impossible to be sure that part of a single skeleton can be representative of mutually infertile species.

What is, however, clear is that in attempts to reconstruct a Darwinian pathway for the evolution of *Homo sapiens*, there are many exceptions to any neat and tidy scheme. What is to be made of the species named by the late Louis Leakey as *Zinjanthropus boiseii* from 1.2 million years ago (now renamed *Australopithecine boisei*), of Java man (800,000 years ago), of Peking man and Boxgrove man (both about 500,000 years ago) and even the Neanderthals, whose often well-preserved remains date from as recently as 30,000 years ago? All these creatures seem to be separate from the line of descent to modern human beings. During the past four million years, it is plain that there were long periods when different forms of hominids were alive at the same time—*H. erectus* and *H. robustus* at similar locations in east Africa, *H. erectus* and Peking man at separate locations and *H. sapiens* together with the Neanderthals in Central Europe as recently as 30,000 years ago.

Confusion can however be dispelled. The human lineage plainly runs from the Australopithecenes (typified by *Australopithecus afarensis)* found in Ethiopia to *Homo erectus* (at 2.1 million years ago), which is known from the anatomy of the skeletons to have walked upright. *Homo erectus* was the genetic source of several forms of hominids, such as *H. robustus* (found alongside *H. erectus* in Kenya) as well as the forms of early man found elsewhere, from Boxgrove in the United Kingdom to Peking and Java in the Far East; presumably they were the first of many waves of emigration from Africa.

The remaining uncertainty concerns the age at which *Homo sapiens* appeared. Wilson's estimate of 125,000 years for the age of the most recent common ancestor of all the human groups is based on the analysis of mitochondrial DNA in modern populations. The stretch of DNA concerned has no effect on people's well-being; it is simply a marker in which mutations signify the passage of time. The implication is that all the people now alive are descendants of a group of people living 125,000 years ago who shared essentially the same stretch of mitochondrial DNA. How many people were in the group? A few hundred, perhaps, or even a thousand or so. The group must have been so small that there would have been no significant variations of the marker DNA among them. And what is the basis on which the group was afterward able to grow so quickly? They must have shared a powerful selective advantage: the faculty of language is the obvious candidate.

There is no proof that the people who came out of Africa 125,000 years ago were the first human beings able to speak to each other, but that fits in well with much else that is known from the fossil record. Why, for example, did the Neanderthals disappear from Central Europe? In the fossil record, they predate the most recent migration out of Africa; presumably they could not speak. Indeed, a remarkable study of mitochondrial DNA recovered from the type specimen of the Neanderthals during 1997 shows that the genetic lineage of these creatures diverged from the lineage of modern man roughly 600,000 years ago.[40] The inference must be that they lost the competition for resources once their articulate successors had occupied their territory.

The best place to look for confirmation of this surmise is not in fossil skeletons, but in the human genome. Soon, there will be a listing of all the human genes. In due course it will be possible to identify those whose chief function is in the development of the brain. Eventually it should be possible to tell which are particularly involved with the development of language and to glean from them something about their origin. It will be a task comparable in difficulty with the old biblical task of making bricks without straw. Every scrap of information from paleontology and ecology will be required to ensure success. But that is how the history of the human race will be reconstructed in the next century or so.

Homo Sapiens?

There are more immediate questions about human evolution to be tackled with the help of the new genetics. What set the process in train, for example? And is it true that the cockpit of change was in Africa, and if so why?

We are not yet sufficiently aware of how far genetics has transformed the business of understanding human life. Take, for example, orthodontics—the branch of dental surgery concerned with malformations of the facial bones and especially of the jaw. The genes involved in the development of the front part of the vertebrate skull are now known and the inheritable human condition known as Saethre-Chotzen syndrome, which includes a misshapen face among its diagnostic criteria, has now been traced to aberrant alleles of a gene known as *twist*, first identified in the fruit fly; normal alleles of *twist* are essential for the normal development of the bones of the face and the head more generally, in both the mouse and in human beings.[41] That is of great interest to orthodontists and, potentially, to their patients.

When people have the time and inclination, they will look in the genomes of Darwin's finches to see whether similar genes are involved in the development of the specialized beaks that distinguish the different species. More to the point, it should soon be feasible to identify in the human genome the genes (probably many of them) involved in shaping the distinctively human junction of the pelvis and the spine that makes bipedalism possible; it may even be possible to tell from molecular chronology when in the past 5 million years it made its first appearance. In the same way, the developmental genes of the nervous system offer a means of learning something about the origin of language in human beings (see Chapter 8).

What set human evolution in train is another open question. Given that the process occupied the geological period called the Pleistocene, during which successive waves of glaciation occupied the surface of the Earth, it has been tempting to suppose that the evolution of *Homo sapiens* was provoked by the stresses (or the opportunities) of a cooler climate. Certainly Africa, where the earliest hominid fossils have been found, was then a cooler and wetter place. But the most ancient fossils seem to predate the onset of the Pleistocene by a relatively small if significant interval. (The Pleistocene period is usually dated at 2.5 million years ago, but little is known of the climatic change that preceded it.) Further study may remove the discrepancy, but it is also possible that the precipitating cause was something different—perhaps even a simple genetic accident, in which case traces of it may be found in the analysis of the "junk" DNA that appears to fill the human genome. That is an important challenge for the years ahead.

Other questions about the origin of the human population also cry out for attention. That the earliest hominids evolved in Africa seems beyond dispute (although common ancestors of the hominids and the great apes have been identified mostly in Central Asia). The importance of Africa as the cradle of human evolution is also borne out by the genetic diversity of the African population now extant: by some measures, the genetic differences between identifiable African populations are greater than the differences between African people and the remainder of the human population. That suggests great antiquity.

So why not apply genetic analysis to the genes in human populations to pin down their evolutionary relationships with each other? That goal has been in the front of many people's minds for a decade, and certainly since Wilson's startling suggestion that modern populations outside Africa are all derived from a "migration out of Africa" as recently as 125,000 years ago. The surprise, of course, is that the fossil record of much earlier times clearly shows the presence of people not markedly different from ancestral forms of *Homo sapiens* in Europe, eastern Asia and the Middle East.

In principle, the occupation of most of the Earth's surface by people could be traced by the analysis of the alleles in modern populations. For the past decade a project called the Human Genome Diversity Project has been seeking approval (and funds) to mount a systematic study of the genetic structure of different populations. The plan is to collect DNA samples from people of recognizably different population groups, and to make those available to approved researchers internationally.[42]

There is no doubt that the claims made on behalf of the project are valid and compelling. It is potentially a way of uncovering in great detail the cultural and even ethnic history of *Homo sapiens*—of telling once and for all when the first people arrived in North America, for example, and where they came from, or how the cultural history of Europe has been shaped by successive migrations from elsewhere in the world. What other new knowledge could as dramatically change our regard for our place in the world and for our history?

Yet the project remains in limbo, and for an important reason. To the extent that the DNA from aboriginal populations is likely to be most informative about the ancient relationships between people, they inevitably top the lists of those from whom DNA would be collected. With some justice, some of the groups concerned insist that they will not collaborate with the project so long as they lack an assurance that the genetic information gathered will not be misused, and to their disadvantage.

There are other difficulties than these. What, for example, would be the process by which an aboriginal (or any other) population would give its informed consent to such an investigation? Who would represent it in the negotiations? And what arrangements would (or could) there be to ensure that DNA collected is used only for the purposes declared?

The consequence is that, for the time being, the project is becalmed. What the world needs is a demonstration, mounted in some developed country, of just such a project intended to throw light on the diversity of a well-established population. Until that is done, there is little chance that one of the greatest opportunities for understanding where *Homo sapiens* came from, and how, will command the support it needs.

EVOLVING EVOLUTION

The recent record reinforces the lesson of the years since 1858: that Darwin's account of evolution remains intact. Hereditable variation and natural selection are indeed the agents that shaped the present richness of life on Earth. The new genetics does not challenge Darwinism but, on the contrary, is the means by which the details of the course of evolution will be unraveled from the sketchy fossil record and the growing accumulation of data about the genetic constitution of animals and plants.

There are several opportunities to be grasped in the years ahead, among which the construction of a reliable chronometry of evolution is conspicuous. It will not, however, be one clock, but several interlocking clocks, all of them founded on a detailed knowledge of several genes and perhaps whole genomes from related organisms. From that, as the decades pass, we shall learn when important adaptations, say sexual reproduction, made their first appearance, and how. An important part of that enterprise will be to match the fossil record of *Homo sapiens* with a genetic account of what has happened in the past 4.5 million years. Then, perhaps a century from now, we shall have a rounded understanding of our place in nature.

It will be noticed that the basis of molecular chronometry is also a way of telling what may have been the structure of DNA in organisms long since extinct. The idea is that related groups of organisms are derived from a common ancestor at some time in the past, and that it is possible to infer the time from the mutations that have accumulated in the DNA, which implies that it should also be possible to infer the structure of the ancestral DNA. Science-fiction writers will appreciate the significance of that; that could put the reconstructed dinosaurs of the film *Jurassic Park* on a firmer scientific footing, among other things. Of course, reconstructing whole genomes would require complete genome sequences for the related organisms now extant, which is not now a realistic prospect. Meanwhile, the replacement of modern by ancestral genes in living organisms could become an invaluable means of investigating their functions, both ancient and modern.

None of this will be accomplished exclusively by geneticists working in laboratories. The fossil record is now even more important, for it is the only way in which it will be possible to guess at the succession of historic accidents that have shaped the present biosphere. Indeed, paleontologists will find that in the years ahead, they will be compelled by the excitement of the new enterprise to be even more painstaking than in the past, seeking more accurate ages for the fossils they find and reconstructing wherever possible the ecology in which the creatures whose remains they collect were once embedded.

With luck, the great controversies of the recent and more distant past will at the same time die away. When it is possible to understand the details of the emergence of novel species, will it not be an academic question to know in precisely what sense genes can be described as "selfish?"

PART THREE

Our
World

Thinking Machines

8

During the twentieth century, people finally came to grips with the problem of the brain. When the century began, the gross anatomy of the human brain had been diligently described and several striking guesses had been made about the functions of some parts. In 1861, for example, Pierre Paul Broca, a Parisian physician, identified a region on the left side of the brain as that responsible for regulating the production of the sounds of speech. More remarkably, in the 1880s, the Spanish physiologist Ramón y Cajal, using a silver stain developed by the Italian Camillo Golgi, worked out that the essential units of the nervous system are cells with richly branched structures that communicate with each other while nevertheless not coming fully into contact.[1] Soon afterward, the first systematic investigation of the functions of the nervous systems of animals were planned by people such as Charles Sherrington (then at the University of Liverpool). Nobody then would or could have guessed how much has since been learned, especially in the past half century.

Yet the cruel truth is that the central objective of the now majestic research program in neuroscience remains beyond reach: there is only the most shaky understanding of how the brain, and the human brain particularly, engenders mind—the capacity to reflect on past events, to think and to imagine.

That is not a scandal. The question is truly perplexing. Since Aristotle's time, people have been asking what mind consists of, but without much success. Aristotle's own opinion was that mind is a combined function of the head, the heart and even the blood. On the principle that the only creatures known to be capable of thought are people, the idea that the inside of the head is occupied by a tiny person, a *homunculus*, gained currency in the Middle Ages, before Copernicus's time. Then and since, the question has been complicated by the notion of the individual soul—that which many people believe to be immortal. If mind is in the head, must not the soul be there also?

For most of the past century, these questions have been considered as distractions from the investigation of the nervous systems of animals and of how they function. Only since the late 1980s, for example, has the question "What is consciousness?" been regarded as a legitimate challenge for neuroscience. Instead, investigations have centered on simpler questions such as how an animal uses its nervous system to direct the movement of its limbs and how signals from the outside world collected by the sensory organs—eyes, ears, noses and fingertips, for example—are processed within the head, usually stimulating an appropriate response. As will be seen, those tasks are difficult enough. Even after a century of remarkable discoveries about the cellular components of the animal brain, their interconnections and the behavior of the species of animals in which they are embedded, not even the nervous system of the simplest creature is yet fully understood. It may have become permissible to ask what and where consciousness is, but it is too soon to expect a detailed answer.

WHAT IS EXPLANATION?

The precise meaning of "understanding" in this connection is all-important. In animals, the brain is usually in the head and is the chief content of the skull if there is one, but it is not the whole of an animal's nervous system. The sensory organs, the eyes for example, are so directly connected to the brain that they are part of it. The same is true of the ears, the nose and even of the specialized cells found in those extremities which have the capacity for transforming mechanical pressure into nerve signals—the sense of touch.[2] The network of specialized nerve cells (called neurons) that penetrates to every part of the body of an animal is an integrated organ, the nervous system.

For the best part of this century, physiologists and psychologists (among other specialists) have provided evermore detailed descriptions of the functions of the nervous system in animals, but these are also matters of everyday experience. Show a cat a mouse, and the cat will begin to stalk the smaller mammal, perhaps eventually killing it; show a cat a dog, on the other hand, and it will arch its back, fluff out its coat and bare its teeth, waiting for a chance to escape safely. These responses, commonplace and universal, are accomplishments of the cat's nervous system. But even the most detailed description of the response of a cat to circumstances like these is not an explanation of how its brain works. It is an explanation only in the sense that the statement, "Rotate the steering wheel clockwise, and the vehicle will turn right!" is an explanation of the complex system called an automobile. With cats, there are further complications; there are many ways of stalking a mouse, depending on the obstacles between the cat and its quarry. The strategy followed in any particular set of circumstances is again determined by the cat's brain, but there are no generalized descriptions of the real-life hunting behavior of a cat. Phenomenological description of this kind, however detailed, is not so much an explanation as a listing of what needs to be explained. The same is true of this century's patient study of animal behavior (now called ethology), most of human psychology and a large part of modern linguistics. From William MacDougall and Sigmund Freud to Noam Chomsky, the people who have seemed to be answering important questions have been asking them instead. The list of what needs to be explained is long.

That statement is not intended to be slighting of psychology and related studies, but is a vivid illustration of how progress in science entails the progressive deepening of the questions we ask about the world. No longer is a description of how a cat responds to the sight of a mouse, or of how a human adult behaves when attracted by a member of the opposite sex, taken as an explanation. With the prospect that there may soon be an explanation in terms of how neurons are organized and behave, it is inevitable that psychology should have largely become a handmaiden of neuroscience (with fancy names such as "cognitive psychology" to mark the change). The interesting question is how soon psychologists will begin to define the questions that will or should concern them when there is a rounded understanding of how the brain works.

At the other end of the spectrum of explanation is the study of the neurons and of their interactions with each other. This is the aspect of the functioning of the animal brain in which this century has won most success. There is a great variety of different types of neurons. Some are chiefly engaged in transferring signals to others like themselves, while others are concerned with keeping the signaling neurons in a functional condition. Different parts of the brain differ in the proportions of the various types of neurons they contain, as do the mechanisms of signaling between neurons.

What has been learned of neurons resembles the knowledge of atoms accumulated in the first two decades of the nineteenth century, and which eventually led to a successful atomic theory. But John Dalton's atomic theory was a framework for explaining the whole of chemistry only when it was supplemented by rules describing how many atoms of one kind could combine with one atom of another to form a stable compound—the rules of *valency*—as they were called. To begin with, they were strictly empirical rules; they were themselves understood only in the 1930s, in the light of quantum mechanics.

So why not keep on gathering more information about neurons until the functioning of a whole brain is understood as the collective response of all its neurons to the stimuli that reach them from the outside world? More information is certainly needed, but there are two reasons why the outcome would not be a satisfactory explanation of the brain. First, even the simplest brain is too complex for its properties to be predictable from those of its smallest constituents. Second, there are conceptual difficulties to be surmounted.

Take the cat's response to the sight of a mouse or a dog. The past century has shown that the visual system stores a distorted two-dimensional image of what the eyes see in the adjacent neurons of several distinct patches of the cortex of the brain—the thin outer layer nearest the skull.[3] Suppose the cat has only two responses: either it begins to stalk (a mouse) or it responds (to a dog) with the most threatening behavior it can simulate. How is the choice between the two strategies made in the cat's brain? What other information than that provided by the sight of the mouse or dog is required for the decision to be made? Perhaps past experience of mice or dogs is relevant, perhaps the sense of smell helps to distinguish between prey and potential predator. Whatever the case, the fact that the cat's brain makes a "decision" and then executes an appropriate sequence of movements is the essence of what needs to be explained.

How is that done? By what means is information gathered by the senses turned into an appropriate pattern of animal behavior, itself brought about by complicated series of muscular movements? It is immaterial whether the cat is "conscious" of what it is about. The question is simply how a collection of apparently autonomous neurons in the cat's brain can conspire to trigger the response that best matches the cat's self-interest in its own survival—catching mice for food, or escaping damage by dogs. That still mysterious step in the working of the brain is the transduction of a sensory image into one of several muscular and behavioral responses.[4]

So far as we know, the human brain is much more complex than that of the cat. For one thing, there is the faculty of language. For another, we know that consciousness enables many decisions to be made by deliberation—the calculation (which may often amount to self-deceit) of which course of action will be most advantageous. But an explanation of the human brain, when it is forthcoming, will be qualitatively the same as that of how a cat responds appropriately when confronted with the sight of either a mouse or a dog. The transduction of sensory information, stored somehow in the brain, into neuronal instructions for the muscles of the body to perform their tricks, will be the nub of it. The nature of consciousness will be something extra.

What form will that explanation take? First, there will have to be an understanding of how the simplest brains store sensory information: there is now ample evidence (from cats as well as primates, usually monkeys) that an external image appears nowhere in the head as a simple snapshot; instead, the several and different two-dimensional visual representations of the image are mixed with information of a quite different character. How these different representations are constructed in the first place and how they are used as a basis for decisions remains profoundly unclear—except that it is accomplished by neurons. Long-term memory, as necessary for a cat's appreciation that "mouse" means "prey" as for a person's rumination on the kinds of mousetraps that may be used to catch the pests, is another puzzle of the same kind. In the past few years, there have been hints as to *where* in the human brain long-term memories are stored, but it is not clear *how* that is done—except that there are only neurons to do the job.[5] Nor is it yet known how information gathered from outside the head is modulated by past experience. Then there is the business of translating into action the brain's analysis (conscious or otherwise) of the problem presented by a visual experience. Neurons are again the only agents in sight, but nobody knows how they do the job. An explanation of the mind, like that of the brain, must ultimately be an explanation in terms of the way that neurons function. After all, there is nothing else on which to rest an explanation.

THE EVOLVING BRAIN

One thing is nevertheless agreed: the animal brain is a product of natural selection—an adaptation to the emergence of multicellular animals or, possibly, the adaptation that first allowed successful multicellular animals to exist. It enables the different parts of an animal, which may be separated by distances very large compared with the dimensions of somatic cells, to adjust their functions to the needs of the organism as a whole. The selective advantage of the animal nervous system is easily imagined; it enables an animal not only to see and thus identify its prey or food, but also to reach out with whatever appendages it may have to grab it. But the nervous system really came into its own in animals capable of moving from one place to another; the nervous system drives an animal's fins, wings or legs in ways that propel it to where its self-interest requires it to move.[6]

A nervous system is not the only way in which the functions of large organisms can be coordinated. Plants (large trees among them), manage without nervous systems; instead, the coordinated response of a whole plant to, say, the arrival of spring each year is determined chemically and physically—by the "rising of the sap" in simple language. And even animal species with elaborate nervous systems, people for example, do not depend on them exclusively for sending signals from one part of the body to another. Hormones produced and secreted by the glands of the endocrine system (as well as by organs such as the kidneys) reach their target organs through the bloodstream, regulating their function decisively.[7] The immune system, which is physically present throughout the body of mammals and which responds in elaborate ways to the presence of foreign molecules, is similarly activated by chemical signals between various kinds of white blood cells to respond by ridding the body of the foreign material—virus, bacterium or whatever.

The decisive role of the nervous system in whole-body communication rests on its speed. The immune system can clear an infection in a day or thereabouts, hormones can influence distant organs in not much more than a few seconds (depending on the hormone and the target organs). But the nervous system can send signals even to distant parts of a body in a fraction of a second. The three systems are complementary to each other and, in vertebrates, have evolved in parallel.

Vertebrates, by definition, have backbones. Indeed, the vertebrate body plan appears designed around the nervous system. Within the stack of vertebrae called the spinal column is a linear array of nervous tissue called the *spinal cord,* perhaps a billion neurons strong, whose origin is traced directly to the way an embryo folds in upon itself in the first few days of its development,[8] forming a cylindrical tube one cell thick running along the length of what will eventually become the mid-line of the back. This structure, the *neural tube,* is eventually pinched off from the external surface of the back and becomes the spinal cord of the adult, with the brain at the upper end. The neural tube is the source of the immature neurons that form the spinal cord and the brain. At about the same time, a separate group of immature neurons is formed in a structure called the *neural crest;* these independently migrate to locations elsewhere in the still-young embryo to lay the foundations of what will become the peripheral parts of the nervous system.[9] Later in development, these peripheral cells grow back towards the spinal cord, making the connections that allow signaling of peripheral sensations to the spinal cord.

There is a strict timetable underlying the proliferation of neurons in an embryo; successive generations of cells are fated to acquire distinct functions when they mature. It is yet another process of differentiation controlled by a hierarchy of genes. The eventual location of neurons is also determined by genes analogous to the pattern-forming genes that shape the development of insects such as fruit flies. Remarkably, many neurons are generated some distance from their eventual locations, to which they must migrate. Yet the precise position and function of a neuron is not exactly specified in advance. A degree of randomness is unavoidable: the vertebrate embryo produces a large surplus of neurons, many of which are removed at an early stage if they are not repeatedly stimulated by use. The extent to which random processes are involved in the full assembly of the brain is obviously a matter of great importance.

Francis H. C. Crick, the co-discoverer of the structure of DNA, has spent the past quarter of a century brooding about the function of the nervous system. His influence on neurobiology internationally has been profound, notably because of the perceptiveness of the questions he has asked. In 1994, he published a book with the title *The Astonishing Hypothesis*[10] deliberately chosen (as the book explains) to dramatize the notion that the brain (and the human brain in particular) acquires its distinctive properties because the neurons appearing in the neural tube assemble into a functional brain by processes regulated chiefly by themselves but also by the environment of the embryo to which they belong.[11] The strong message of the book was that the time has come for "serious" neurobiologists to study consciousness, and that the study cannot be simple hypothesizing, but must be tested by experiment and observation. Mind, Crick says, is one of the collective properties of an appropriate collection of neurons. The details of the apparently autonomous self-assembly of the brain are now being worked out, in all their intricacy, by the techniques of molecular genetics.

The spinal cord in every vertebrate is almost a self-contained organ. It enables animals to do what most clearly distinguishes them from plants: to move about. The muscles that move the limbs are controlled by bundles of neurons leading off projections from the spinal cord and whose activity can be modulated within the spinal cord itself by signals gathered by sensory nervous tissue in the limbs. Put a hand too near a fire and the spinal cord will arrange to pull it back before the brain has had time to reflect on what is happening.[12] But the movement of an animal's limbs must also be modulated by the evidence of other senses, sight and hearing for example; otherwise animals would forever be colliding with objects in their path. Memory is also a selective advantage for an animal, enabling responses to be modulated by past experience. The spinal cord is not, therefore, an autonomous nervous system, but is repeatedly overridden by signals from the brain.

The vertebrate brain is not one organ, but has several recognizably separate parts, among which the sense organs are conspicuous. Its most remarkable feature is that it consists of two halves which are almost mirror-images of each other and which are linked together (in human beings) by a robust bundle of about a billion neurons forming a structure called the *corpus callosum*. Famously, bundles of neurons from the sense organs, the eyes and ears for example, mostly (but not exclusively) project to the opposite sides of the brain. (Thus the nerve fibers from the right eye are routed primarily to the left-hand side of the cerebral cortex—or to those parts of it responsible for vision, and called the "visual cortex"—but through a cascade of relay stations in the brain at which signals from right and left can be exchanged.) This crossing over of the neural connections does not, however, apply to the sense of smell; neurons from the right and left nostrils project to the right and left hemispheres of the brain respectively. That circumstance is almost certainly an evolutionary fossil—a sign that the sense of smell is the most primitive, and consistent with the observation that invertebrate animals, earthworms for example, do not have divided brains.

The identification of the left-hand side of the brain (behind the temple of the skull) as the region primarily concerned with the production of human speech shows that the two hemispheres of the human brain are not identical. In reality, the region, called the *language center*, is more complicated than originally described by Broca and in some people is on the right, not the left; those affected are often *left*-handed. The emergence of the language center in the course of evolution is a matter of great interest. Broca's discovery was based on his study of a patient with brain damage in the left temporal region, and who could not speak. So one inference is that the language center is necessary for vocalization. More recent observations, using the relatively new technique of magnetic resonance imaging (MRI, or "brain scanning," which can be used to tell where activity is concentrated even in the brains of living conscious people), have shown that the same structure is used when merely thinking about the meaning of words. Many questions arise. Is the language center involved in the grammatical and syntactical work that the brain must do when producing speech? (And if not, where is that done?) Where (and how) is the brain's lexicon stored?

Little is yet known of the neural arrangements for vocalization in the brains of animals other than people, yet dogs bark and birds sing. No doubt the years ahead will reveal structures in the brains of these animals that are used for making noise, raucous or song-like as the case may be; most probably there will be similarities of some unknown kind between those still unidentified structures and Broca's region, in the left templar region of most people's heads. Then it will be possible to attempt to reconstruct the evolutionary history of the human language center.

The question of when (and how) what is now the human language center became a unilateral or one-sided structure has particular interest. So far as is known, human beings are unique among animals in their capacity for using language, which is a means of communication far more stylized than the sounds that other animals make to or at each other. Noam Chomsky has argued persuasively, on linguistic grounds, that the rules of grammar are innate: sentences without verbs are ambiguous and even unintelligible. But the distinctive feature of our language, whichever we use, is its flexibility. Old words can acquire new meanings, new words are repeatedly devised to convey new and often abstract meanings, within limits novel syntactical tricks can be used to give a spoken sentence a special effect on those who hear it. Naturally it is tempting to suspect that the process in which the language center became unilateral (usually on the left) has something to do with the development of distinctively human language.

The evidence is equivocal. A very recent study has shown that, in 17 out of 18 chimpanzees (examined after death), there is an asymmetry that may betoken a precursor of the human language center; interestingly, the left hemisphere is larger than the right.[13] But chimpanzees, however intelligent they are, cannot speak, while the general principle of natural selection is that an evolving lineage of animals does not acquire adaptations that confer no benefit. It is therefore crucial to the understanding of how human language evolved to discover what value the slight asymmetry of the chimpanzee's brain may be (or have been).

Meanwhile, the one-sidedness of the human language center provokes other questions. Of all the distinctive characteristics of *Homo sapiens*, the faculty of language seems to be recently acquired. (The diversity of the languages still in use is compelling evidence in that direction.) It is tempting to speculate—but this is mere speculation—that the faculty of language gave the modern version of *Homo sapiens* the competitive advantages over early contemporaries such as the Neanderthals, in which case language may precede by only a little the last migration out of Africa perhaps as recently as 125,000 years ago. That possibility has important consequences, even for the understanding of the high prevalence in modern populations of psychiatric diseases such as schizophrenia. Timothy J. Crow, an Oxford-based researcher, thus argues that schizophrenia would long since have been eliminated by natural selection were it not a consequence of some important selectional advantage, and that language is the obvious candidate for that role. This has led him to the hypothesis that the genes that presumably control the development of the brain's asymmetry during childhood and early adolescence may not always yield the usual result, and that susceptibility to schizophrenia may be the result.[14]

Fortunately, there is now a prospect that the understanding of all these processes will be illuminated not merely by further studies of the anatomy of the brain, but by the study of the genes that guide the development of the nervous system. It should soon be possible, for example, to tell which genes are crucial to the development of the mammalian as distinct from the vertebrate brain more generally. Given some knowledge of what the genes concerned accomplish, it should then be feasible both to reconstruct the recent evolution of the mammalian brain and to understand in greater detail that process by which "a bag of neurons" (Crick's phrase) is assembled into a working brain.

Units of Thought

At one level, the nervous system is but a machine—a self-assembled collection of neurons with a complicated function. As with much other living machinery, the molecular engineers have been dutifully taking it apart, naming its parts and describing their linkages with each other. Although the rudiments of the neural connections were accurately described by Cajal more than a century ago, only in the past two decades has the intricacy of the neuron been fully appreciated.

Axon

Dendrite

Figure 8.1 Signaling Neurons. A signaling neuron from the human cortex, after Ramón y Cajal; the input end of the cell (the dendrite) is below and the output end above. The total length of the cell is 2 millimeters.

There are neurons proper, concerned with signaling, and cells with ancillary functions, many of which are probably yet to be identified. Some, called *glial* cells, provide a kind of scaffold upon which migrating neurons travel from their origin to their final destination (at least in the cortex). Others have the special function of wrapping themselves around the signaling processes of neurons proper, coating part of the target cell with a helical coating of the protein *myelin,* which enhances the speed and fidelity of the signaling process.[15] Still others have housekeeping functions, reprocessing materials secreted into the spaces between neurons. It will be no surprise if the years ahead show that some of these ancillary cells are more important than they now seem.

Understandably, however, neurons proper have been the principal objects of study in the past quarter of a century or so. It is helpful (if sometimes dangerous) to liken a neuron to a tree, with a complicated root system, a trunk (which may be branched) and an elaborate canopy of branches supporting the leaves that collect sunlight. In the nervous system, the roots (called *dendrites*) collect signals from other neurons, the cell as a whole transmits a composite signal upwards along the trunk (called the *axon*) and the composite signal is passed onto other neurons through the branches of the canopy. Like trees, neurons have different habits, or shapes, and have acquired different names; those with the most ramified structure are found in the spinal cord and, otherwise, in the mammalian cortex—neurons that may span several millimeters or, when (as many do) they communicate laterally with regions elsewhere in the cortex, several centimeters or more.

What, in such a structure, constitutes signaling? The wonders of the modern telecommunications industry, with high-frequency radio waves carrying images of moving pictures, are not a model for how neurons communicate with each other. So far as is known, neurons communicate chiefly "on/off" messages to cells lying downstream in the communications pathway; the message says simply whether the neuron sending a signal is active or inactive. But the message's strength is determined by the speed of the "on/off" repetition; the faster the repetition, the stronger the signal. To a first approximation, the communication is one-way: signals are transmitted along the axon in a direction that carries them away from the dendritic end of the neuron, along the axon and toward the connections with succeeding neurons in the pathway.[16]

By contrast with the simplicity of the messages, the signaling system in neurons involves a complicated electrochemical device that appears as if it has been stitched together from cellular mechanisms from other kinds of cells (which is almost certainly how natural selection has fashioned them). Neurons are outstanding for being electrically sensitive. As an indirect consequence of keeping excess water out of the interior, there is a small permanent voltage between the outside and the inside of most cells amounting to between one-tenth and one-twentieth of a volt, but the membranes of neurons are equipped with molecular devices allowing this voltage to be temporarily reduced for a fraction of a millisecond or so.[17] If that should happen at, say, the base of an axon, a wave of electrical disturbance will then propagate along the structure

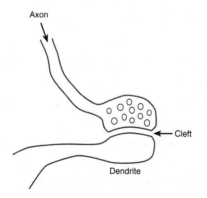

Figure 8.2 Synaptic Connections. A highly schematic diagram showing the connection between an axon and the dendrite of a downstream neuron.

like the shock wave traveling ahead of a supersonic aircraft (but at a smaller speed).[18]

The connections with downstream neurons, called *synapses*, are indirect, as Cajal observed in the nineteenth century. It is as if the tip of one of the branches leading from the axon extends the palm of a hand towards a lump grown on one of the dendrites of the downstream cell, almost but not quite touching it. That is a *synaptic junction*. Note that it has two parts: the *up*stream or *pre*synaptic junction and the *down*stream or *post*synaptic junction; it is not yet known what holds them so closely together. What then happens when one cell signals to another? Simple. The presynaptic junction will already have secreted into roughly spherical bags of membrane a quantity of one of the chemicals called *neurotransmitters;* the arrival of an electrical wave will discharge that into the synaptic gap. The chemicals are then absorbed at the postsynaptic junction, affecting the electrical behavior of the dendrite of which it is a part and perhaps of the whole cell to which it belongs; another wave of electrical disturbance in the downstream cell may be the outcome.

That account of interneuronal communication is laughably oversimplified and does not identify the points at which the present corpus of knowledge is growing quickly. But this is not a book about neurons, which are of interest for the present purpose only because they may shed light on how we shall eventually understand the working of the human brain. It is important, therefore, to remember that neurons are the strangest cells in the whole animal kingdom. First, there is their shape, with slender dendrites and even more ramified axonal processes at the end of a long and possibly branched stalk, which is adapted to insinuate the influence of individual cells over a much greater volume than their substance justifies. Second, there is their curious permanence: although neurons often have to migrate to their fated position in the brain, once in place they are there for the rest of an animal's life (unless they die). Third, they appear to have the function simply of telling downstream neurons with which they communicate whether or not they are active, and to what degree. How can the exchange of such simple messages between neurons yield the sophisticated images we are all familiar with in our heads—the recollection of a Beethoven symphony or the remembered spectacle of the Grand Canyon?

It would be ridiculous to suggest that the process of remembered imagination is yet understood, but part of the explanation must lie in the great number of neurons active in the head and in the complexity of the connections they make with one another. A human infant's nervous system probably contains many times as many neurons as our galaxy contains stars—100 billion of them. The complexity of the connections between neurons is similarly gargantuan; cells of the cortex may be in a position to signal to as many as 10,000 others, either in the cortex or elsewhere in the head. The combinatorial possibilities are immense, as can be told by multiplying 10 billion by itself 10,000 times.

We forget that what we know about neurons and their connections with each other has been learned in a very short time. The first studies of the interaction between neurons and other cells were begun only in the 1930s, using the junctions between motor neurons and muscle cells as a model of what might happen in the brain. The electrical properties of neurons became accessible only after the Second World War, with the help of electronic equipment originally developed for military purposes. (It was not then a simple task to measure a voltage change of some thousandths of a volt lasting for a thousandth of a second or so, especially when that occurs within the space of a few thousandths of a millimeter.) The fine structure of the neuron had to wait on the development of the electron microscope during the 1960s and on novel techniques for making individual neurons visible under a microscope by injecting them with enzymes. During the 1970s, it became possible to make electrical recordings from single neurons in the heads of anesthetized animals and to tell which parts of the brain were active under chosen conditions without interfering with the skull at all. In the 1980s, two German scientists developed a technique for excising a small piece of membrane from the external surface of a neuron[19]; that is invaluable for studying the channels in the external membranes that regulate their electrical behavior. That branch of neuroscience has become an ingenious amalgam of electronics, chemistry and the watchmakers' art. The grandeur of the goal—that of understanding how the brain works—makes it worthwhile.

MAKING CONNECTIONS

Meanwhile, the question of how neurons in the brain make appropriate connections with one another has become accessible. The problem has three parts. Many neurons have to migrate from where they are formed to the places they are destined to occupy. The most remarkable migrations occur in the formation of the cortex, whose six visibly distinct layers of cells are formed in succession from the inside out. The cells destined for the successive layers are formed in sequence, and must migrate through the layers of the cortex already formed to reach their destinations. That is done with the help of the scaffolding of glial cells already in place and by chemical clues provided by the cells themselves. The migration of the cells of the neural crest to the periphery of a young embryo is probably arranged similarly, although that remains to be proved.

Setting up communication channels between neighboring neurons involves both an axon and a dendrite. The tips of axons can and do grow if they are not already anchored in a synapse, in part attracted by chemical signals from potential target dendrites. But there is also evidence that protein molecules in the intercellular spaces between neurons can also affect the direction taken by growing axon tips, although little is known about the sources of those chemical clues and their placement. When the two halves of the synaptic junction are in place, each is adjusted in shape and biochemical character to ensure some degree of permanence.

It is a triumph that the details of this process have been described, at least in outline, but there are several important questions still not answered. What, for example, determines when and how neurons make connections with others relatively far away, say elsewhere in the cortex? And how malleable are the connections anyway? One general rule, for the nervous system as a whole, seems to be that a vast excess of interneuronal connections is established early in development, followed by the atrophy and death of those which are not used. It is also likely that infant human beings, whose brains continue to develop after birth, establish many new connections while gathering early experiences of the real world,[20] but it is an open question whether new connections are formed in the brains of adults when they learn new skills, say the capacity to speak a foreign language. And while it is known that axons whose tips have been damaged are capable of regrowing them in favorable circumstances, it is not known how often this happens in real life. Issues such as these are evidently of great importance for those who would assist the maintenance of the human brain in good working order.[21]

There is, however, one sense in which the human brain is supposed to change constructively during adult life: the continuing capacity of adults to remember novel experiences. The most generally accepted theory of long-term memory is that it entails a strengthening of the synapses linking intercommunicating neurons; the more efficient they are, the greater will be the influence of one neuron on its downstream partners. There is experimental evidence that animal brains function in such a way, but the idea is still to be firmly established—and understood.[22]

Otherwise, the general opinion is that, again to a first approximation, an adult brain does not change in the course of life—except that many neurons die either through disuse or damage (by alcohol, for example). It is the common experience of neurosurgeons and of those who care for the victims of strokes that a damaged brain remains permanently damaged; fresh neurons are not formed to make good whatever damage may be done.[23] But some kinds of neurons are routinely renewed, notably those of the sensory organs of the nose, where everyday attrition takes a toll. (By contrast, the cells of the lens of the eye appear to be unchanged from birth.) It is a matter of great practical importance to know whether some parts of the human nervous system are in principle renewable; vertebrates such as salamanders have retained that faculty, after all.

In reality, the experience suggesting that the adult brain is incapable of renewal could well be a consequence of the complexity of the communications network established in early life; it is far from clear how a newly arrived and immature neuron would be able to fit into such a network long after the largely unknown influences guiding early development have disappeared. There may be a similar explanation of the poor results so far found in attempts to improve the treatment of Parkinson's disease by grafting appropriate cells from embryonic tissue. The question of whether the nervous system is in principle renewable will be settled only when there has been a systematic search for stem cells (such as are found in many other organs of the human body) from which neurons might be regenerated. But there is no prospect that the outcome will ever be a technique for replacing people's brains: everything now known about the architecture of the brain and of its neurons describes an integrated structure in which the "hardware" inextricably embodies the "software"—the neurons on the one hand and memory (which may be determined by the strength of synapses) on the other.

Computing by Neurons

That statement implies that the human brain may be likened to a computer, which is a contentious but unavoidable question. The fact that neurons signal to others chiefly the information that they are "on" or "off" is suggestive, for do not electronic computers store information in just such a form? The sheer complexity of the networks established in the head is also suggestive—a sign that the brain may be sophisticated enough to store the sheer volume of information involved in our conscious imagination.

With the development of electronic computers on the heels of the

Second World War, the optimistic designers of those machines nursed the belief that they would be quickly led to an understanding of the brain. Even as the first electronic computers were entering into commercial service in the 1950s, groups of designers were enthusiastically devising electronic circuits to simulate known functions of the human nervous system—the recognition of simple visual patterns such as printed characters and even the corrupted replicas of them that constitute handwriting. (The term "artificial intelligence," or AI, was in common use by the 1960s.) Inevitably, many of the schemes then devised appear in retrospect to have been naïve and even unrealistic. But the metaphor of mind as computer is not to be scorned because of the early follies.

Early (and some latter-day) practitioners of AI made mistakes of two kinds. First, they let optimism get the better of realism, setting out to design computer programs that would, say, enable computers to hold real conversations with people (via a keyboard). In the minds of the enthusiasts, the invariably stilted character of the exchanges was excused by the difficulty of endowing the programs with the whole of human wisdom. To outsiders, it seemed instead to be a proof that machines are incapable of associative thinking of the human kind.[24]

The early practitioners were also muddled about their goals, yet dogmatically arrogant as well. Was AI a way of learning about the functions of the brain by computer simulation? Or for exploring the nature of abstract faculties such as intelligence in isolation from what happens in the brain? Or was it a means of developing devices that would be useful in the real world, borrowing some of the tricks with which natural selection has endowed the vertebrate nervous system?

The sometimes fierce disputes about the usefulness of the metaphor of mind as computer resemble the controversy following Darwin's publication of *The Origin of Species*. To suggest that machines may be capable of intelligent behavior seems to many people to diminish the faculty of human intelligence in exactly the way that Darwin's belief in the "Descent of Man" (from the great apes) diminished the standing of

Homo sapiens. But as always on these occasions, the issue in dispute is misstated. The two kinds of devices have quite different origins and functions, which only occasionally overlap.

Electronic computers are made in factories of materials such as silicon; they are usually much faster than human brains at simple tasks such as arithmetic, but their life is no longer than the duration of electricity supply that keeps them working (and they tend to become obsolete every few years). Human brains, by contrast, are the products of a long period of adaptation driven by natural selection and come equipped with built-in input/output devices as well as appropriate software (some of which can be modified by experience); by the test of the complexity with which their units are linked together, brains are much more intricate than any electronic computer yet manufactured, while their lifetime (so far as is known) matches that of their principal user. And what electronic computer could invent without conscious thought an innovation as remarkable as the faculty of language? The self-esteem of *Homo sapiens* is hardly dented by the electronic simulators.

Meanwhile, there are practical reasons why the metaphor of mind as computer is necessary: it is difficult to see how it will be possible to make sense of the information now being gathered in the laboratory about the properties of neurons and their interconnections without using the language originally devised for the description of computer components and the elements of communications networks. Moreover, exactly similar language will be necessary to account for what is the central question about the working of animal brains in general—making decisions about the appropriate response to a sensory input and then executing that response.

The value of computer language is evident in attempts to simplify the standard description of the neuron. Each cell has several inputs (through the postsynaptic junctions on its dendrites), but the signal along its axon is either "yes" or "no," modulated by the emphasis it is given by rapid repetition. How are the conflicting inputs to a single neuron integrated into a single output? To the extent that the arrival of a neurotransmitter at a postsynaptic junction will trigger electrical changes in the cell membrane, the response of the downstream neuron may be simply determined by whether the electrical changes reinforce each other when they reach the base of the axon, or whether some signals are cancelled by others. No doubt the postsynaptic junctions strengthened by learning (if that is how learning is accommodated in the brain) would tend to dominate the output of the neuron, perhaps because their electrical connections with the remainder of the cell are more efficient.

Remarkably, cells that integrate together disparate inputs in such a manner can be shown to be capable of simulating all the logical operations performed by individual transistors in electronic computers.[25] None of that *proves* that neurons function in such a manner, of course, but it makes the metaphor more persuasive—and suggests investigations of the properties of neurons that might usefully be carried out.[26]

The metaphor is further strengthened by the recognition that the animal brain routinely makes what would be (for human beings) quite complicated calculations. Consider what happens when a bird of prey has a small mammal on the ground within its sights. Success requires not simply that the hunter knows where its target is, but that it should be able to estimate where the target will be even if it happens to be moving. That implies a calculation of a trajectory that will carry the hunter to its prey at an appropriate time. The calculation does not, of course, have to be made consciously or even explicitly—say in numerical language. It is sufficient that the tracks of the hunter and its prey should coincide in a manner that makes capture possible.

A cat's choice of an appropriate response to the sight of a mouse or a dog is also the result of a calculation—not a numerical calculation, but a symbolic one. How this is done may be unknown, but computer language makes it possible to describe what may happen. Some particular part of the cat's brain will be involved, which must involve connections with the visual field (which is where the image of the mouse is stored). If smell is relevant, there will also be connections with the olfactory centers and no doubt with whatever parts of long-term memory may be relevant as well. The cat's appetite (for mice) may also come into the calculation.

How will the result of the calculation be recorded in the brain? Nobody yet knows, but one strictly hypothetical way would be if there were a neuronal circuit (not necessarily predetermined) somewhere in the network of the brain that would transfer "yes" signals from cell to cell repeatedly if the balance of the information received justified the conclusion that the animal in sight was prey rather than predator. That would allow both for time delay in making a decision and for the possibility that an unambiguous decision cannot be reached on the information at first available. If that same circuit also included connection with the parts of the brain driving the muscles of the cat's skeleton, the result could be the appropriate hunting behavior (of a mouse by a cat) without the cat having given the matter a moment's conscious thought.

It would be wrong even to suggest that decisions between prey and predator are made this way in the heads of real cats. There is as yet no evidence from cats or other animals to show that repetitively excited neuronal circuits have special significance. Nevertheless, the essence of the task carried out must be as described. The brain must somehow bring together all relevant sensory inputs as well as memory so as to make a judgment, unavoidably a matter of probability, of whether the object concerned is prey or predator, and must then execute an appropriate response *via* the muscles. The purpose of this strictly hypothetical scenario is merely to illustrate that describing what happens in the brain in terms of neuronal circuits and of the neurons which are their elements makes it possible to convert abstract notions such as "to decide" and "to execute" into more tangible models which, if they were persuasive in other ways, could even be tested.

In reality, it is unlikely that the brain has a single mechanism for making decisions, or for executing them. Did not the senses evolve separately?

What lies ahead is the patient task of telling which structures in the brain are linked to others by long axonal projections and making educated and testable guesses about the neuronal circuitry that may be involved.

The present state of the exploration of the neuronal circuitry in animals is nicely illustrated by the way in which a leech will bend away from a stimulus applied to its skin. This is strictly automatic reaction, so that the brain is not involved, which means that experiments can be carried out with single segments of the worm, each of which has four neurons specialized in detecting an external stimulus. How does that activate the muscle fibers in each segment of the animal? Two scientists from the University of California at San Diego have now been able to show that the signals from the four sensory neurons in each leech segment communicate with the ten neurons that regulate the muscle fibers through an intermediary group of at least 17 other cells, and that the outcome (which has been simulated by computer) is a bending of the leech segment in almost exactly the required direction. No doubt similar mechanisms will be found to account for other responses of

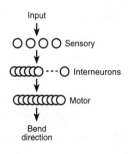

Figure 8.3 **Bending in the Leech.** The neural circuit consists of three layers of cells. The output of the four input cells is found to be a good representation of simple trigonometric functions which specify where a light stimulant has been applied to the skin of the animal. The function of the intermediate layer of cells appears to be to "calculate" which fibers must be relaxed or contracted to ensure bending away from the stimulus.

what is called the autonomic nervous system, in which the brain is not necessarily involved.[27]

But progress in this field will not be quick. One difficulty is that the detailed arrangement of neurons is not copied faithfully between one animal and another of the same species: inferences about which neurons do what must rest on statistical interpretations. But the most serious obstacle to gathering of information about the human brain is that human beings are able to talk about the functioning of their brains (in interviews with psychologists, for example), but crucial studies of, say, the fine structure of neuronal circuitry must be carried out with animals.

There is now at least reason to hope that the obstacles to the direct investigation of the living brain may be removed in the years ahead through the further development of imaging techniques. Devices for the visual examination of the working brain have been in use since the early 1970s. The first of these involved injecting radioactive compounds into the bloodstream of people, so that active areas of the brain could be recognized as the sources of radiation characteristic of the materials used.[28] Now there is greater promise in the technique called MRI (for magnetic resonance imaging), which has the advantage of being noninvasive and which is already widely used in neurological examinations (for the diagnosis of brain tumors, for example). The technique is capable of revealing how different areas of the brain are used for processing information of different kinds. A neat illustration of this is a recent study showing that distinct but overlapping parts of the language center are used for processing speech in a native and a foreign language.[29] What remains to be seen is whether the detail that can be identified by these means will be fine enough to answer the questions being asked about the function of the brain as a computer. Probably not.

Yet the metaphor of mind as a computer has come to stay, and deservedly. The justification is that it is partly true, and that it has already proved of great practical and theoretical value. To be sure, none of that implies that the brain resembles a PC of the kind that can be purchased in the stores. If anything, it is a parallel computer (capable of carrying out several different tasks at the same time) and an asynchronous computer (meaning that there is no central clock to time the passage of the signals). But it is also indispensably part of the body to which it belongs, in which it has the utilitarian function of actuating the limbs and the other movable parts of the body. To expect that it will eventually be understood in terms of neurons and their connections, much as computer engineers seek to understand their competitors' products by "reverse engineering," is no shame. But it will take a long time.

Simulations of the Brain

Meanwhile, simulated brains appear to be thriving. Despite initial disappointments, AI has borne fruit in unexpected ways. At the outset of the first wave of enthusiasm for simulation, in 1954 Frank Rosenblatt, a U.S. engineer, devised a machine called "The Perceptron" intended as a device for recognizing printed characters. The essence of the device was a flat array of photodetectors onto which the image of a character would be projected, each detector was linked to a transistor itself linked into a network with all the others. Instead of working out, no doubt in complicated mathematics, the specification of the patterns of illuminated and dark cells expected from an "A" or a "Z," the network could be "trained" to recognize which patterns corresponded to which letters of the alphabet by altering the connections between the transistors in the network; the cleverness of the machine lay in the simple rule for telling how the connections should be adjusted at each training step. Enthusiasm for the innovation was immense; in Britain, a conference was organized at the National Physical Laboratory (then one of the principal government laboratories) and an electronics company (Solartron Ltd.) vowed that it would soon have working devices for reading text on the market.

Sadly the Perceptron in its original form proved a failure,[30] but the idea that networks of simple electronic elements can simulate some functions of the brain was not forgotten. After an interval of 20 years, John Hopfield of the California Institute of Technology published a persuasive account of how such a network could indeed be built so as to store "memories" of several different patterns of inputs and then tell to which of them an unknown stimulus most closely corresponded. Hopfield also specified the learning rules that would make the network perform designated tasks more efficiently. The devices are called *neural networks*.

The idea that networks of electronic elements should be capable of storing a memory (or indeed several) seems magical, but is not. Imagine that the elements of the network are little computer memory elements each capable of storing either "0" or "1" and that there are four elements altogether. Suppose also that each is connected to every other (making 12 connections altogether). Suppose also that the effect of the connections, whether weak or strong, is to influence the elements they link to carry the same digits. Then there are only two stable structures, those in which all elements carry "0" and all carry "1." But now suppose that the connections between one pair of elements and the remaining pair are made much weaker than those within the two pairs; then the arrangements "0,0;0,0," "0,0;1,1," "1,1;0,0" and "1,1;1,1" are all stable.[31]

As neural networks go, that is not a particularly useful device, but it would be relatively easy to train it to recognize the pattern "1,1;0,0." One must suppose that the network is given this pattern as repeated inputs to its elements, that its four elements interact with each other briefly (trying to force the others into their own mold) and that it then reads out the contents (either "0" or "1") of its memory cells. The result will differ at first from "1,1;0,0" because of the interactions between the cells, but the trainer will then weaken the interactions between the first and second pairs of cells to reduce the error at each successive step. And the end result will be a set of four elements capable of recognizing each of the fourfold patterns "0,0;0,0," "0,0;1,1," "1,1;0,0" and "1,1;1,1"; in the idiom of the Perceptron, if these symbols stood for four letters of the alphabet, the appropriate letter could be printed out.[32]

However valid the analogy between devices like these and the way in which some parts of the brain function, there is no doubt that neural networks will become important elements of the technology of the years ahead. Inevitably, the military are intensely interested in neural networks as means of recognizing predefined patterns (targets for cruise missiles, for example), but the civil applications of these techniques are likely to be even more important. The archetypal application may be a device[33] built at the Salk Institute in California by Terrence Sejnowski, which has the facility for turning printed text into recognizable spoken (English) speech; this neural network, called NETtalk, really works as advertised (if somewhat slowly). Its achievement is to be able to tell when the same character must be pronounced in different ways—in "psychology," for example, the "p" is silent, the "ch" is pronounced as "k" and the two "y"s are respectively an "i" and an "ee."[34]

How can a mere device become so smart? The training process consists of a person reading aloud an English text, whereupon the device compares the sequence of sounds with the authentic printed version. (NETtalk is both an audio recorder and a speech synthesizer.) Several English characters are handled simultaneously, so that the context in which they have to be pronounced can be identified. The connection-correcting instructions are generated from the difference between the authentic and the synthesized sounds. A few hundred units of the network are sufficient for a workmanlike output. This, and devices like it, are part of the shape of things to come.

NETtalk is not just a clever gadget, however, but may be suggestive of the way in which the brain works, at least in one small but significant detail. It is built from three layers of network elements—one (the largest) to handle the input text, a second that effectively serves as an intermediary with the output layer, which generates the sounds. The intermediate layer of network elements seems to perform an intermediate analysis of the problem to be solved (turning text into meaningful sounds), dealing separately with vowels and consonants, for example. But in the brain, the cortex is a notably layered structure, with six distinct layers of neurons. Is it possible that the construction of neural networks, simulations of brain function, have to copy nature's architecture to function successfully? The mere question lends substance to the metaphor of the mind as computer.

WHAT IS CONSCIOUSNESS?

There remain the questions of what and where (in the brain) is consciousness. For that matter, are animals other than *Homo sapiens* "conscious" in any sense? Until the publication of Crick's book, the topic was widely regarded as beyond the reach of objective neuroscience, yet an immense amount has been written on the subject. What follows is not the careful assessment of the many discussions of the subject that might have been expected, but a statement of some of the obvious features of the phenomenon that need to be explained.

First, whatever consciousness may be, it must be an adaptation brought about by the usual forces of natural selection. How else have we acquired any characteristics? In this case, the selective advantages are easily imagined. To have acquired the ability to reflect on a course of action before embarking on it would have been, in a rational world, a great advantage. So, too, would have been the capacity to invent courses of action that are not habitual, calculating the benefits and risks: "If I go that way, the walk will be longer, but there are fewer hills so that I should get there more quickly," for example. A conscious animal would be a fitter animal in Darwin's sense, and would derive great advantages therefrom.

In that sense, human consciousness is on a par with the human faculty of language, which is almost certainly a recent evolutionary acquisition, probably not much older than the last migration out of Africa. That, too, must have been a huge selective advantage. The benefits of being able, say, to plan hunting expeditions in advance, arranging where to meet other members of the group at the end of the day, may even have been decisive in the competition between *Homo sapiens* and preexisting hominids. But it is difficult to believe that language can be used effectively by creatures who are not conscious, at least in the rudimentary sense that noun phrases such as "end of the day" and verbs such as "to meet" (or earlier versions of vocabulary) mean the same thing to different individuals. And how is that to be done except by evoking similar pictures in the heads of the individuals concerned, presumably derived from their common experience? On the other hand, the benefits of consciousness do not wait on language; to be able to make considered decisions privately is in itself advantageous. In short, it is likely that consciousness preceded the ability to speak.[35]

But what is consciousness anyway? One is conscious of what is going on in one's own head—or at least of some of what goes on. At one level, consciousness is merely the monitoring of the information gathered by the senses and stored, temporarily at least, in short-term memory. The use of the passive verb "monitoring" merely indicates that the brain does not pay equal attention to all sensory inputs. Walking along a city sidewalk, a person might pass 100 people in five minutes without "giving them a thought," but then pay close attention to a man in civilian clothes carrying a machine gun; or be indifferent to most of the odors encountered on the same journey, but then be given pause by the distinctive smell of rotting meat. The monitoring of the sensory inputs must evidently be continuous; it follows that there must be mechanisms for drawing them to conscious attention when their emotional content is significant—and for interrupting standard responses pending *thought* when something is out of the ordinary.

The computer metaphor has a name for this kind of function: *operating system.* That is the part of a computer's software that arranges where, on a disk or in active memory, a file should be stored, but which obtrudes on the operator's attention when something goes wrong, as when the memory is full. Precisely such a model of consciousness has been advocated by Philip Johnson-Laird, originally a British cognitive scientist now working in the United States.[36] In the present climate, when the need is for a coherent account of what the faculty of consciousness may be and how it is provided by part of a "bag of neurons," the simplicity of this idea does not count against it.

There is, however, one sense in which the model of the PC operating system is not mirrored by the properties of human consciousness. Computer systems can be designed to yield whatever degree of detail about the internal workings of the machine the user may wish for. "Precisely which parts of extended memory are being used on this task?" for example. No doubt the provision of this information would slow down normal operations, but the information could be provided. The human brain's monitoring of its internal activities appears, by contrast, to be limited. No amount of introspection can enable a person to discover just which set of neurons in which part of his or her head is executing some current thought-process. Such information seems to be hidden from the human user. One of the incidental benefits of an understanding of mind complete with an account of how it is represented in neuronal circuitry is that it will then be possible, at last, to test whether Freud's concept of the *unconscious* represents a normally hidden set of neuronal processes or whether it is a figment of analysands' (and Freud's) imaginations.

The human brain, however, does more than monitor a person's sensory inputs. What, for example, of the capacity for deliberate consideration of a problem—that of a man carrying a machine-gun on a city sidewalk, for example. A cat's confrontation with a dog may well be familiar to the cat, while its response is likely to be instinctive, but people with machine guns are not often seen in cities. How does a person respond? Call the police? Hide? Engage the man with the gun in conversation? In principle, the need for a decision is no different from the cat's need to decide between prey and predator. Again there could be a neuronal circuit for each alternative course of action, but the inputs to the neurons repetitively stimulated would mostly be from long-term memory reflecting each individual's experience of previous courses of action like the alternatives that suggest themselves. In other words, there is no reason why the mechanisms of conscious decision making should differ radically from those required of a cat confronted with either a mouse or a dog. But that will be known only when there is an understanding of where in the brain and in what kinds of neurons the two kinds of thought processes are carried out.

Imagination is probably, in principle, no different. Conceiving of objects or other entities not previously encountered is plainly possible. The value (to the person concerned) of particular imagined entities presumably depends on their purpose—a novel solution to a problem, perhaps, or an imagined device that generates a pleasing aesthetic affect. If neuronal circuitry allows the assessment of the utility of alternative behavioral patterns (as with the cat and the mouse or dog), why should not similar circuitry allow the assessment of the value of imagined thoughts?

That line of argument is the justification for the claim earlier in this chapter that the eventual explanation of consciousness will be qualitatively no different from an explanation of a simple behavioral choice by a cat's brain. That statement has the status of a guess, not even of a hypothesis. But it is a reasonable guess, given the common tendency for natural selection to convert existing mechanisms for new tasks. Unfortunately (or perhaps fortunately), there will be no quick answers to the question of the neural correlates of consciousness.

What little is at present known is itself uncertain. Crick argues that the lower (innermost) layers of the cortex may have a special role in this respect, while the frontal lobes of the human brain have been known for a long time to play a part in the making of decisions. It will be good luck if the careful tracing of neuronal connections on which hundreds of neuroanatomists are now engaged quickly leads to an account of the circuitry of consciousness. Probably new techniques will be required before the goal is truly within sight. But that the goal is attainable seems now to be plain.

The Numbers Game 9

When David Hilbert, from the University of Göttingen in Germany and one of Europe's outstanding mathematicians, addressed the International Congress on Mathematics in 1900, he had every reason to tell a cheerful tale. Had not the century then closing been that in which mathematics had been restored to what it had been in the time of Pythagoras, 2,500 years earlier: the engine of discovery in the whole of physical science? Hilbert and his contemporaries were also impressed that mathematicians seemed to have regularly devised techniques that turned out, years or decades afterwards, to be precisely those required by physicists for handling problems in the real world.

In 1900, for example, Hilbert could not have known that non-Euclidean or Riemannian geometry (in which "parallel" lines may intersect each other) would prove to be the basis for Einstein's theory of gravitation, or that his own construction of abstract spaces with infinitely many dimensions by the use of simple mathematical functions would prove, 25 years later, to be the enduring mathematical basis of quantum mechanics. (This abstract space is known, even now, as "Hilbert space.") Newton, in the seventeenth century, had been compelled to devise for himself the differential calculus that made the conclusions of the *Principia* persuasive, but it would have seemed to Hilbert that such drastic remedies had been made unnecessary by the efforts of mostly European mathematicians. Mathematics and physics had become mutually stimulating, with mathematics (in Hilbert's opinion) at least a little in the lead.

Hilbert's appraisal of the situation was correct. Beginning early in the eighteenth century with the work of Leibnitz, whose friends had disputed Newton's claim to be the originator of the differential calculus, European mathematicians had made the calculus into a powerful tool for solving physical problems, making it into a system of great interest in itself to mathematicians.[1] During the nineteenth century, Euclid's arithmetic had been generalized to include entities (corresponding to numbers) that could be combined together by rules that do not correspond to arithmetical addition and multiplication; among other things, that led to the development of *group theory,* which is of great value in handling physical problems in which symmetry of some kind plays a part.[2]

So what should Hilbert tell his audience at Paris at the beginning of the new century other than that mathematics had become a central tool in understanding what the world is like? He also took the daring step of listing what he considered the unsolved problems in mathematics. One of them was the issue of "Fermat's conjecture," more commonly known as "Fermat's last theorem," which has been proved correct only in the last decade of this century. There were 26 other problems in Hilbert's list, one of which was breathtaking in its ambition. He was talking soon after two British philosophers, Bertrand Russell and Alfred North Whitehead had published their monumental work *Principia Mathematica,* which was an intricate exercise in symbolic logic intended to demonstrate that mathematics in its entirety can be derived from a finite set of axioms (or assumptions) about the elementary properties of mathematical objects. Hilbert's ambition was to turn the argument around, using this or some other formal system of logic to prove the truth or falsity of all propositions in mathematics. Hilbert invited his fellow mathematicians to devise a scheme that would, on the face of things, automate the drudgery of mathematics, allowing the practitioners to imagine new mathematical propositions, but to relegate proof and disproof to a formal procedure that could in principle be carried out by a machine.[3]

The Worm in the Bud

Hilbert's dream remained a challenge for just three decades, when there came a great surprise: the dream was proved to be logically unattainable. This was a remarkable development, not so much because proofs that some intellectual task is impossible are inherently difficult to construct, but because there had previously been no reason to suppose that Hilbert's goal was unreasonable—difficult it might be, but not beyond reach.

The applecart was overturned by Kurt Gödel, an Austrian then working in the United States, who was able to move to the Institute of Advanced Study at Princeton University on the strength of his achievement. What he demonstrated is that no formal axiomatic system of mathematics can be at once consistent and complete. Consistency means that the formal system should not yield inconsistent results—both a proof that 123,456,789 is a prime number and that it is not, for example.[4] Completeness means that the formal system for assessing the truth or falsity of propositions in mathematics should be capable of testing all possible propositions. These criteria, of course, are the least that could be required of a formal system of mathematics satisfying Hilbert's demand. Gödel showed that such a system cannot be constructed.

Abstract though the conclusion may be, it drives a coach and horses through the idea of mathematics as a comprehensive logical system.

Gödel's theorem, as the conclusion is usually called, forced mathematicians to accept that there are some questions that cannot be decided. For if there is constructed a genuinely comprehensive system of logical argument, some propositions will be shown to be both true and false; they will be "undecidable" in common language. On the other hand, a system free from contradictions can never be applied to every proposition.

Gödel's conclusion raises several questions beyond mathematics. Roger Penrose, the Oxford mathematician and physicist, describes it as "one of the greatest achievements of the twentieth century."[5] Gödel has also something to say about the relationship between mind and brain. If, in 1900, there seemed to have been for the previous two centuries an uncanny parallel between the development of mathematics and the understanding of the physical world, how can it be that there seem to be limitations to the degree to which formal mathematics can represent reality? In any case, what is the biological basis for the capacity of the human mind to practice mathematics? On the face of things, there can be little in the past history of *Homo sapiens* to prompt the evolution, by natural selection, of such a faculty. And what, if any, are the consequences of undecidability for mathematics?

CAN COMPUTERS HELP?

By a curious coincidence of timing. Gödel's theorem was quickly followed by another intellectual *tour de force* whose implications were afterwards recognized to lead to the same conclusion. Although electronic computers were then still a decade off, a small band of mathematicians had begun to anticipate how they would function. One was Alan Turing, a mathematician then at the University of Cambridge who is best remembered for his work during the Second World War in decoding intercepted messages from the German High Command.[6] In 1936, Turing developed many of the concepts afterward embodied in the machines that now sit on our desks. In the process, he outdid Gödel.

Already in the 1930s, it was plain that electronic computers would be useful tools only if they could be provided with sets of instructions (now called *programs*) by which they could reliably operate on the data with which they were supplied. A computer program corresponds to what mathematicians had learned to call an *algorithm*—a set of procedures for operating on *input* data so as to yield the desired *output*. To be useful, a program (which may be very large) should nevertheless be finite; otherwise, the process of instructing the computer would never be complete, with the consequence that it would never get around to operating on real data.[7]

Even in 1936, it was plain that the best programs would be *recursive* programs, in which the machine would be instructed to follow the same small set of instructions repetitively. So much is clear from how we have all been taught at elementary school to add numbers together. The sum of 1,234 and 5,678, for example, is obtained by first adding together 4 and 8 (to give 12); we enter 2 as the right-most digit of the sum, recognize that we must add the remaining 1 (which represents the number 10) to the sum of the second pair of numbers (3 and 7), which gives a total of 11; so we enter the right-most 1 in the second place (from the right) of the sum and add the left-most 1 to the pair of numbers in the third column from the right. And so on. At all stages in this process, exactly the same instructions are followed; that is a recursive algorithm. There is no limit to the size of the numbers that can be added in this way, except that the sum of indefinitely long strings of digits will take indefinitely long to calculate. And there is no limit to the number of numbers that can be added together; adding the first two strings of digits together, then adding the third string to that intermediate sum, and so on, is also a recursive process. Moreover, there is no need of mathematical insight in carrying out the rules: computers can perform them perfectly well.

If this elementary-school procedure were the basis for a real computer, it would be found to have one potentially serious defect. There is a sense in which it runs back to front. It yields the least significant digits of the sum of two numbers first, and the most significant digits only when the whole addition is complete. That would mean that the user of a computer might have to wait inconveniently long before he or she had an idea of how large the result will be. So why not start from the left, not the right? That is easily done. To form the sum of 1,234 and 5,678, first add 1 and 5, putting the result (6) in the left-most place. Then add 2 and 6, giving 8 to put in the second place (from the left). The third step is more tricky; adding 3 and 7 gives 10, which is not a valid entry for the third place (from the left), but the algorithm can easily be amended; simply enter the 0 in the third place and go back a step to add the spare digit 1 to the 8 already there, making 9. And then the same thing happens when one comes to the fourth place; one must enter a 2 there and go back a step to add in the 1. The result of the addition is the same as in the first procedure, but the calculation may be halted at any stage to give an approximation to the result. Evidently the algorithm is also recursive. But the digits of the sum at intermediate stages of the calculations may have to be changed by the outcome of later stages, sometimes quite radically. Try adding 899,999,999 and 100,000,001 by starting from the left! Algorithms that are normally efficient ways of reaching a result can evidently be thoroughly inefficient when faced with particular input data.

Turing recognized the far-reaching implications of that state of affairs in 1936. The essence of his work was a generalized concept of the ingredients of a computer. A machine must have means of taking in data and putting out the result, correcting the latter to allow for amendments of provisional results arising in later stages of a calculation (as in the second algorithm above). The machine must also be capable of existing in several internal states such as would be required to "remember" that certain digits must be "carried" to the left in each of the procedures specified above. Again it is essential that the number of internal states should be finite; otherwise the machine could never be built.[8] In Turing's argument, "machines" that satisfy these conditions are defined by the various algorithms (the "software" in modern language) with which the hardware can be supplied. Turing called his abstract machine a *universal* machine, and showed that more elaborate specifications are logically equivalent to the rudimentary machine. The computers now on our desks are all "Turing machines"; the status of the computer in our heads is, for the time being at least, less certain. Only now is neuroscience unraveling the algorithms used by the simplest neural circuitry (as with the example of the leech described in the previous chapter).

What does this have to do with the bombshell Gödel had sprung a little earlier in the 1930s? Evidently it was incumbent on Turing to analyze the faults as well as the capabilities of his universal computer. (Remember that nobody had yet built a real electronic computer.) Turing anticipated that his universal computer might cease to print out numbers at some stage in its operations. The lights would keep flickering on the (hypothetical) box, suggesting some kind of activity inside, but the calculation would appear to have halted. Is there something wrong with the design of the algorithm? Or is the calculation of the next digit unusually difficult, perhaps because it requires carrying a spare digit arising in an arithmetical operation over several decimal places? Or is it worse than that? Or is the problem insoluble?[9] Telling which of these is the correct explanation Turing called the "halting problem." He went on to show that there is no way of identifying the cause without direct examination of the machine's internal states.

Another way of putting the dilemma is to ask what kinds of algorithms are suitable for use in computers. Evidently they must be finite, for otherwise, however fast the machine, there will never be even an approximate result. So suppose that somebody has the ambition to compute all the possible numbers between 0 and 1 in the ordinary decimal notation, which means numbers such as 0.25, 0.992 and even 0.123,456,789,101,112,131. These are all *real* numbers, a technical term meaning that all the usual operations of arithmetic can be carried out with them. Plainly, there is an infinite number of them—indeed, there is an infinite number of real numbers beginning with each of the three finite real numbers above, even that with 18 decimal places. So is it possible to devise a finite algorithm that will compute all of them?

That is a pointless exercise, but in principle it could be simply done. First print out the numbers 0.0, 0.1, and so on up to 0.9, then take each of those numbers in turn to generate new numbers with a second decimal place in which the digits are, successively, 0, 1, 2 and so on up to 9, giving numbers such as 0.19. And then repeat the process to form numbers with three decimal places, and continue indefinitely. The algorithm is finite, as a Turing machine requires, even if it is modified to eliminate duplicates (such as 0.5 and 0.50), but the output from the computer would take a long time to appear and then would say nothing about the numbers themselves, not even that 0.333,333 is a good approximation to $1/3$ for example.

Is there a more meaningful way of generating the real numbers? One obvious class of real numbers that can be computed by finite algorithms are those that are the ratio of two digits, say $3/12$, $5/7$ or $11/17$ (with the numerator less than the denominator). Technically, these are the *rational* numbers. Elementary schools have taught us all the algorithms for computing them. For example, $5/7$ is 0.714,285,714,285, ... in which blocks of six digits repeat indefinitely, or *recur*. In principle, the algorithm could be modified so as to recognize this block structure, and then truncate the computation.

Plainly, the number of rational numbers is infinite, but they have two remarkable properties, established only towards the end of the nineteenth century. First, although there is an infinite number of them, they can be put systematically into order in such a way that it is possible to say which is the third, the fourth, the fifth, ... the tenth, ... the thousandth ... the millionth and so on.[10] Second, the rational numbers are only a small fraction—indeed, a vanishingly small fraction—of the totality of the real numbers—those represented by all possible decimal fractions.

There are, of course, other ways of computing real numbers by finite algorithms. For example, finite algorithms (learned at school before pocket calculators came on the scene) can be used to extract square roots. There are also rules, based on complicated algebraic formulas, for calculating the successive decimal digits of so-called *transcendental* numbers such as π. But they also are countable, like the rational numbers, even if they are infinite in number, and they constitute a vanishingly small fraction of all the real numbers. Collectively, these are what Turing called *computable* numbers—the essence of the case is that even the algorithms should be finite, for otherwise even the fastest computers would be unable to reach a conclusion. They are also a vanishingly small fraction of all the real numbers. The others are the numbers for which there is no finite algorithm, and which can be defined only by specifying the individual digits which, for a real number of indefinite length, means an algorithm with infinitely many steps.

These abstract arguments have remarkable consequences for the capabilities of the universal Turing machine that might well have discouraged less stoical people than those who, in the 1930s, were planning the electronic computers not yet built. For one thing, most real numbers are not computable numbers in the sense in which Turing defined the concept.[11] In practice, of course, this is not usually a serious difficulty. The numbers people want to compute are usually defined by a formula that can be translated into machine form, while the precision required of the calculations is usually a finite number of decimal places. But it is a further sign that even a universal computing machine has limitations.

Turing's argument, intended to explore the concept of computability, came upon a profound obstacle, which he called the *halting problem*. Suppose you have provided your computer with an algorithm appropriate to the task you want it to carry out (which would not be called a computer program), and suppose that it begins the job (which you can tell because the lights flicker on the front of the box), putting out some numbers, perhaps the first few digits of the decimal form of a computable number. And then suppose that the supply of digits stops. The lights on the box flicker as before, but computation seems to have halted. What can be the explanation? There may be several. The algorithm may be faulty, the computer may have a logical fault in its design—attempting to divide a finite number by zero is an elementary recipe for trouble of this kind—or the task the computer has been given may involve so much repetitious computation that there has not yet been time to produce a result. Turing's subversive conclusion seems trivial when stated in plain language: there is no way of diagnosing the cause of the halt in computation except by taking the machine and the algorithm (program) apart, looking for logical flaws that may provide an explanation. This again is a case where it is decidability that matters. Turing's achievement, in 1936, was to show that this is logically the equivalent of Gödel's theorem.[12]

What are the consequences of this remarkable outcome of the quest that Hilbert began in 1900? While there is no sense in which the power of mathematics has been diminished, it has certainly not reached that state of grace in which the correctness or otherwise of mathematical propositions are questions that can be relegated to machines. Most mathematicians and all scientists may have carried on doing what they would have been doing anyway.

Mathematics and Reality

Nevertheless, some mathematicians have been persuaded to make friends with philosophers. With good reason. It is an illusion that mathematics necessarily reflects the real world. Rather, mathematics is a means by which conclusions are drawn from stated assumptions, including the rules of inference and the axioms (defined in Euclid as "that which is self-evident"). If the axioms are changed, so is the system of mathematics and the conclusions that can be reached within it. Modern "algebra" is, for example, a generalization of classical arithmetic to allow its elements to be combined in ways that Euclid never dreamed of. So in mathematics as in the rest of science, truth has been discovered not to be absolute—but, in this case, to be relative to the assumptions, which are fortunately (in mathematics) usually explicit.

A familiar illustration will prove that point. Euclid's geometry, mostly concerned with figures that may be drawn on a flat sheet of paper, takes as one of its axioms the notion that parallel straight lines lying in the same plane never intersect.[13] The converse is also true: straight lines lying in the same plane which never intersect are parallel to each other. Then, in the 1840s, the rigors of that axiom were relaxed, allowing (among other things) that two lines might be counted as parallel if they are each perpendicular to some other line, as the meridians of longitude are all perpendicular to the Equator of the Earth (at the Equator). That turns out to be a very different kind of geometry from the geometry of the flat plane. The three angles of a triangle do not add up to 180° on the surface of a sphere such as Earth.[14] Changing an axiom can evidently radically change the mathematics based on it.

The uncanny success of mathematics in anticipating the needs of nineteenth century physics is not, in that light, surprising; over the best part of two centuries, mathematicians built their logical systems on axioms molded by commonsense physical principles. At the outset, with Newton, the elements of the calculus had a geometrical interpretation; the quantity called an *integral*, for example, which is easily recognizable in most modern mathematics texts by the symbol $\int$, began life (thanks to Leibnitz in the early eighteenth century) as representing the area beneath a curve.[15]

The dilemma thrown up by the Gödel-Turing conclusion remains an open question. Why should it be impossible to devise a formal system of mathematics that is at once consistent and comprehensive (or "complete")? Systems of mathematics are, after all, entirely in the head, so that the dilemma must point to a quirk of some kind in our way of thinking. In short, what are the neural correlates (as neuroscientists would say) of Gödel's theorem? Roger Penrose believes that the mathematical undecidability reflects a hitherto neglected role for quantum mechanics, with all its uncertainties, in the operation of the brain on the scale of neurons. A remarkable but virtually unreadable book by Douglas R. Hofstadter canvasses other possibilities.[16] The plain truth is that nobody has at present a persuasive idea of what the implications are, except that they are philosophical rather than practical.

UGLY MATHEMATICS

Gödel and Turing are not the only ones to have cast doubt on the general impression (shared by mathematicians) that mathematics is one of the constructs of the human imagination which, because of its elegance and explanatory power, is intellect perfected. Some problems are undecidable, others are just ugly to solve. Whatever the explanation, the decades since the Second World War stand out in the history of mathematics for their preoccupation with problems whose solutions are not elegant but, rather, are untidy.

The remainder of this chapter is about the ways in which mathematicians may contribute to the soluton of problems in physics and biology, and thus to a better understanding of the world.

There is, for example, the matter of *nonlinear* equations. By half a century ago, the physical sciences basked in the knowledge that a whole class of problems could be systematically solved by methods developed by mathematicians in the previous two and a half centuries. An oscillating pendulum? Easy. The wavelike oscillation of an electron bound by electrical attraction to the nucleus of an atom? More difficult, but the methods had been nearly codified by mathematicians and turned into algorithms that even machines can use.

The problems soluble in this way are *linear* problems, which is a technical term with a specific meaning. Imagine that there is some set of equations which must be satisfied by any solution of the problem; usually the equations will be written in the language of the calculus, but not in this book. Consider the oscillation of a weight suspended from a spring hanging vertically. This a standard problem for students in secondary schools, who are asked to find how the displacement of the weight changes in the course of time. The fixed data are the mass of the weight, the strength of the spring and the downwards gravitational pull, which between them fix the frequency of the vibration (so many oscillations per minute, say). Then it is an instructive problem to find an algebraic way of describing how the displacement of the weight varies with the time. That "solution" of the equations is easily found, at least with the assumption that the force exerted by the spring is directly proportional to the displacement. Unsurprisingly, the displacement varies rhythmically with time. And having found this one solution, it is easy to find a host of others: simply multiply the first expression describing how the displacement varies with the time by any real number. That is a sign that the underlying problem, as stated, is a *linear* problem. The meaning is simply that the frequency of the oscillation does not change as its amplitude is increased or decreased.

But what if the underlying assumptions are false? It could be, for example, that the elastic behavior of the spring is not that specified. Perhaps the spring is hung in such a way that the weight collides with an obstacle if the amplitude of its oscillation exceeds some fixed amount; that could be arranged by hanging the spring through a hole in a fixed piece of wooden board whose diameter is large enough to accommodate the spring, but not the weight. That is not the kind of problem high-school students are given for homework, for it has become a *nonlinear* problem. That is simply seen. So long as the oscillating weight does not collide with the board, the rhythmic solution of the problem will suit well. But if the amplitude is doubled, and if necessary, doubled again and again, the point will come when the weight and the board collide, whereupon quite different things happen. A simple linear problem has become a much more complicated *nonlinear* problem.

Difficulties of this kind first thrust themselves on mathematicians in the eighteenth century, when people tried to use Newton's theory of gravitation to make more accurate predictions of the position of the Moon in its orbit about Earth, allowing for the disturbing effect of the Sun's gravitational pull on the Moon's orbit. By then the Newtonian equations for the simple problem of an object such as the Moon or a planet moving around its attracting center (the Earth or the Sun respectively) had become familiar. Any ellipse with the attracting center at one of the two foci is a valid solution of Newton's equations (witness the vastly different orbits in which the known planets move about the Sun). All of them are solutions of a linear equation. The mass of the orbiting object does not affect the shape of the orbit or the time taken to complete one revolution in it.

So why not extend Newton's argument to the joint motion of, say, the Sun, the Earth and the Moon? That turns out to be a very different problem. The Sun's influence on the Moon's motion around Earth depends on where the Moon is in its orbit. Moreover, the Moon does not exactly repeat its tracks as each orbit is completed; instead, it spirals outwards under the Sun's influence. Mathematicians early in the century were at first unsure whether the difficulty they found in treating this problem was a consequence of their still new mathematical techniques; by the end of the century they had convinced themselves that the problem did not have exact solutions, but that numerical approximations would be necessary. (Joseph Louis Lagrange synthesized a century's argument in his great work *Analytical Mechanics,* published—with a French title—in 1788.) People seeking to publish accurate nautical almanacs were forced to construct approximate methods for calculating what is still called "the equation of the Moon."

In reality, most of the practical problems arising in the real world are nonlinear problems arising in a similar fashion. Quantum mechanics, for example, gives a prescription for describing the motion of a single electron attracted electrically to an atomic nucleus. That problem is linear in the sense that there are exact solutions to the appropriate equations for each allowable energy. So what about real atoms or even, more interesting, molecules, where there are usually several electrons? Sadly, there are no exact solutions. The difficulty is that the electrostatic repulsion between the electrons is comparable with the attraction of the nucleus for each of them; it is the three-body problem in a novel context. Even in the case of the helium atom, with two electrons and a single nucleus, the motion of the electrons makes an intractable problem that can be treated only by approximate methods. Notoriously, Einstein's theory of gravitation (the "general theory") is inherently nonlinear. Superficially, Einstein's equations look simple enough, but the essential trap is that the elements of Einstein's gravitational field determine how much mass or energy there is at some place in space time in terms of complicated derivatives of the same elements at other places. In the circumstances, it is a triumph that a number of exact solutions of Einstein's equations admittedly describing idealized universes, have been found.

So what has mathematics done about the general nonlinearity of the real world? At least since the Second World War, the problem has commanded the attention of many of the world's outstanding mathematicians, yet there is no general method for solving these problems except computation. Nor is it possible to enumerate all possible solutions, meaning that potentially important solutions may be missed altogether. The result is that nonlinear problems abound with surprises, pleasant or otherwise.

There is, for example, the matter of the waves that can be generated on the surface of a body of liquid.[17] The surprise is that solutions to these particular nonlinear equations can represent not only familiar repetitive waves, but also the motion of a single distortion of the surface of the body of the liquid—it is as if there is a hump on the surface that propagates as a whole in one direction, losing speed (and changing shape) only slowly as it does so. This remarkable phenomenon, called a "solitary wave" or "soliton" (because there is only one of it) turns out to be more than a mathematical curiosity.

Quite quickly, the particle physicists realized that solitons are potentially models of particles moving through empty space; in 1962, T. H. R. Skyrme, a British physicist, devised a scheme for describing the particles of nuclear matter by means of solitons and, true to the whimsical traditions of the field, particles called "skyrmions" are now frequently referred to in the scientific literature. More generally, the shock waves generated by explosions are usefully represented in this way, with the result that the expanding spherical shell of energetic radiation and particle motion caused by the explosion of a supernova is generally accepted to be a soliton solution of the appropriate equations. So too do similar solutions represent electrons trapped near the surfaces of solids, which is an important problem for those who manufacture and manipulate semiconductor devices. And nobody will be surprised if it turns out that the electrical signal traveling the length of a neuron's axon is an example of the same phenomenon.

It is intriguing—and somewhat chastening—to read an observation reported to the British Association for the Advancement of Science in 1845 by one J. Scott Russell, who described just such a soliton wave on the surface of a narrow English canal caused by the sudden halting of a barge:

> I followed it on horseback, and overtook it still rolling on at a rate of some eight or nine miles an hour, preserving its original figure [of] some thirty feet long and a foot to a foot and a half in height. Its height gradually diminished, and after a chase of one or two miles, I lost it in the windings of the channel.[18]

The phenomenon also occurs in many tidal rivers at neap tide, when the large quantity of water forced into the estuary can generate a soliton wave that travels several kilometers inland.[19]

The unexpected behavior of solutions of nonlinear equations has become a serious impediment to the understanding of the natural world. The most serious difficulty is that often it is impossible for those who have found several solutions of them to be sure that they have found them all. Yet the practical importance of these systems is yet another sign that Hilbert's contentment with what mathematics had done, by 1900, to throw light on natural phenomena was a little premature, to say the least.

CHAOS EVERYWHERE

As it happens, nonlinear phenomena have become even more obtrusive since the 1970s, with great potential importance both for mathematics and for science generally. What has emerged is that entirely well-ordered systems of equations (and the systems in the real world they describe) can yield solutions that have exceedingly disordered behavior. This has spawned a whole new subfield of mathematics called *chaos theory* and a host of explanations of phenomena in the real world as different as the irregular motions of asteroids and the capacity of groups of cells in the heart to behave aberrantly, in a disorderly fashion.

The name "chaos" is potentially misleading. There is no suggestion, for example, that the origin of disorder in chaotic mechanical systems is a consequence of the physical uncertainties that have proved unavoidable in quantum mechanics. The mechanical and other systems covered by the label "chaos" are strictly deterministic: in a system that changes in the course of time, for example, the equations specify exactly how the state of a system at an initial time determines the state of the system at all later times. But in the real world, it may not seem like that. Strictly speaking, chaos should be called "deterministic chaos."

The origins of this new idea are, as often, confused, partly because it was infiltrated into science and mathematics in several different ways at about the same time, in the early 1960s. With hindsight, as so often, it soon became clear that many others had flirted with the idea much earlier. One of these was Henri Poincaré, the French physicist who may well have anticipated Einstein's special theory of relativity, who studied in 1893 the peculiarities of systems now recognized to be capable of chaos. Later, but long before the 1960s, several Russian mathematicians had demonstrated properties of nonlinear equations that proved afterwards to be of great value: they include G. N. Kolmogorov, A. S. Lyapunov, Lev Landau and Georg Slaslavskii.

In the event, chaos was put on the map of mathematics by two separate developments. The more conventional was an attempt in 1963 by E. N. Lorenz to solve a set of equations devised (in a highly simplified form) to describe the onset of turbulent flow in a liquid heated from below.[20] The second, due to the U.S. mathematician E. B. Mandelbrot, was more startling because of its simplicity and more appealing because one of its by-products was a series of computer-drawn patterns on a sheet of paper that were bizarre in their complexity and, on the face of things, accessible to anybody with a home computer and a little knowledge of how those devices work.

What follows should not be mistaken for mathematical prose, even though it includes a number of symbols which are merely labels for nouns it would be too cumbersome to define each time they are used. Suppose some point on a circle is used to represent the position of a particle, and that the objective is to describe where it will be after the lapse of successive fixed intervals of time. Suppose that the particle initially is at a point on the circle called x_0, and that its successive positions after 1, 2, 3, etc. intervals of time are x_1, x_2, x_3, etc. If the particle is moving with uniform speed (say 10° in a counterclockwise direction every second), meaning that it takes 36 seconds to complete one circuit, then the successive positions around the circle are easily foretold. Each successive x is 10° greater than the previous number in the sequence, or $x_1 = x_0 + 10$. The general rule is that $x_{n+1} = x_n + 10$. Then there arises a minor complication. After 36 steps, the particle will be back where it started, or $x_{36} = x_0$. Now suppose there is a second particle, initially at x_0, that moves in the same way. After 36 steps, that too will return to its original position, and the angular separation between the two particles will not have changed. No sign of chaos there.

But matters can easily be more complicated. Suppose that "law of motion" for these particles is that, at every time-step, their angular position (in degrees) is doubled. This is not nearly as straightforward. Particles that begin at a small angular position will move only a small amount at each time-step, but those beginning at just under 90° will have been carried to nearly 180°; indeed, after one time-step, all particles starting between 0° and 90° will be spread out between 0° and 180°. And what about the particles starting from a position between 180° and 270° (which is just one right-angle greater)? A little reflection will show that they are also uniformly distributed between 0° and 180°, mixed up with the particles originally in the opposite quadrant of the circle in the figure. What this shows is that the law of motion described has the effect of carrying particles that were originally close together into positions that are very far apart. That is deterministic chaos—indeed, this particular illustration is a grossly simplified case of what is now called "Mandelbrot's set," which in its more general form can yield remarkably intricate patterns.

Why had this strange behavior not been discovered earlier? For three centuries, since the publication of Newton's *Principia*, physicists had bent their minds to the refinement of the laws of motion while mathematicians were busy with better ways of calculating the trajectories of moving objects such as planets under the influence of gravitational forces. That was also the spirit of Lorenz's calculation of heat convection. But Mandelbrot turned the question inside out, as if to say, "What kinds of motion are there, anyway, and what can be learned from their variety?"[21]

These developments would be mere curiosities were they not so common in nature. The case of the asteroids is a good illustration. These are meteoritic objects in orbit about the Sun, but at certain times they come close enough to Jupiter and even to Earth for their motion to be affected. In principle, this is an exactly soluble problem (with due allowance for the approximations needed in the three-body problem). But the precise strength of the interaction with Jupiter turns out to be exquisitely sensitive to the exact positions of the asteroid and of Jupiter in their respective orbits when they are in gravitational range of each other. Most asteroids are not so well known that the interaction can be calculated with sufficient precision that the orbit of the asteroids can be predicted for more than a few hundred years in advance.

That, however, is a rarefied illustration of the role of chaos in the natural world. The limited time-span of accurate weather forecasting is a more common inconvenience, reflecting the way that initially small disturbances in the atmosphere can grow to be dominant in about 15 days or so. The changes in the Earth's pattern of magnetic fields, whose origins lie in turbulent motions of liquid rock in the Earth's core, are similarly chaotic, as is the slight rocking of the Moon on its axis of rotation (called *nutation*). But chaos is also common in various evolutionary contexts, as when a small increase of the fitness of an evolving species leads to a very large increase of its population—perhaps to be followed by a rapid decline as food supplies are exhausted.

All that is certain at this stage is that chaos has come to stay. Already, mathematics has done much to explain the details of chaotic systems—and even to suggest how the disturbing effects of chaotic signals on, say, communications networks may in some cases be eliminated. But much remains to be done.

Problems Unsolved

Meanwhile, there are at least three fields in basic science crying out for the kind of partnership with mathematics of which Hilbert boasted in 1900. There is, for example, the matter of *complex systems*—problems that can be accurately defined, but which are literally too complex to be solved by known methods.

Problems of this kind are typified by this mathematical puzzle-designer's favorite tease: "here is a map of a dozen cities that a traveler has to visit in any sequence; what is the shortest route that will accomplish that?" The distances between the cities are provided with the specification of the problem. Provided that the number of fixed locations is small, the solution can be reached by the tedious method of working out all possible itineraries and adding up the distances involved in each of them. But the magnitude of the problem grows very rapidly as the number of locations is increased. So quickly does it grow, indeed, that it must at some stage exceed the capacity of even the largest super-computers.

It is not, of course, the end of the world that traveling salesmen may have to work out their routes empirically, but the problem has analogs in fields of science that are important. There is, for example, an obvious similarity with the question of how large protein molecules, comprising perhaps several thousand amino acids strung together, fold so as to form functional proteins. There, again, the huge variety of possibilities increases rapidly as the number of amino acids in the chain is increased. Without a systematic way of solving these problems, there will be no coherent way of tackling one of the rudimentary problems in the new biology. Meanwhile, it is worth noticing that people with an interest in the mathematics of complex systems have expanded their ambitions to embrace the working of the world's stock markets.

The idea that some problems in science, as now formulated mathematically, are beyond the capacity of existing and even foreseeable computers does not mean that their usefulness is limited to mere computation. So much is clear from the way in which, in 1994, what had been called "Fermat's last theorem" was eventually proved to be correct. The French mathematician Pierre de Fermat, in the seventeenth century, had conjectured that there are no whole-number solutions of the equation $a^n + b^n = c^n$, where a, b, c and n are all integers. The truth of the proposition had been verified with the help of computers for all possible combinations of a, b and c for all values of n up to about 100,000, which was sufficient to persuade most people that Fermat must have been correct. But verification by computer does not amount to a mathematical proof. The years ahead are certain to engender great interest in the question when computer demonstrations can be counted as tests of the truth of mathematical propositions. The second crying need of mathematics now is for computational schemes that are better suited to emergent needs. In particular, given the certainty that cell biologists will soon be driven to computer models to make sense of the huge amounts of data accumulating, user-friendly ways of carrying out these tasks will evidently be required.

The third demand of mathematics now is more taxing, stimulated by the current interest in fundamental physics in the idea that the structure of space-time may be more complex than is now supposed. If, for example, particles are to be represented as submicroscopic strings or even membranes which are inseparable from space and time, how are they to be used to modify present conceptions of what space and time are like? Will it be necessary to modify the way of describing space in mathematical language that has served the world well since the time of Descartes and how will that, in turn, change emerging views of how matter and space are interrelated? If Hilbert were giving his address in 2000, that would be one of the most important unsolved problems on his list.

Avoidance of Calamity 10

The purpose of this book is to demonstrate that science, far from being at an end, has a long agenda ahead of it. Many of the questions identified will no doubt soon be answered; others will still puzzle people for centuries ahead. By then, still other questions not yet cogently asked will clamor for attention. The record of the past shows that novel conundrums are forever treading on the heels of those that still perplex. The 500 years of modern science are a good beginning, but only a beginning.

Since the early 1960s, there has been a substantial body of opinion that science has occasioned novel man-made problems that may in themselves be threats to human survival. In its most modest form, this complaint amounts to the assertion that the pace of change is now so rapid that social and political institutions cannot adjust quickly enough. Others remark on the problems occasioned by pollution of the environment and disturbance of the biosphere, claiming that they are potential sources of calamity for the human race. While the more extreme versions of these arguments do not command general respect, the cumulative effect of their general repetition has engendered the widespread opinion that it would be simpler for the human race if the pace of discovery were somehow abated, perhaps by restricting the funds spent on research.

That is a serious error. Among those who advocate the more vigilant "care and custody—even love—of a small planet," [1] there has recently grown up a tendency to equate Earth with *Homo sapiens*, in phrases such as "the interests of the planet." The usage overlooks the extent of human interference with the biosphere in past centuries, and depersonalizes arguments that might otherwise seem anthropocentric. The usage also begs an important question: even if the interests of the planet and of the human race are at present coterminous, will that always be the case?

It is a matter of perspective. This book has so far been concerned with the present and the near future. Gaps in understanding now apparent are regarded by researchers as opportunities to be seized. And who would dispute that it would be a great boon for the human race if, for example, there were a speedy remedy for psychiatric diseases? But there are also, on the horizon and beyond it, threats to the survival of the human race even more serious than the dangers inherent in an exchange of nuclear weapons between the major military powers (which have at least temporarily receded) and the continued degradation of the environment. What, in the interests of the planet (so to speak), should be done to guard against such threats?

The short list of potential dangers that follows can readily be misunderstood. To be sure, the idea that the accumulation of carbon dioxide in Earth's atmosphere may change the climate permanently is so familiar after the Kyoto conference in December 1997 that it needs no special pleading. Nor is the idea that novel and resurgent infections are a threat to human health, even survival, surprising; has there not been more than a decade's anxiety about AIDS, after all?

There are more serious if more distant dangers. It now seems clear that meteoritic impacts on the surface of the Earth at the transition from the Cretaceous to the Tertiary Periods 64 million years ago were the cause of a major transformation of the biosphere. Dinosaurs, then the most conspicuous forms of life, simply disappeared. What can and should be done to guard against the risk that something similar may happen again? And what if accumulating knowledge in genetics shows that the human genome is inherently unstable? Although Darwinism gives a favorable bias to the fitter species, it cannot guarantee that they will survive indefinitely. On the contrary, the fossil record is full of the relics of organisms, animals and plants, that no longer exist, apparently without external cause. Unlike other species driven to extinction by some inherent fault, human beings would be able to appreciate what was happening if there were an inexorable increase of the burden of genetic handicap. Would they then seek to avoid what seemed to be their fate, and how?

The first two hazards are immediate, and will preoccupy the century ahead. The hazard of meteoritic impacts is remote in an interesting sense; a calamitous impact could happen at any time, but nothing could be done at present to mitigate the consequences. The hazard of genome instability, on the other hand, is not so much a present danger as a challenge to contentment. Even if the human genome is accumulating unwanted sequences of nucleotides—and the evidence one way or the other is insubstantial—hundreds of generations would have to pass before anxiety became immediate. Yet the possibility that the genome may be unstable is a stringent test of what people mean when they insist that *Homo sapiens* should find its proper place in nature. Would enthusiasm for what is called "sustainable development" persist if it entailed the extinction of our species?

A PLAGUE OF PLAGUES?

It is no mystery that we have been afflicted by several novel infections in recent decades. Better detection is largely responsible; in prosperous countries, people are no longer allowed to die of unknown causes. People also travel more and further afield, and so transfer tropical infections to temperate climates (as, in the old days, European infections were carried to the New World). In many places, human population growth has greatly reduced the habitat available to other animal species and so has helped the transfer of infectious agents from other vertebrates to people; propinquity makes transfer easier. People, meanwhile, have created novel evolutionary selection pressures on microorganisms, notably by the use of antibiotics against bacteria. Who can wonder that known organisms have become more virulent and that organisms previously confined to other species have adapted to make a living off *Homo sapiens*?

The list of recently novel infections is nevertheless impressive. Legionnaires' disease surfaced in July 1976 in Philadelphia and was eventually linked to a bacterium adapted to life in water-cooled air-conditioning systems. AIDS followed on its heels; it was first recognized as an infection in 1981; now, the World Health Organization (WHO) reckons that more than 20 million people worldwide are infected with the virus responsible, known as HIV. Despite novel drug treatments, most of them will die of the disease.

Over the same period, there have been repeated alarms about a series of outbreaks of virulent viral infections with their origins in Africa: Lassa fever (1969), the similar infection caused by Ebola virus (1976) and the Marburg infection (named after a German, not an African, city when a shipment of green monkeys from Uganda infected 30 German laboratory workers and killed seven of them). At about the same time, Lyme disease (named after the town of Old Lyme in Connecticut) was diagnosed as an infection. More recently, the United States has been preoccupied with hantavirus (named after the river Haanta in Korea), which killed more than a score of people in the U.S. Southwest in 1993.

In rich countries, agents that cause food poisoning are a continuing cause for concern. In the past decade, the United States, Japan and Britain have in turn been afflicted by outbreaks of food poisoning due to a novel virulent strain of the ubiquitous gut bacterium *E. coli*. The resurgence of malaria in India and Africa and of the drug resistance and renewed virulence of the bacteria responsible for tuberculosis are a general anxiety.

So who can be sure that the largely unknown world of microorganisms does not harbor a bacterium or a virus with the potential to kill off a large proportion of the human race, putting its way of life, perhaps even its survival, in hazard? Bubonic plague came close to causing such a crisis, between 1300 and the late seventeenth century, when successive epidemics affected most countries in the world, reducing many populations by a quarter or a half their number. That was a consequence of the rise of international commercial shipping; the ships carried the rats that carried the fleas infected with the bacterium *Yersinia pestis*. And, thanks to the longevity of modern populations, it is at the upper range of living memory that the influenza epidemic of 1919–21 killed 20 million people throughout the world, more than were killed in battle during the whole of the First World War.

The threatening balance sheet has another side of course. Did not the long effort to control smallpox infection culminate, in 1994, in the eradication of the disease?[2] WHO has now set its sights on the eradication of poliomyelitis, which still causes endless damage in poor countries, but which can be effectively controlled by vaccines. There are also vaccines effective against measles and rubella ("German measles") and effective drug treatments of West African river blindness, otherwise *onchochersiasis*. And yellow fever, a restraint on travel to the tropics as recently as half a century ago, has been contained, at least, by the heroic development of a vaccine by the U.S. Army in the late 1920s.

This list of pluses is not as long as that of the minuses of the infections recently sprung to prominence, but mere listings are no way to assess the impact on public health. The chief reason why the death rate in the early years of life has fallen dramatically in the rich countries, and is also now falling quickly in the poor countries (whence the frequently rapid growth of population there), is that infants and children are now less often killed than formerly by infectious diseases. The benefit to human populations is a great reduction of death among young people that seems to strike at random, and which is preventable, and the sense of security and civility that can bring to whole populations.

Notoriously, the causes of the general increase of the expectation of life of the young have less to do with the technology of modern medicine than with more old-fashioned stratagems: the treatment of public water supplies to remove bacteria and improved personal hygiene, the rationales for which are founded on nothing more than the general principles of Pasteur's germ theory of disease.

Only a century after Jenner's empirical development (at the end of the eighteenth century) of a technique for immunizing people against smallpox (by the use of living viruses of the closely related cowpox) did vaccination become important in human medicine. We forget that most of the vaccines now widely used in the protection of infants and children against infection have been developed since the 1930s and that the dramatically effective polio vaccines were first tested in human beings only in 1956, by which time antibiotics such as penicillin had been in use for a decade or so. Even rich countries' investments in the defense against infection has taken the best part of a century to bear fruit. Ignorance (until recently) of the details of the human immune system is only part of an excuse.

What of the emergence of resistance to antibiotics of some bacteria and the resistance of malarial parasites to the drugs developed for use against them (or of the mosquito vectors to insecticides such as DDT)? In the halcyon 1950s and 1960s, when novel antibiotics appeared on the markets every year, the phenomenon was not anticipated. Warnings by a few microbiologists that plasmids in bacterial cells could transfer genetic material between different species of bacteria were often derided as misanthropic overcaution.[3]

Nevertheless, the continued medical benefits of antibiotics probably still outweigh the difficulties caused in medical practice by resistant bacterial strains. Yet if it turns out, as seems probable, that the novel virulent strain of *E. coli* has become prevalent as a result of the widespread use of antibiotics in animal husbandry, and particularly so as to accelerate the growth of cattle, it will seem very strange to future generations that such a trivial use should have been made of such powerful medicines. The rapid spread of *E. coli* infection, which first appeared in the United States in the 1980s, to at least two other continents (Asia and Europe) is a dramatic proof, if one were needed, that novel infections are an international problem.

The spread of AIDS is perhaps the most sensational illustration. Within a decade, countries such as Thailand and India were more heavily infected than the United States, where the infection was first recognized. But there is also the case of bovine spongiform encephalopathy (BSE), apparently engendered in British cattle by rearing them on high-protein feed containing processed sheep offal, and from which 14 people in Britain and one in France had apparently died by the end of 1996. Not merely is the infection novel, but so is its mechanism. The infectious agent seems to be neither a virus, a bacterium nor a protozoan (as in malaria and syphilis), but a naturally occurring protein molecule whose aberrant shape can induce the same aberrant form in other such molecules. It remains to be seen whether this novel infection can and will be eradicated before it spreads beyond Britain and the few other cattle herds in other European countries in which infected cattle have been found.

What is to be done to keep these novel infections at bay? The first need is a general recognition that those recently come to light are but the tip of an iceberg yet to become visible. Lassa fever, for example, a great source of alarm in the 1970s, is probably an old infection in Central Africa, probably maintained in some unrecognized animal reservoir, that was first identified only when it infected Western travelers to Africa. Significantly, the cancer of the human superficial tissues called Kaposi's sarcoma, which is a common accompaniment of AIDS, was found in 1997 to be caused by "Herpes-8," one of a large family of herpes viruses of which the most common is that responsible for non-cancerous "cold sores" of the mouth.[4] In reality, the one certainty about the catalog of conventional microorganisms that can damage people is that it is far from complete.

By definition, newly emergent organisms are more difficult to deal with, but the case of AIDS is not the discreditable proof of indolent science that it is usually represented to be. On the contrary, it is a marvelous example of how modern techniques and substantial resources can solve difficult problems with exemplary speed.

Within three years of the first proof (by the U.S. Centers for Disease Control, or CDC) that AIDS is caused by an infectious agent, the HIV virus had been characterized as a retrovirus (with a genome made of RNA, not DNA); a year later, the whole nucleotide sequence of its genome had been worked out. Because the virus attacks the particular white blood cells (called "T-cells") normally involved in recognizing and killing other cells harboring virus particles, causing a profound immune deficiency in those infected, talk of developing a vaccine was always somewhat speculative. But by 1989, a drug (AZT) designed to inhibit replication of the virus was found to decelerate development of disease. A further five years of energetic research were required to uncover the intricacy of the interaction between HIV and the cells in which it replicates; the genetic structure of the virus (published in 1985) has since led to the design of drugs that seem to promise indefinite postponement of overt AIDS. At any earlier time, it would have been astounding that so much could have been learned of so complex a virus in a mere 15 years. That is a testament not only to the fears initially engendered by AIDS but to the vast resources put into the investigation of the disease.

How can any part of that tale be discreditable? Because the hunt for HIV and for remedies against infection brought out the worst in the research community's innate competitiveness—not so much for riches as for fame. Robert C. Gallo, then head of the virology branch of the U.S. National Institute of Allergy and Infectious Diseases in the early 1980s, first sought to diminish the discovery of the virus by Luc Montagnier, a professor at the Pasteur Institute in Paris, and then to give his own comparable work the ring of authenticity. By the spring of 1984, he and Margaret Heckler, then U.S. Secretary of Health and Human Welfare, were promising a vaccine in a few years. This corrosive incident has reflected badly not only on those concerned, but on the research community as a whole.

It is nevertheless the plain truth that the great effort to understand the mechanism of AIDS and to find at least a palliative for the disease seems to be yielding the intended results. The total spent on AIDS research in the United States alone probably amounted to $20 billion by 1996 which, by coincidence, is not very different from the total cost of the Apollo Program by which the first U.S. citizens were landed on the Moon. At the outset, whatever improved treatments emerge, they will be expensive, well beyond the pockets of the poor countries where AIDS is now most prevalent. It is striking that AIDS research, however, has also uncovered a great deal about the working of the human immune system that was not suspected when the first AIDS cases were being treated in makeshift clinics.

There is no reason why inquiries of the same intensity should not be mounted against other truly novel infections. Indeed, that would be the prudent strategy to follow. The need is for a detailed understanding of the mechanism by which new strains or even species of microorganisms have emerged. It is not merely that a knowledge of the genome of a novel bacterium or virus can point to the design of effective drugs (which has happened with AIDS and even influenza[5]), but that modern knowledge of cell biology and molecular genetics can lead to other ways of treating infected people, while the ecology of novel infections may often suggest even more economical and effective ways of controlling them. The difficulty is that thorough investigations of this kind are expensive. National governments skimp on them. WHO has only a meager budget for research. (Its successful Program of Research and Training in Tropical Diseases, begun in the early 1970s, has been supported mostly by private foundations.)

Yet the threat of infectious disease to the health and well-being of human populations, although now blunted in the rich countries of the world, is a present danger for us all. To be sure, our health and well-being has been greatly improved by the general use of antibiotics, vaccines and drugs that inhibit infectious organisms, but AIDS is probably simply the first of the several novel infections that will arise in the decades ahead. What if the next novel infection is a further cancer-causing virus?[6] It is ironical that when, for the first time in human history, there are intellectual tools available for understanding the causes and the course of infectious disease, the world has not been galvanized to deploy them effectively.

BACK TO METHUSELAH

The idea that continued industrial activity could radically alter Earth's climate goes back to 1896 and to the Swedish chemist Svante August Arrhenius, who then made a simple calculation to show that carbon dioxide (CO_2) accumulating in the atmosphere would increase the temperature at the surface by roughly 5 degrees Centigrade. Half a century ago, the issue was taken up by Roger Revelle, a geophysics professor at Harvard University; his case was based on a celebrated series of measurements of the concentration of CO_2 above the Hawaiian volcano of Mauna Loa (which is far enough from fuel-consuming industry to represent the well-mixed atmosphere). The measurements have continued since the 1950s and are paralleled now by observations from dozens of laboratories around the world.

The outcome is unambiguous. The CO_2 concentration is indeed increasing steadily, by approximately 0.5 percent a year. The amount of CO_2 has increased by a quarter since the early eighteenth century, two-thirds of that increase has been within the twentieth century. The quantities involved are huge: the weight of CO_2 now being added to the atmosphere each year comes to just under 14 billion tonnes.[7] Large though the quantity may seem, it is just over a half of the CO_2 produced each year by the known consumption of fossil fuels throughout the world.

It is natural to attribute the measured quantity of CO_2 added to the atmosphere each year to the burning of fossil fuel, but a little caution is in order. The total quantity of CO_2 in the atmosphere is almost exactly 1,400 billion tonnes. The rate at which atmospheric CO_2 is exchanged with the oceans on the one hand and the biosphere on the other is much greater than the rate of production from burning fossil fuel; each year, some 340 billion tons of atmospheric CO_2 is dissolved in the oceans and an almost equal quantity is returned.[8] There is a comparable exchange of CO_2 between the atmosphere and the biosphere, whose vegetation and soil humus are reckoned to store the equivalent of more than 8,000 billion tons of CO_2 or its equivalent (as other chemicals). The atmosphere cannot therefore be likened to a gas-filled balloon into which CO_2 from burning fuel is discharged and is there inevitably and permanently retained. Indeed, the reason why only half the CO_2 known to be produced by burning fuel remains in the atmosphere is that its exchanges with the oceans and the biosphere are unequal; more is lost to the atmosphere than is returned. The "missing" half of the CO_2 remains either in the oceans or locked up in trees and humus. So much seems now to be agreed.

There is also agreement on the mechanism by which CO_2 can affect the climate, and the surface temperature in particular. CO_2 is transparent to visible radiation, which is the bulk of the radiation from the Sun, but it absorbs infrared radiation of the kind emitted by Earth's surface. The result is that the more the CO_2, the more difficult will it be for infrared radiation from the warmed surface of Earth and from the atmosphere lying above it to escape from near the surface. But the energy that comes in (as visible radiation) must eventually go out again (as infrared radiation). The response must therefore be an increase of temperature near the surface. Aptly, the phenomenon is called the "greenhouse effect"; CO_2 in the atmosphere affects the temperature beneath in exactly the way in which the glass covering a greenhouse increases the temperature inside, and for the same reasons. It is important that the greenhouse effect is not always a malevolent phenomenon. Without greenhouse gases in the atmosphere, the temperature on Earth's surface would be below the freezing-point of water, so that the Earth would not then be inhabitable. But water vapor plays a more important part than CO_2 in allowing life on Earth.

So what are the likely consequences of the increased amounts of CO_2 now in the atmosphere and of the further increases there will be as the consumption of fossil fuel continues? For half a century, people have been building models aimed at predicting what will happen. Like the cell-division cycle, this is a dynamic phenomenon: solar radiation comes in and terrestrial radiation goes out, after being transmuted into infrared radiation by the atmosphere and the Earth's surface. The predictions of the first models, in the 1950s, were scary but insubstantial, suggesting increases of average temperature of 2 or 3 degrees Centigrade by early next century. The early model builders readily agreed that they had neglected many important physical processes in the balance between the incoming and outgoing radiation.

There are several complications. CO_2 is not the only greenhouse gas; so too are methane (produced naturally, in agriculture and by waste dumps), the refrigerants known as CFCs,[9] nitrous oxide (from wetlands, tropical forest soils and vehicle exhausts) and water vapor. The link between the atmosphere and the oceans cannot be neglected—it is crucial, if only because the oceans are a potentially huge reservoir of heat energy and because the death of single-celled organisms in the oceans is a means of turning organic carbon into carbonate rock, so permanently locking it away. Even the amount of sea-ice in the polar regions matters; it reflects incoming solar radiation unchanged. One of the most obdurate difficulties is the presence of clouds in the sky; their upper surfaces act like sea-ice, reflecting solar radiation directly, but they are also impediments to the escape of infrared radiation from below.

Meanwhile, the modeling has been improved enormously; it will be a long time before there is a model of the cell cycle that is anything like as elaborate. What began as a back-of-an-envelope calculation has grown into a set of parallel computer-based models at a dozen centers across the world. The models are called general circulation models (to allow for the inclusion of the oceans and the biosphere). Their common purpose is to study the evolution of Earth's climate. Climate, it will be appreciated, is not the same as weather, but rather is weather averaged over times comparable with the duration of seasons. The word "average" conceals a trap; the most delicate step in constructing climate models is to know what features of the constantly changing weather at any place can be safely represented in the climate models by their averages and other statistical properties (such as their variation in time).

The least secure feature of existing models is the treatment of clouds —the models usually deal with a quantity called "average cloudiness," whereas the edges of real clouds (especially thunder clouds) are plainly important means of transferring heat energy vertically in the atmosphere. But there is also the need to take account of the slowly acting phenomena affecting climate, notably the interaction between the atmosphere and the oceans. Given the complexities, it is remarkable that the models of climate change function as well as they do.

The threat of global warming has already been a boon in one respect: the behavior of the Earth's atmosphere and oceans is now much better understood than could have been expected half a century ago. The composition of the atmosphere and especially its content of the minor greenhouse gases is known in detail and is closely monitored, as is the output of energy from the Sun (which varies with the eleven-year sunspot cycle and which, in principle, could fluctuate more markedly over longer times). Earth satellites have been designed and launched from the United States to make direct measurements of the radiation entering and leaving Earth's atmosphere, which is a breathtaking exercise: although the total amounts should ordinarily be the same, the distribution in wavelength of the outgoing infrared radiation should yield important information about the processes by which solar becomes terrestrial radiation. By contrast, the oceans are less well known and understood.

And the outcome of this gigantic effort? Like weather forecasting, the forecasting of climate changes has become a big business for computer modelers. The models now concur that the average temperature will have increased by 2050 by just about 1 degree in 60 years (since 1990), whatever is now done to restrict the emission of greenhouse gases.[10] Moreover, even with restrictions on greenhouse gases, there would be a further increase of temperature by the end of the next century that, depending on the severity of whatever constraints may be placed on the consumption of fossil fuels, could amount to a further 1 or 2 degrees. (In 1990, the unrestrained model calculation suggested a total increase of between 3.5 and 4.0 degrees by the end of the next century, but did not then allow for the aerosol dust in the atmosphere—see below). Even if (by some means not now foreseen) CO_2 emissions were promptly halted, the temperature would continue to increase after 2050 as heat previously stored in the upper layers of the oceans is delivered back to the atmosphere.

Small though the projected increases of temperature are, their consequences—if realized—would be important. It is, or should be, irrelevant that natural fluctuations of the surface temperature involve changes from year to year that exceed the amounts predicted by the computer models, for the natural fluctuations are both upwards and downwards. A *consistently* increasing trend of temperature, even as little as one-fifth of a degree each decade, would be a very different matter. It is relevant that the melting of the ice at the end of the last Ice Age took more than 1,000 years, during which the average temperature of Earth's surface increased by roughly 6 degrees[11]; that is the equivalent of 0.06 degrees each decade, or less than a third of the lowest rate of temperature increase now predicted by the climate models.

The effects of a one-degree increase in average temperature in the first half of the next century would be significant and economically important. The balance of natural vegetation would be changed, together with the animals dependent on it. At least in some parts of the world, especially where the natural ecology is already delicately balanced as in the near-desert regions of the subtropics and the subpolar tundra, land that now supports vegetation of a kind would support nothing and, given the erosion that would follow, might become incapable of regenerating itself naturally.

Agriculture would be most seriously affected. One of the predictions is that the temperature of the land mass of North America would increase roughly twice as quickly as would the average temperature of the whole of Earth's surface, and that precipitation would decline. A 1991 study[12] of the outcome for farmers concluded that some (but not all) crops could yield what they do at present or even more—maize could benefit from the extra CO_2—but only if they were irrigated to a degree that is now considered uneconomic. This implies that food production could be maintained, but that there would be a substantial increase in real cost. In the poor regions of the world, where there is now hardly any irrigation, it is improbable that similar adaptations of practice would be feasible.

Sea level would probably also rise, both because the increase of the temperature of the oceans would cause the water near the surface physically to expand; and because of continued melting of glaciers and of the ice caps surviving from the last glaciation (in Greenland and Antarctica). But the model builders acknowledge that it is difficult to predict sea-level changes partly because of uncertainties about the long-term behavior of ocean currents but also because increased precipitation in the polar regions would add to the bulk of the permanent ice. With no restraint on CO_2 emission, sea level is reckoned to be 50 centimeters higher by the end of the next century (but the uncertainties would accommodate half or twice as much as that). That compares with a well authenticated increase of sea level of between 10 and 20 centimeters over the past century.[13] Nearly half of that increase is believed due to the melting of terrestrial glaciers, the remainder due to the expansion of the warmed surface layers of the oceans.[14]

Although the magnitudes of the predictions are again small, their consequences for many parts of the world could be serious. Sea level is the base from which the rise and fall of the twice-daily ocean tides are measured. It also determines how far inland will be the reach of the waves raised by oceanic storms such as hurricanes (or typhoons), which have historically done much to shape the topography of coastal regions. Regions such as the Bay of Bengal, already repeatedly exposed to typhoon damage on a huge scale, would be seriously affected; for example, the exposed coastal area of Bangladesh lying around the common delta of the Ganges and the Bramaphutra, would be greatly reduced in area and the region of the country exposed to regular inundation would move far inland.

In this litany of unwelcome happenings, there is also a recipe for catastrophe. Since radar measurements in the 1970s first revealed the underlying topography of Antarctica,[15] it has been clear that some parts of the Antarctic ice cap could be unstable. There is particular concern about the West Antarctic ice shelf, which faces from Antarctica to the Pacific. This mass of ice, reaching for several hundreds of kilometers out to sea from the solid continent, is partly grounded on the sea bottom and supports hundreds of cubic kilometers of ice. If this block of ice should become detached from Antarctica, perhaps because melting ice lubricates its junction with the rock beneath, the whole mass could become the world's largest iceberg. In that case, there would be an increase of some 5 meters (16 feet) in global sea level. The map of the world would change quickly[16]: in northwestern Europe, the only capital cities left unscathed would be Berlin, Berne, Brussels, Luxembourg, Madrid and Vienna; in the United States, most major cities on the eastern seaboard would be flooded (but parts of Manhattan would remain as islands). On the West Coast, Los Angeles would sink beneath the waves. Nobody suggests that the slumping of the West Antarctic ice sheet is imminent, but merely that it is likely to happen at some stage (after the end of the next century) if global warming is a reality and continues unchecked.

So what is the reality? The fact that computer models make similar predictions is not decisive; they are all constructed along similar lines, using similar methods and the best data available; may they not all be making the same mistakes? The two acknowledged inadequacies are the treatment of real clouds by the computer models (and, more generally, the treatment of water vapor, which is a greenhouse gas) and the still imperfect knowledge of the long-term behavior of the oceans, but there are others. The processes in the oceans by which organic carbon (locked up in living plankton) is converted into carbonate deposited in the deep-ocean sediments are only poorly understood. If (as might be expected) the process becomes more rapid as CO_2 concentration increases, the proportion (now one-half) of CO_2 from burning fuel remaining in the atmosphere could progressively decline, reducing the long-term concentration of CO_2 below the values now assumed.

There are now more basic difficulties with the models, which use a grid of points in latitude and longitude to plot the climate at the surface and in the atmosphere above it. The inevitable result is that each point in the grid represents a substantial surface area, with dimensions of about 300 kilometers. This amounts to the assumption that the climate is uniform in patches of the world's surface comparable in size with Ireland or the U.S. state of Virginia. Quite what this means for the calculation of quantities such as the average temperature is far from clear.[17] And it is striking that only in the past few years have the computer models been modified to include the effects of atmospheric aerosols[18] (which have a cooling effect). Only that step has made it possible successfully to reconstruct past climate—"postdiction" as it is sometimes inelegantly known. Unfortunately, the hurried incorporation of new knowledge like that seems like an afterthought, which does not inspire confidence.

The reality or otherwise of global warming therefore remains unproven. Given that estimates of the degree of global warming have steadily declined since the first predictions in the late 1940s, and have declined between 1990 and 1996 by about a half, skeptics can and do argue that more scrupulous attention to the details of the modeling might show that the threat of global warming is an illusion. On balance, as will be seen, that conclusion seems both unreasonable and imprudent.

It is true that Dr. James Lovelock, an independent British scientist, has argued that there is something special about the resilience of the Earth's surface in the face of external pressures—the switch from a reducing to an oxidizing atmosphere 2,700 million years ago or earlier, for example, or the climatic ups and downs caused by long-term oscillations of the Earth's orbit. Lovelock has exercised great ingenuity in identifying processes that tend to make the match between Earth's changing atmosphere and the evolving biosphere hospitable to the latter. But his "Gaia hypothesis" has been widely misused in claiming either that Earth is somehow immune from environmental insults or, more mystically, that it has a life of its own. The truth is that the laws of physics and chemistry will not be suspended in favor of species whose only claim is that they are now extant. Nor does it imply that the successful adaptation of the biosphere to the slow geochemical changes of the distant past will necessarily apply in the future, when man-made geochemical changes represented by the accumulation of CO_2 are relatively more rapid. The Gaia hypothesis, wish fulfillment of a kind, is a flimsy basis for asserting that global warming will not happen.

It would be different if there were a plausible mechanism to explain how natural processes of some kind might prevent what the computer models predict. There have been several ingenious attempts to produce one, but none has surmounted the scientific community's unique test of what to take seriously: is a new idea so appealing as to suggest that effort spent on it will yield an interesting discovery about the world?

Almost by default, the argument has gone to the modelers. That, one might guess, is how it must have been in Galileo's time. When Galileo proposed that force and acceleration are equivalent, nobody had a better idea and the equivalence principle won the day. On global warming and climate change, victory has gone to an organization created in 1988 by two United Nations agencies—the Environmental Program (UNEP, based in Nairobi) and the World Meteorological Organization (Geneva); the organization is what its name says, the Intergovernmental Panel on Climate Change (IPCC).

Originally the intention was that IPCC should make a scientific assessment of the likely effects of global warming on the climate, similarly to assess the effects of changed climate on human activities (from habitation to the practice of agriculture) and to recommend policies that might ameliorate or avoid the changes in prospect. The first assessment, intended as raw material for the U.N. conference held in Rio de Janeiro in August 1992 (and widely described as the "Earth Summit"), was a mixture of naiveté (in the second and third parts of the exercise) and of careful analysis and explanation (in the first, scientific component, of the project). IPCC has been retained as one of the organs of the Framework Convention on Climate Change adopted at the Rio meeting and which, now ratified, has the force of an international treaty.

The continuing scientific work of this organization does not carry the conviction it might, partly because IPCC has become part of a political process—that of administering a treaty whose usefulness is disputed by many of its signatories (the United States, Saudi Arabia and China, for example[19]) and whose most burdensome (because expensive) provisions have not yet come into effect. The scientist members of IPCC's committees are distinguished researchers whose objectivity is mostly unquestioned, yet the organization's avowed objective is to "build a limited consensus" on the science of climate change.[20]

That is a misguided enterprise. It is not so much that scientific consensus, as during the 1,200 years when Ptolemy's account of the motion of the planets was assumed to be the truth, often ends in tears; current investigations of possible climate change involve no great imaginative leap. Instead, they are founded on sound atmospheric physics and a great wealth of observational data. But IPCC's scientific assessments (of which there have been two so far, in 1996 as well as 1990) would have been more persuasive if they were plainly the products of an independent organization. The distinction of IPCC's members is such that it could issue its views of the likely evolution of the Earth's climate on its own considerable authority. The consensus that is needed, after all, is not a *scientific* consensus, but a *political* consensus—what, if anything, to do about the threat of global warming, and by when.

In an unrelated but important environmental field, there is an independent scientific organization that has been able decisively to shape public policy internationally: since 1918, the International Commission on Radiological Protection (ICRP) has periodically issued recommendations about the standards required to protect people from potentially damaging radiation (X rays and radioactivity).[21] The recommendations have almost invariably been followed by national governments. The commission's independence from governments has been an important guarantor of the integrity of its advice.

In radiological protection, there is also a truly intergovernmental body, the UN Scientific Committee on the Effects of Atomic Radiation (UNSCEAR), which was set up in 1955 (at the request of the government of India) by the United Nations. The secretariat of UNSCEAR (originally in New York, now in Vienna) collects data and reviews published evidence bearing on its problem. It does not seek consensus about the hazards of ionizing radiation in any context, but publishes opinions in technical language (which governments are nevertheless competent to evaluate).

IPCC has followed a different path. Its summary opinions are negotiated at meetings at which governments are represented.[22] Its published assessments are written in quasi-technical language; all the intricacies of the problem are mentioned, but most data are presented graphically rather than numerically, so that readers cannot easily take their slide rules to them. The organization has also grown sensitive to criticism, whether technical or procedural. IPCC seems not to have appreciated that the prospect of climate change is bad news for most people and that its own method of inquiry would be more persuasive if it were demonstrably open to the possibility that the bad news may not be true. Such a stance would even lend weight to IPCC's authority.

In the pursuit of scientific consensus, IPCC has neglected some aspects of its huge problem that might make its conclusions more palatable. One source of confusion, for example, is that the succession of years with high average temperature in the 1980s is not the only one on record. The 1930s were also warmer than expected. On the other hand, the decade from 1940 onward included so many below-normal temperatures that many climatologists openly canvased the possibility that the Ice Age was returning. In other words, it seems that there can be fluctuations of climate not simply from year to year, but for decades at a time. In that case, may not the high temperature of the 1980s have been the consequence of such a fluctuation?

The obvious place to look for an explanation of these long-term fluctuations is in the oceans; the remarkable equatorial current in the Pacific linked with the climatic phenomenon called el Niño, first recognized (and named for) currents flowing along the coast of Central and South America at intervals of three or four years, is known to influence the climate of the whole South Pacific basin and probably that of the Earth's as a whole.[23] IPCC's consensus building would be easier if it found explanations for the recent major fluctuations of climate still fresh in the records and in the minds of skeptics.

There is also the matter of "technical fixes," ways of ridding the atmosphere of CO_2 or other greenhouse gases without the inconvenience and the economic cost of restricting the consumption of fossil fuel worldwide. Given that the prospect of global warming is bad news for the world at large, schemes for avoiding some of its consequences should fall squarely within IPCC's remit; little attention has been paid to them.

There has already been one experiment of this kind. The growth of oceanic plankton, the source of carbonate rocks on the deep-sea bottom, is limited by the availability of soluble iron. So why not "fertilize" the oceans with iron, with the objective of accelerating the conversion of atmospheric CO_2 into carbonate? An experiment in the Antarctic Ocean in 1991 yielded negative results, one in the North Pacific in 1995 proved more encouraging, but is not yet a basis for practical action. As things are, there is no mechanism by which such devices could be investigated internationally.

By contrast, the reforestation of land not needed for other purposes, widely advocated, is not a permanent way of avoiding global warming, but merely serves to increase the amount of CO_2 locked up in the biosphere; it has the effect of postponing global warming to some degree; it would be more effective if genetic engineers had bred trees that will turn CO_2 into wood more quickly than even eucalyptus does. The hope of finding chemical reactions that will remove CO_2 from the atmosphere economically and safely is remote, but the removal of other greenhouse gases may be a worthwhile challenge for the imaginative in the years ahead.

What, meanwhile, should those who are not climatologists consider to be the likely course of events? Despite the uncertainties of the models, global warming is on the cards. What remains in doubt is its degree and timing. Because of processes not included in the models, the present estimate that the temperature will rise by one degree by 2050, and will continue to rise in the second half of the next century, may well be an overestimate, perhaps by a factor of two.

The truth will eventually come from a comparison of the computer predictions and observation, especially the measurement from Earth satellites of the temperature near the surface of the Earth and its variation with altitude above the surface. (The computer models will undoubtedly be improved, but the obstacles are formidable and will not be quickly overcome.) By the end of the century, there will be three decades worth of reliable satellite observations. Interestingly, the satellite observations so far give an average rate of increase of temperature of 0.09 degrees (just under one-tenth of a degree) each decade, which is just about a half of the rate of increase predicted by IPCC for the half-century ahead.[24]

The essence of the dilemma presented by the threat of global warming is simply stated. The artificial greenhouse gases in the atmosphere are capable of increasing the temperature at the surface and the evidence agrees with the expectations (of the computer modelers) that warming is already under way. If so, there is every likelihood that the warming trend will continue for several centuries ahead. At some point, there will have to be a halt to the emission of greenhouse gases. Long before then, the sheer cost of adapting to the changes (the cost of irrigating crops, for example) will exceed the cost and inconvenience of abating the emission of greenhouse gases.

When will that time be? Only two answers make sense: "Now!" and "Yesterday!" Luckily the world acted yesterday. To be sure, the Rio treaty is not an agreement on steps to abate global warming and its effects, but a framework within which signatories must decide what steps to take, and when. The treaty includes an exhortation that governments in industrialized countries should restrict the emission of greenhouse gases in 2000 to the levels of 1990. At a conference of the signatories in Kyoto at the end of 1997, it was agreed that total emissions in 2010 should be reduced to 8 percent below those in 1990; some have said they will comply, others that they will think about doing so while developing countries, including China, are exempted. Diplomacy moves glacially, more slowly even than global warming.

Deciding what to do is complicated by the lack of an economic analysis of the alternative courses that might be followed. What, for example, is the trade-off between adaptation to changing temperature and the cost of reducing the consumption of fossil fuel? (If global warming is a reality, adaptation would not serve indefinitely.) Or, what would be the cost of switching from rice production (which is the source of perhaps half the methane in the atmosphere) to other cereals? Fuel efficiency is evidently the most effective way of abating CO_2 emissions, but how are governments to encourage—let alone enforce—best practice? What would be the economic cost of replacing electricity generating plants by nuclear power stations, given that the capital costs of nuclear plants are more than four times greater? (Unconventional sources of energy, although free from CO_2 emission, are physically incapable in the next century of substituting for any but a small part of present energy consumption.) What of the cost and complication of replacing gasoline with liquid hydrogen as a fuel?

So the world, or "the Planet," faces a dilemma. Although there is ample room for argument about the construction and predictions of the computer models, there is no serious doubt that global warming will occur, if the addition of greenhouse gases to the atmosphere continues unchecked. Moreover, the end point would be global catastrophe. So there is a need to make an early start on the formal restriction of emissions of greenhouse gases. At this beginning stage, modest targets should be set for reducing emissions, but on principles that are likely to endure for the centuries during which a control regime will have to endure. We shall no doubt see whether those who administer the Rio framework are able to steer a course between the political correctness that the planet comes first and the need to generate the resources that will make a transition to a low-CO_2 regime politically feasible.

DODGING PLANETARY MISSILES

In July 1994, an object called Shoemaker-Levy-9 after its co-discoverers plunged into the surface of the planet Jupiter. Although the impact was just beyond the illuminated limb of the planet, on the far side of Jupiter as seen from Earth, debris from the interaction between the object and the gaseous atmosphere of the planet was thrown high enough above the surface to be just visible. For several weeks afterward, dark spots on the surface (some several times the diameter of Earth) were seen at what were presumed to be the sites of impact. The object had been discovered only in 1993, and was classified as a comet on the strength of sunlight scattered from gas and dust surrounding it; its most remarkable feature, though, was that it was not a single object, but several pieces, spread out over several million kilometers along an orbital track. And the orbit made it clear that Shoemaker-Levy had become captive to Jupiter's gravitational attraction; the several pieces of the object visible are presumed to have been produced, at an earlier close passage of Jupiter in 1992, by the gravitational forces of the planet on the much less massive comet, much in the way in which the Sun and the Moon distort Earth's shape by raising tides in the oceans.

The occasion was not merely the first authenticated observation of the impact of a comet or any other object with the surface of a planet in the solar system.[25] For many, it was also a demonstration of one of the hazards to the long-term survival of *Homo sapiens* on Earth.

There is now abundant evidence that the hazard is real. The Moon's surface has been shaped by meteoritic impact. Some of its craters have been dated to 4,200 million years ago (by the radioactivity in rocks returned by the Apollo missions) or to the very earliest stages of the Moon's existence. Others are much more recent. The same is true of the surface of Mars, where impact craters are mixed with ancient volcanic cones. The surface of Mercury, the innermost planet, is similarly pock-marked. How could Earth have escaped events like that?

The record shows that Earth has not been immune. Duncan Steel, an astronomer at the Anglo-Australian Telescope at Coonabarabran in New South Wales, reckons that 130 impact craters have so far been found and surveyed[26]; erosion and geological processes ensure (on Earth) that only the most recent craters survive. Meteor Crater in northern Arizona, a deep pit in the surrounding sandstone which is more than 1.2 kilometers in diameter is all that is now left of the impact 50,000 years ago of a metallic object, mostly made of iron and nickel. The crater left in Arizona is the equivalent of the explosion of a 40 megaton nuclear weapon.[27] The most recent incident of this kind is the event on 30 June 1908, when a vast tract of northern Siberia was devastated by the explosion of an extraterrestrial object at an altitude of about 10 kilometers above Earth. Thousands of square miles of forest were blown over by the blast and then set on fire. The most likely explanation of this happening, called the Tunguska event, is that a stony meteoritic object perhaps 30 or 40 meters in diameter exploded as a result of friction in Earth's atmosphere, releasing energy estimated to have been the equivalent of a 20 megaton nuclear weapon.

The Chicxulub crater of the Yucatán Peninsula in Mexico, on the western seaboard of the Caribbean, is more vividly related to the question of what should be done about the danger of meteoritic impacts. The crater is at least 180 kilometers across, and is almost certainly the site of the impact of the object whose arrival 64 million years ago marked the end of the Cretaceous period—and the end of the dinosaurs. The object that excavated this pit is reckoned to have been at least 10 kilometers in diameter. The energy release on impact is counted as the equivalent of many millions of megatons of nuclear explosives—more than the explosive power of the world's present stock of nuclear weapons.

The trail leading to the Chicxulub crater is one of the romantic tales of recent science. The late Luiz Alvarez, the distinguished particle physicist from Berkeley, California, and his son Walter Alvarez, a geologist, first startled their colleagues by arguing that the extinction of species at the end of the Cretaceous period was caused by a meteoritic impact; the evidence was the discovery of unusually large (but still small) proportions of the rare metal iridium in deposits in a thin layer separating the Cretaceous from the Tertiary rocks that succeeded them 64 million years ago. The original finding came from Gubbia, in the Italian Alps, but then was recognized to be a worldwide phenomenon. How did the excess iridium come to be so widely scattered? The impact of the extraterrestrial object would have filled the atmosphere with dust for months on end, hiding the Sun and so killing off vegetation and herbivorous animals (such as several species of dinosaurs). The eventual recognition that Chicxulub is the likely impact site was inspired by the discovery that the iridium anomaly is especially well-marked in the Caribbean. Skepticism of the original proposal of the Alvarezes was almost banished in a mere 15 years.

Agreement has not always been reached as easily. As recently as the 1920s, people were still disputing whether Meteor Crater was caused by an extraterrestrial impact. The cratering of the Moon and the other inner planets (as well as that of the planet-sized satellites of Jupiter and Saturn) has decisively changed opinion. Chicxulub shows that impacts can cause global calamity. Tunguska shows that events such as this have happened in the recent past. Now it is also known where in the sky to look for the objects capable of causing disaster—among the asteroids and comets.

Asteroids are really minor planets traveling in orbits around the Sun, of which more than 6,000 are now known. Most are confined to orbits between Jupiter and Mars and thus pose no immediate danger, but others travel in orbits that carry them within Earth's orbit around the Sun and so are potentially capable of colliding with Earth. Objects in this class as much as 20 kilometers across have been identified.

On the face of things, comets are different; for one thing, they are routinely detected by the light reflected from the clouds of gas and dust which they emit under the influence of sunlight, while the detection of asteroids relies on reflection from their more compact solid surfaces. There is circumstantial evidence that comets derive from a cloud of objects tenuously captured by the Sun's gravitational field and lying at the margins of the solar system. (The cloud of "comets-in-waiting" is named after the Netherlands astronomer J. H. van Oort who first guessed that it may exist.) By the gravitational influence of passing stars or molecular clouds, the argument goes, individual objects will be deflected from their orbits into paths that carry them into the solar system proper, when they may again be deflected by the gravitational pull of one of the massive outer planets, Neptune or Uranus, into the gravitational influence of Saturn or Jupiter. And the result of that may be an orbit that eventually carries the comet well within Earth's orbit, raising the possibility of a collision.

The transfer of a comet from the Oort cloud to an orbit that happens to cross that of the Earth is a matter of chance and is necessarily slow, perhaps occupying millions of years. Because comets are visible only when they are sufficiently near the Sun for them to have a visible coma of gas and dust, a census of the cometary orbits that lie partially within, say, the orbit of Jupiter, is not at present possible.

Nevertheless, the known comets that cross Earth's orbit seem not to be destined to collide with it. The orbit of Halley's comet, for example, which last approached the Sun in 1986 and which returns at intervals of 87 years, is now so well-known that its position can be predicted confidently for the next 2,000 years or so; it will not cause trouble during that period. There is less certainty about a comet called Swift-Tuttle (after its discoverers in 1862), which was observed again in 1992 and whose next apparition is set for 2126. The object is big enough to cause an explosion comparable with that which ended the Cretaceous Period, but the orbital calculations suggest that the comet will miss the Earth by two weeks in August 2126. No doubt there will be anxious recalculation of the orbit as the date approaches.

The direct impact of a comet is not the only hazard, however. The disruption of Shoemaker-Levy by the gravitational field of Jupiter illustrates one source of danger; comets passing near the Sun or one of the major planets may be disrupted in just that way into still-large fragments that convert a single projectile (the comet) into a discharge of shot as from a blunderbuss. Indeed, the showers of shooting stars, or meteorites, which can be seen in the night sky at particular times of the year (in June, October and November, for example) consist of nothing but fragments torn from comets in this way whose orbits have drifted away from the original orbit of the comet with the passage of time. Each shooting star is a fragment no heavier than a gram or so which is vaporized in the upper atmosphere of the Earth, but the Tunguska calamity in 1908 appears to have been caused by a much larger lump of rock that was originally part of what is known as Encke's comet, lying wholly within the orbit of Jupiter around the Sun and which crosses Earth's much smaller orbit. In the past few years, a number of lumps of rock (technically asteroids because they do not glow) have been found in the Encke track, but they are probably only a proportion of the total. On that showing, nobody can tell when the next Tunguska event will occur.

So what is to be done? The case for knowing what projectiles are in near-Earth orbits is compelling. If great damage is about to be done to some region on the Earth, does not our dignity require that we should know ahead of time? There is even the possibility that something might be done to avoid the calamity.

Although none of the objects whose orbit is yet known is thought likely to hit the Earth in the next few centuries, it would be rash to suppose that more than about 10 percent of the large asteroids have yet been found. Smaller objects, perhaps only a kilometer in size but with an explosive power equivalent to a million bombs of the kind dropped on Hiroshima in 1945, are necessarily less completely cataloged. Still smaller objects a few tens of meters in diameter (in the Tunguska or Meteor Crater class) can be spotted within a million or so kilometers of the Earth, but detection should improve as the decades go by. It will be a long time before people will hope consistently to detect and track still smaller objects, those just 1 meter across; in any case, most of those are at present broken up high in Earth's atmosphere, which is a powerful shield against all but major disaster.

So what are the chances of impacts that cause serious damage in the coming years? Given the still very large uncertainties about the nearby population of asteroids, not to mention that of the degree to which individual objects will be disrupted in the atmosphere,[28] accurate estimates of the chances are not possible (although they have frequently been attempted). Nor is there much foundation for past attempts to demonstrate that the frequency of meteoritic impacts depends on the oscillation of the solar system as a whole above and below the plane of our galaxy.[29] But there is little doubt that 10 kilometer impacts will be much less frequent than 1 kilometer impacts, that those will be less frequent than impacts by Tunguska-sized objects and that they will be in turn much less frequent than the arrival of the meteors that shoot across the sky in showers at the right season. Crude (and statistically incorrect) though it may be, it is not unreasonable to suppose that the interval between impacts of different degrees of seriousness is that between the last known event of a particular size and the present; that would imply that there will be one Chicxulub impact every 50 million years, one Meteor Crater (caused by a large lump of metal) every 50,000 years and one Tunguska-like event every century.

The precise probabilities do not matter, although the idea that Tunguska events may happen once a century may seem overstated, given that there is only one record of such an event. But near-misses by objects of this energy are indeed frequent: they are represented by the newspaper accounts (from inhabited regions of the Earth) of fireballs flashing across the sky for thousands of kilometres before disappearing from sight. The data collected in the United States by the Spacewatch project suggest that there is a near-miss of the Earth by an object capable of causing a 20 megaton explosion once a week, and that objects in the Meteor Crater class or larger pass within the orbit of the Moon around Earth roughly once a year, on the average. Sooner or later, one such object will cause serious damage on the surface of the Earth—it can be only a matter of time. Yet the impacts that matter for the survival of *Homo sapiens* are those in the class that produced the Chicxulub crater. Would not advance warning, at least, be worth having?

The collision of Shoemaker-Levy with Jupiter appeared to electrify governments as well as the rest of us. Within a few days, the U.S. Congress was calling for a plan to identify all objects of this kind, while the Parliamentary Assembly of the Council of Europe was considering a similar resolution.

Now that the research community accepts that impact by meteoritic objects are as likely to happen in the future as in the past, its curiosity has been aroused; tracking asteroids and comets has since become fashionable at the world's optical observatories. On present performance, there is a high chance, perhaps even as high as 90 percent, that objects in the Chicxulub class will be detected in advance of their collision with

Earth. How far ahead? Asteroids moving in known orbits about the Sun and returning every three or four years can be tracked with great accuracy, so that there may be several years of warning. Comets approaching the Sun from the outer regions of the solar system, on the other hand, may never previously have been seen and may be recognized only a few months ahead of impact (although, again, there might be several years' warning). Of necessity, there will be only a short warning of impact by Tunguska-sized objects, perhaps none at all.

The dilemma for the world at large is what to do—and how to do it. The United States has taken the lead in the de facto improvement of the arrangements for detecting meteoritic objects in space; it is best equipped to do so. But there is every reason why the task should be shouldered internationally. For one thing, if the objective of a systematic search for large meteoritic objects is to consider whether something might be done to avoid global calamity, it would be essential that the data gathered should be publicly available to people of all nationalities; in equity, those with an interest in the data should also contribute to the cost of gathering them.

The need for an international framework is further reinforced by the awkward truth that the only known way of avoiding cataclysmic impacts would involve military technology of the most advanced kind. Obtaining accurate orbits for previously unknown extraterrestrial objects requires the use of radar installations of the kind tightly regulated by the Anti-Ballistic Missile Treaty of 1972. (The most appropriate installation of this kind is that built by the former Soviet Union at Krasnoyarsk in Siberia in the 1970s.) Avoiding impacts, on the other hand, requires that an object on course to collide with Earth should somehow be deflected from its course, which in practice can be done only by exploding a large nuclear bomb at a precisely determined point near the object.[30] Of necessity, the further away it is, the smaller will be the nudge required to arrange that it does not collide with Earth. It is unthinkable that one of the few powers able to command the appropriate rocket and space technology could build and deploy asteroid interceptors without raising hostile suspicions among rivals. Even winning consent for the inevitable practice exercises that would be necessary would be a major diplomatic exercise.

Yet there is no avoiding the truth that the presence of millions of lumps of rock in orbits carrying them near to Earth constitutes a hazard for the continuation of life (or at least that of *Homo sapiens*), that such a collision could happen at any time, possibly with only a few months' warning. And Tunguska events could happen without warning of any kind—tomorrow, perhaps. The case for a properly constituted international organization for collecting, maintaining and analyzing data on the orbits of extraterrestrial objects is unshakable. And the world's grave diplomats, occupationally wedded as they are to the belief that the solution of their current problem will leave the world safe for the rest of time, will soon find themselves caught up in what now seems to them to be science fiction—making arrangements to defend the world from an impact that threatens the survival of people. The next Tunguska event will make that inevitable. Let us hope that it comes before the next Chicxulub crater is excavated.

INSTABILITY INHERITED

Deeper understanding of the genetic apparatus of human beings has been one of the most conspicuous achievements of the past quarter of a century. The understanding is not complete, of course, but the outcome is surprising: the human genome appears to be little different from the genome of the chimpanzee or even the mouse. All mammals have more or less the same genes, but in the human genome they are differently arranged, apparently in a way that accounts for the characteristics that distinguish human beings from the great apes—being able to stand erect, to use language for communication and even to be able to reason. So far as is known, people are the only creatures yet evolved with these particular characteristics. So far as can be told from the fossil record (which is far from complete), the turbulent sequence of human evolution has been compressed into 4.5 million years during which little has happened to change the attributes of the great apes (see Chapter 7).

Darwinism does not guarantee that the products of natural selection will survive for the rest of time. On the contrary, the fossil record is full of evidence of other apparently successful species that disappeared from the strata for no apparent reason. What can account for the unexplained disappearances? What if the selective advantages that sustained the disappeared species in the short run were accomplished by genetic machinery that was physically incapable of replicating itself faithfully in the long run, over hundreds or even thousands of generations? And what if *Homo sapiens* is such a species? Then, in due course, *Homo sapiens* will become extinct.

There is nothing that could be called evidence of such a danger. There are only straws in the wind, but they may be illusory. The roughly 500,000 copies of the *Alu* nucleotide sequence in the human genome may just as well be a proof that much of our DNA is really junk as a sign that the human race has an unstable genome.[31] Even neurological diseases such as Huntington's disease, which appears to involve a systematic error of the DNA-replication machinery of the male reproductive cells yielding, ratchetlike, disease conditions but not getting rid of them, are not proof that the human genome is unstable. Further investigation may even show that the ratchet effect is an artifact of the first few observations, and is unreal. In any case, the disease is self-limiting in the sense that those who inherit it are likely to die at earlier and earlier ages. Nevertheless, it is a cause of mild anxiety that there are half a dozen other neurological diseases stemming from an apparently identical fault of the replication machinery.

These scraps of information are not proof that the human genome is intrinsically inadequate, but are simply a reminder that there is no guarantee of long-term stability. The somber speculation in this supposedly sober text is meant simply to provoke two questions: How would people respond if there were solid evidence that there is systematic error in the working of the human genome? And how would that danger affect our evaluation of the other sources of calamity—infection, global climate change and meteoritic impact? In contrast with the threat from meteoritic impacts, there would be ample warning of such a calamity, hundreds of generations at least. However, the warning would not be clear-cut, but rather would consist of the gradual, even insidious, accumulation in the human population of people with impaired faculties— a slowly but steadily growing proportion of people born with "condition X," say. What would we then do?

The simplest option, that of doing nothing, has much to be said for it. For one thing, it is an uncontentious course to follow. For another, human societies have a long tradition of caring humanely for disadvantaged people. The challenge of shouldering that social task with compassion and imagination is even widely held to be one of the spurs to the improvement of the moral condition of modern societies; if only the precept were more often matched by performance, we should have even more to boast about. There is nothing to suggest that the present incidence of congenital handicap is insupportable or that any but a small part of it could be avoided by techniques based on prenatal diagnosis.

But what if, after some hundreds of generations, the incidence of hypothetical condition X had risen to 10 or 20 percent, and showed signs of increasing steadily from one generation to the next?[32] So far into the future, it is reasonable to suppose that the genetic basis of condition X will be fully understood. Would it not then be tempting to wonder whether the human genome could be corrected in some way, in the interests of the survival of the human race? And would not the temptation be irresistible?

That amounts to the supposition that the practice of eugenics would have become acceptable. It is therefore important that condition X is strictly hypothetical. Even if it were a condition whose genetic basis is now understood, as is that of Huntington's disease, there is at present no known way of manipulating the human genome so as to avoid its occurrence, sporadic as well as inherited. This is strictly a dilemma of the future, but it is a dilemma that will inevitably arise at some point.

Even if the prospect of real genetic calamity does not arise, it is likely that the deeper knowledge of the working of the human genome now being won will suggest ways in which the design of Homo sapiens provided by 4.5 million years of natural selection could be decisively improved upon by genetic manipulation. After all, people are now manipulating the genetic structure of genes so as to make plants resistant to infections. Why not manipulate the human genome to the same end? It is a reasonable guess that Homo sapiens will not always disclaim such opportunities.

CONDITIONS FOR AVOIDANCE

The long-term threats to our survival are more easily identified than the means by which they can be avoided. That is the nature of the world and always has been. The fossil record is a vivid reminder that species are perishable. Even during the emergence of Homo sapiens in the past 4.5 million years, there have been several occasions when species of hominids disappeared for reasons not yet understood (see Chapter 7). Indeed, present dangers are probably—even certainly—less immediate than those of the past. The difference is that we know what the dangers are, or *believe* we know.

How should we respond? Knowledge is half the battle, of course, but is not sufficient in itself. Preparedness is also necessary. How is that to be arranged? The long-term threats so far identified have sprung from scientific discovery and understanding; on the principle that it is often convenient to blame the messenger carrying bad news, science is often blamed that the world now seems such a dangerous place. Yet the avoidance of calamity requires continuing scientific investigation and innovation. Yet even the well-studied threat of global warming is ill-posed, while the likelihood that the future will see the emergence of more virulent microorganisms is not matched by a knowledge of what form they will take.

Is the planet (in the anthropomorphic sense), and are national governments in particular, prepared to accept the implications that small armies of scientists will be required to remove persisting uncertainties and to devise effective strategies for the avoidance of calamity? The experience of recent years in rich countries is not encouraging. Research communities are everywhere under pressure to be more "relevant," usually understood to mean that they should assist with national competitiveness in the production and sale of tangible traded goods. By contrast, most rich national governments have skimped on funds for strengthening the public health organizations whose efforts should rightly be enlarged to meet the still unknown needs of the years ahead. Similarly, the monitoring of the surface temperature of the Earth by instruments mounted in satellites is too often regarded as part of supposedly effete "space research" rather than as a means by which an economical strategy to combat global warming could be devised. Yet preparedness requires that these activities should be supported on a scale, and with a vigor, that is likely to meet the unknown needs. That is one sign that the rhetoric of survival is not matched by resolution.

Another is social and political. By definition, threats to what is called "the planet" are international in their implications, but the past decade has shown all too clearly that the pace at which international agreements on matters such as global warming can be negotiated are not noticeably quicker than the times over which such threats become real. Worse, the strategies eventually adopted internationally are usually burdened by the compromises required to override the vested interests of many of the participants. Governments are ready enough to commend themselves to their electors with protestations of regard for "the planet," but appear not to appreciate the urgency required of them.

In reality, there is no reason why any of the potential calamities now foreseen, even the most scary among them, cannot be avoided. But avoidance requires vigilance and courage. Will we have enough?

Conclusion: What Lies Ahead

This book is about science now and in the future. What stands out is that there is no field of science that is free from glaring ignorance, even contradiction. Loose ends, which historically have been the origins of quite novel understanding of what the world is like, persist and abound. Who, in those circumstances, can believe that science a century ahead will be a tame extrapolation into the future of what it is now?

The situation in the physical sciences now closely resembles that of a century ago, when the doctrine of the æther was firmly established in people's minds. For the first time in human history, there is now an account of how the universe came into being and, moreover, a theory of matter that is to some degree constructed on the principle that matter derives its properties from the event in which the universe as a whole made its appearance. That is a great achievement, especially because until 1929 nobody knew that the universe is expanding, and because ways of dealing self-consistently with quantum particles were worked out only in the early 1950s.

But it is an interim achievement only. There are many defects in present understanding. Continuing difficulties about the rate of expansion may well melt away when there is a usable map of the nearby universe; the fact that the amount of matter obdurately falls short of expectation is a more serious difficulty; the possibility that the clumping of galaxies into clusters will falsify the assumptions on which present cosmology is based is real enough for discomfort; while the status of the inflationary universe as a way of smoothing out Gamow's big bang remains obscure. As with the æther, serious people hunt for the constituents of the "missing mass" without acknowledging that the whole idea may be no more than a sign that present understanding of the universe is incomplete, as was Maxwell's electromagnetism without relativity. In the understanding of matter, the classification of particles as either hadrons (quarks and things made from them) or leptons (electrons and neutrinos) has probably come to stay, but it will be some years before it is known whether the so-called standard model of particle physics is complete. My hunch is that the future will follow the past in revealing a new nest of Russian dolls to be unscrewed.

The outstanding contradiction in fundamental physics is the failure, so far, to reconcile Einstein's theory of gravitation with quantum mechanics—perhaps the two outstanding intellectual achievements of this century. Not for want of trying. Many who work in this recondite field are convinced that the reconciliation will eventually come from the model of particles as tiny vibrating strings or membranes. The obvious question that arises is whether the strings or other structures must be regarded as parts of the structure of empty space, the vacuum, or as structures imposed upon it. The likelihood is that the distinction is semantic only. Bridging the gap between quantum mechanics and gravitation seems destined to endow space-time with a structure, in contrast with the view of what space is like on which science has relied since the time of Descartes. Such a development might have the benefit of accounting naturally for the intrinsic properties of the elementary particles of matter, such as their built-in angular momentum, or spin. In principle, there is no reason why space should not have structure, but nobody can yet imagine what it is like. That will be a discovery to look out for.

The present situation in the life sciences is more complicated. As a direct consequence of the structure of DNA, both the internal working of the cell and the mechanism of inheritance in plants and animals (including the development of embryos from fertilized eggs), has become susceptible to laboratory investigation. Any well posed question can apparently be answered. The consequences for human health and well-being are indubitably immense; the harvest of novel drugs or techniques, such as the regeneration of organs, that will flow from these developments is not yet appreciated. We (or our children) will all live longer healthy lives. Yet even in confident molecular biology, there are clouds on the horizon. The working assumption that single genes (or their products) have a predominant function in the life of a cell is unlikely to persist for long. The complications that will follow will add to present difficulties of comprehending the intricacies of cellular life. Molecular biology is being driven inexorably toward the mathematical modeling of these processes. And while molecular biology has proved marvelously adept at identifying genes in large numbers, that is merely the naming of the parts of living organisms. The promise of a full listing of the human genome a few years from now will not in itself tell us how the human genome functions or how natural selection made it what it is.

There are also two outstanding problems in biology that are virtually untouched. The origin of life on Earth remains a mystery, despite the accumulation of evidence that modern life has sprung into being within the past 4,000 million years from molecules fashioned from inorganic materials that were capable of stimulating their own replication. Whether pure thought or laboratory investigation is the best way to understanding remains to be seen. What is clear is that it is a matter of time only before the design of a plausible self-replicating organism will be demonstrated in some fortunate laboratory. But that will not happen soon.

The more recent evolution of living things will also enliven the years ahead. It is now known that cells whose genes are arranged in a manner more complicated than in bacteria (which have a single circular chromosome) date from at least 2,500 million years ago. That raises questions about the origin of multicellular organisms and about the origin of sexual reproduction. The dramatic discovery, reported at the beginning of 1998, of fossilized animal embryos in Chinese rocks dating from before the Cambrian Explosion suggests that both evolutionary adaptations were firmly established much earlier than has previously been supposed. And there is now the exciting prospect that the detailed course of human evolution will be established by the combination of genetics and classical paleoanthropology. We shall soon have an understanding of the whole of human history.

Understanding how the brain works is, unfortunately, a more distant goal. The past century has marvelously described and cataloged the properties of the cells in the working brain as well as its several separate structures. The proposition that the neurons organize themselves into effective machines for converting sensory information into a form that evokes an appropriate response from limbs and other organs of the animal body is well established. What remains is to identify the neural circuits in the head that make these responses possible. Even for the simplest responses, that will take a long time. We are a long way from the time when the neural circuits responsible for the so-called higher functions of the human brain—remembering and reflecting on past events, choosing between imagined courses of action and imagining more generally—have been identified. Yet only then shall we be able to claim that we understand how the brain works.

This brief listing of the foci of our present ignorance is incomplete, of course. These are all fields in which new knowledge will be gathered in the years ahead. But what the historical record shows is that progress in science, even though it may consist of asking familiar questions of nature more perceptively than has been possible in the past, is not free from surprise. The historical record shows that clearly enough. A century from now, people will be occupied with questions we do not yet have the wit to ask.

The record also shows that the conceptual foundations of science are repeatedly remodeled. On the long view, the present dilemma in the physical sciences about the structure of empty space is an echo in our own times of the argument about Newton's notion of "action at a distance," at which Christiaan Huyghens protested loudly. Maxwell broke with the Newtonian tradition in formulating his theory of electromagnetism in terms of the direct interaction of electrical charges with electrical and magnetic "fields," which was a step towards realism in the philosophical sense: the electromagnetic field, which can in principle be measured, permeates the whole of space (or of space-time), so that its effect on charged particles is determined in the here and now. All the modern theories of particles follow Maxwell's in this regard (with appropriate adaptations to account for quantum mechanics and special relativity). But these theories also imply that there is more to empty space than the name suggests: particles (or pairs of particles) can be conjured from it, for example. But that is not a case of, "The æther is dead, long live the æther!" The quantum vacuum is a different kettle of fish from Maxwell's. So, too, would be space-time with an internal structure such as may be necessary to accommodate the strings or membranes that may be necessary to reconcile quantum mechanics and gravitation.

Realism has also become a guiding principle in the life sciences. The rise of molecular biology has ensured that. One molecule can influence another in ways that can be predicted, if not from first principles then from empirical rules that steadily become more sophisticated—to match the increasing intricacy of the molecular behavior uncovered in the laboratory. In many ways, the situation resembles that in the early nineteenth century, when people knew for sure that matter is made of atoms and had to rely upon empirical rules for telling how atoms combine with each other to form molecules. Now it is accepted that the components of the machinery of life are molecules, but the ways in which they interact with each other are still described by rules of thumb.

That statement can be, and will be, greeted with the cry, "Reductionism!"—a term widely used, among biologists especially, as a polite synonym for abuse. The truth is that the conflict between reductionism and its supposed opposite, "holism," is both arid and meaningless. The conflict springs from confusion about the character of explanation in science. Even in the physical sciences, where there is no dissent from the proposition that the properties of large objects are consequences of the properties of the atoms of which they consist, people do not routinely calculate the properties of objects in the real world from the properties of their atoms. Even now, the boiling point of water is much more easily and reliably measured than calculated. The belief that the calculation *could* in principle be carried through is sustained by many decades of experience showing that the predicted interactions between atoms are a sufficient basis for calculating the properties of even quite complicated molecules. Moreover, as methods of calculation are refined, the calculations are found to be better and better approximations to the measured properties of the systems they supposedly describe. That is the intellectual basis for supposing that the properties and behavior of large objects can be derived from a knowledge of their atoms.

Since the death of vitalism, the belief that living things have an intrinsic quality that distinguishes them from inanimate objects, there has been no choice but to suppose that living things or tissues are similarly the products of the molecules, and ultimately of the atoms (and even of their components, electrons and nuclei) of which they are made. Indeed, the whole of molecular biology is predicated on the belief that this is so. One of the themes of this book is that too little has yet been done to put the assumptions to the test. It is not simply that a more quantitative approach to the behavior of living things would strengthen the assumptions, but that it would in itself be a source of novel understanding. One of the nasty surprises of the eighteenth century, when it was found that the simple Newtonian problem of three objects interacting with each other could not be calculated exactly, is in retrospect a discovery rather than a blot on the fair face of physical science. Who knows what surprises of that kind lie in wait for molecular biology?

Realism rather than reductionism has thus become the guiding principle of science. Certainly its opposite, idealism, is thoroughly out of fashion. That is so even though one of the great practitioners of this way of regarding problems in science was Albert Einstein in his later years. Unwilling or unable to accept the seemingly paradoxical behavior of single particles, such as electrons moving through both of two slits at the same time, for example, he sought instead a set of equations whose elegance and symmetry would command respect, and by which even paradoxical phenomena would be explicable. Einstein's quest was no doubt impelled by his great success with the general theory of relativity (otherwise the theory of gravitation), which first won attention through its elegance. As the world now knows, it was a fruitless search. Quantum phenomena are often wrongly described as paradoxes for no better reason than that they conflict with the expectations of common sense, which themselves spring from human senses that have been honed by natural selection for telling what the macroscopic world is like. It is disconcerting that phenomena on the small scale are at odds with expectation, but there is a wealth of experimental data for which no other explanation is possible. How else than by experiment can reality be described?

In the century now ending, realism has given observation and experimental measurement the whip hand in their continuing interplay with theoretical explanation. Although the landmarks of the past 100 years are often described by reference to novel theories—two theories of relativity, quantum mechanics and the structure of DNA, for example—there has never previously been a period in the history of science when the correctness of explanations is more rigorously tested by experiment. Indeed, there is now even a philosophical basis for that procedure. That is what the late Karl Popper asked of science, with his repeated plea that hypotheses about the world, or potential theories, should be falsifiable. What that implies is that an explanation has no value unless it can be tested by observation or experiment (which is obvious) and unless some of the predictions made with it may be either true or false: truisms (such as "all objects have mass") do not make useful theories.

This austere doctrine is now generally accepted by the research community, at least in principle. Practice, unfortunately, often falls short. Bright ideas are too quickly regarded as explanations of the real world. Take, for example, Einstein's relativistic theory of gravitation, whose immediate claim on public attention was its novelty and breathtaking scope, bringing gravitation within the idiom of realistic field theories, subsuming electromagnetism with gravitation and providing for the first time a framework for describing the evolution of the universe as a whole. Two predictions were quickly proved correct—anomalies in the motion of the planet Mercury in its orbit and the bending of distant starlight passing near the edge of the Sun during the solar eclipse of 1919. Yet the mathematically unshakable prediction that sufficient concentrations of mass or energy should lead to the formation of black holes remains in limbo, and will remain so until there is an accommodation between Einstein's theory and quantum mechanics. Much the same is true of Gamow's big bang universe, a respectable hypothesis in 1947, with several successes in its support. Then, in the 1970s, the big bang was falsified and then "fixed" by the device of the inflationary universe, implying parallel universes which are, by definition, unobservable—so that the "fixes" are not themselves falsifiable hypotheses. There is no reason why Popper, who was not a scientist, should absolutely dictate the current agenda of research; *his* hypothesis about the nature of science may also be mistaken. But it is mystifying that a large part of the community of astronomers and astrophysicists around the world should regard the big bang model as a good approximation to something called "the truth" when they are aware of the empirical problems crying out for attention. Can we no longer live with the knowledge that we are ignorant of many things?

Impatience seems also to have infected the life sciences. Gene hunters are forever announcing that they have found "the gene" for this or that human trait, male homosexuality for example, implying that they have an explanation for the trait. But what they mean is that a particular allele of some gene is linked with identified cases in which the trait appears to run in families. Especially because the functions of the more common alleles of the same gene are often unknown, to offer the offending allele as an "explanation" is like offering hunger as an explanation of obesity.

The truth is that the sheer success of science in the past half-millennium has engendered a corrosive impatience. We too easily forget how recent are the empirical and theoretical foundations of present understanding. Prudence, or merely good manners, would dictate a seemly recognition that they may also be incomplete. Yet practitioners invest their bright ideas with the attributes of tested truth; the rest of us demand that science should prosecute its agenda energetically, so curing the sick, enriching the impoverished (and even the well-to-do), forfending calamity and giving us all something to hope for as well. It is an awesome and an impossible agenda. Can it be carried through?

The history of this century is again instructive. Soon after the Wright brothers first flew, Marconi spanned the Atlantic with radio waves and Henry Ford was planning the first people's motorcar. These three innovations are the well from which a large part of this century's contentment and prosperity has flowed. But now we know that the future will be different. The understanding of life that has followed from the structure of DNA ensures that the century ahead will be transformed by engineered forms of plants and animals, and by different and more effective human medicines. Proselytizers lyricize about the potential of genetic manipulation (and "investors" pour money into seemingly bottomless pits). But the technology of the decades ahead will be dominated by genetic manipulation—and by those who know how to manipulate a gene. They will feed the hungry and cure the sick of simple ailments; it is a matter of time only before they can list the kinds of cancers that are curable this year, and those that are on the cards for cure next year or sometime soon afterwards. Many of us will then fret at the novel dilemmas that will be created. Do we wish to live to 110? Or (more probably), why can they cure *his* cancer but not *my* Alzheimer's disease?

As will now be evident, this book has been mostly concerned with the question of what science can do to help us understand the world we live in. (Always pretentiously, that is called academic, or basic, or fundamental or pure science.) Yet the science that has dramatically changed and improved the lives of people in the past century is applied science. The principles of aeronautics were established soon after the Wright brothers flew their still-evocative mile in North Carolina at the beginning of the century. The aircraft in which we now fly are designed by engineers who are no less worthy of our regard because the reasons why aircraft can stay in the air (if not always) have been understood for the best part of a century. Indeed, that kind of science is more difficult to practice than the other kind; notoriously, engineers and their ilk must pay attention to the practicability of their machines and devices ("Can they be manufactured?"), their reliability ("How safe are they?") and the overriding question of whether people want them ("Will they sell?"). Nothing in this book, concerned as it is with how we understand the world, must be taken to imply that the science and the craft of changing the world with practicable, reliable and marketable innovations are anything but a crucial element in our common struggle to survive.

Almost certainly, the scale of the effort that will be required to carry through the wished-for agenda for the years ahead is nowhere appreciated. In genetics, for example, the task of understanding the functions of all the 100,000 human genes will require a much greater effort than that involved in their identification, and by a factor of 10 or more. Turning the knowledge thus gathered into usable medicines will be an extra. Testing whether the dream of exploiting the "paradoxical" properties of quantum mechanics for the construction of still faster computers than those now on sale will require a great effort in fundamental science, but will surely be followed through. Devising the systems that will safeguard human populations against novel infections is more than a matter of science, of course, but requires the evolution of novel technically based social institutions as well. Other potential sources of calamity will create comparably gargantuan demands on skill and science. There is no question that science and its application will be even more conspicuous in the centuries ahead than they have been in this century.

That, of course, will simply be an extrapolation of the trends apparent over the past five centuries. Since the time of Copernicus, when the modern idiom of science took root, the interplay of observation, experiment and imagination has greatly deepened our understanding of the world. Yet the questions that we ask of nature do not radically change. Aristotle asked how the universe is constructed. Copernicus answered the question as best he could by putting the Sun at the center. By Hubble's time, in 1929, it seemed that Aristotle's legitimate curiosity about the world had been satisfied, but then the question was deepened (and made more difficult): Gamow set out to explain not just what the universe is like, but how it began. That has been the course of events in most parts of science. The idea that life on the surface of the Earth began spontaneously is a daring and modern assertion; the hope of discovering precisely how it began is a question that could not have been asked confidently much earlier than this. The working of the brain is still a closed book only because the practitioners now demand an explanation, in terms of cells and their interactions with each other, that would have been self-evidently out of reach only a decade or so ago.

Thus progress in many fields of inquiry is measured not by mere discoveries (however enumerated) but by the deepening of the questions people ask of nature. It is no scandal that many of the questions now in people's minds are extensions of questions asked by Aristotle and his contemporaries. They have become more interesting questions, and they are more taxing. Quite apart from the clamant demand for more applications of science, there is also not yet an end in sight to the process of inquiry. The problems that remain unsolved are gargantuan. They will occupy our children and their children and on and on for centuries to come, perhaps even for the rest of time.

Notes

Introduction

1. In the absence of friction due to the air, the square of the velocity is proportional to the height.

2. The potential energy is not reflected in the internal structure of the object; the word "potential" refers only to the way in which the energy spent in lifting the object can be recovered by letting it fall back to its starting-point.

3. In *Encyclopaedia Brittanica.*

4. Pythagoras, the Greek, is best-known for his theorem about the length of the sides of a right-angled triangle $a^2 = b^2 + c^2$. Several whole-number solutions to this equation are known, such as $3^2 + 4^2 = 5^2$. Fermat's theorem (more properly a conjecture) is that there are no whole-number solutions of the equation $a^m + b^m = c^m$, where m is any integer larger than 2.

5. See O. Lodge in *Nature* 46 (1893): 165.

6. Einstein is (wrongly) best known for his formula $E = mc^2$, where E is energy, m is mass and c the velocity of light.

7. Real hot objects do not all behave in this regular way; some may emit extra amounts of radiation at particular frequencies according to the atoms of molecules of which they are made. Planck's argument succeeded because he

defined an idealized "black body" as a source of radiation; that is now recognized to be the equivalent of empty space, or the vacuum, at the temperature concerned.

8. In *Proceedings of the Royal Institution* 15 (1897): 415.

9. Several physicists in Britain, Germany and Scandinavia were involved, but the general principle seems first to have been spelled out by Gustav Kirchhof of Heidelberg. Earth satellites have now made possible the direct investigation of solar material, which is continually flung off the surface of the Sun as what is called solar wind; but there is no way of telling, as yet, how biased a sample this may be.

10. Rydberg noticed that the frequencies of the spectral lines in the so-called Balmer series are all proportional to $(1/2^2 - 1/n^2)$, where n is an integer greater than 2.

11. The value of the gravitational field at any point in space-time is not a single number, but a set of ten numbers, each of which can vary from place to place and from time to time; technically, it is known as a tensor field. (There are ten components of this field because there are exactly ten ways in which the four dimensions of space-time can be selected in pairs.)

12. Also unresolved is a more particular problem: that of the *cosmological constant*, introduced by Einstein in 1918 to accommodate a solution of the general equations due to his German colleague Wolfgang de Sitter in which even a region of empty space would expand indefinitely. Believing the universe to be static, Einstein supposed the presence of the constant in the equations would counter de Sitter's implied expansion; now there are some who hold that the cosmological constant is in part responsible for expansion observed.

13. Strictly speaking, the gametes, or the sperms and ova of animals.

14. Systematic ways of doing this waited on the development of the ultra-centrifuge in the 1940s.

15. The formal name is *deoxyribonucleic acid*.

Chapter 1. Beginnings Without End

1. Gravitational energy alone would give the Sun a lifetime of only 10 million years, less than the age of rocks on the surface of Earth. Some of these rocks were already known, at the beginning of this century, to be older than a billion years. The case that the solar system is off-center was argued principally by the U.S. astronomer Harlow Shapley; the modern estimate is that the solar system is roughly a third of the distance between the galactic center and the periphery.

2. This simply means that light pulses from a distant object take time to reach Earth.

3. Astronomers measure distances in terms of a unit called a "parsec," which is the distance at which the diameter of Earth's orbit about the Sun subtends an angle of 1 second of arc; one parsec is equal to 3.26 light-years.

4. Although a nuclear physicist, Gamow harnessed his enthusiasm to genetics after the discovery of the structure of DNA in 1953, devising a genetic code for specifying the structure of proteins in terms of nucleic acids and founding what he called the "RNA Tie Club" to identify like-minded scientists.

5. By Dirac; see Introduction.

6. The temperature is lower than that at which helium gas turns into a liquid at 4 degrees above absolute zero; the fact that the radiation can be detected at all is striking proof of the sensitivity of radio telescopes.

7. Alpher, personal communication, October 1995.

8. The belief that Gamow and Alpher had not explicitly forecast the microwave background radiation is supported by Steven Weinberg in *The First Three Minutes* (Andre Deutch, London, 1977), a popular account of the big bang.

9. Hoyle was the first to christen Gamow's theory the "big bang," intending the reference to be derisory.

10. In the original theory, matter was supposed to appear as hydrogen atoms. Later versions of the theory suppose that matter appears as so-called *Planck particles* (see Chapter 2). In both the steady-state and the big bang theories, there is not necessarily a conflict between the simultaneous appearance of matter and radiation and the principle that energy cannot be conjured out of thin air. The energy of a dense agglomeration of matter as in a big bang is negative relative to the condition in which it is dispersed, while that of the radiation is always positive.

11. The measurement of the redshift, accomplished by instruments called spectroscopes, usually entails measuring the frequencies (or wavelengths) of identifiable spectral lines.

12. Astronomers usually refer to H_o in terms of kilometers per second (of recession speed) for each extra megaparsec (or million parsecs) of separation.

13. The self-attraction of matter in the universe decelerates its expansion, meaning that it would have been faster in the early universe than it is now. That affects the estimated age, and the more so, the greater the density of matter. At the extreme, the ages quoted in the text would be reduced by one-third, to 13.5 and 7.5 billion years respectively.

14. The old stars lie in compact groups called "globular clusters" which are mostly found in the central bulge of the Galaxy.

15. Terry D. Oswalt et al. *Nature* 382 (1996): 692–694.

16. Some supernovae (called Type II) originate in stars much more massive

than the Sun and explode when they have consumed all of their potential nuclear fuel (which may include helium and carbon as well as hydrogen), becoming gravitationally unstable as a result; the amount of energy released varies with the mass of the star, so that they could not be used as standard candles. Supernovae of Type I, on the other hand, are white dwarf stars that adventitiously accrete interstellar material. Luckily, the two types can be distinguished from each other.

17. This prediction of Einstein's theory of gravitation has been confirmed; the case that it might account for quasars was made chiefly by F. Hoyle and G. R. Burbidge, in *Astrophysical Journal* 144 (1966): 534–552.

18. That means that the light output is many billions of times the output of radiation from the Sun.

19. Precisely how far away the quasars are requires a knowledge of the degree to which the expansion of the universe has been decelerated by its self-gravitation in the period since the big bang.

20. Temperatures in molecular clouds range upwards from that of the cosmic microwave background radiation (2.7 K) up to 80 K, and may be much greater near incipient stars.

21. The best-known brown dwarf, known as Gliese 229B, is estimated to be between 30 and 55 times the mass of Jupiter. A member of a binary system, it has a surface temperature of less than 1,000 K. (See *Science* 272 (1996): 1919–1921.

22. Pythagoras, the Greek, is best known for his theorem about the length of the sides of a right-angled triangle which are related by $a^2 = b^2 + c^2$. Several whole-number solutions to this equation are known, such as $3^2 + 4^2 = 5^2$. Fermat's theorem (more properly a conjecture) is that there are no whole-number solutions of the equation $a^m + b^m = c^m$, where m is any integer larger than 2.

23. Assuming that H_0 is 80 km per second per Megaparsec, or near the upper end of the allowable range.

24. Willem de Sitter, a Dutch astronomer, produced a solution of Einstein's equations in 1917, the year in which Einstein also tackled the cosmological implications of his theory. De Sitter's universe was empty of matter, but appeared to be static. Einstein's own model universe included matter in the form of dust, but was prevented from expanding only by his introduction of what he called the "cosmological constant," whose effect is that of a general repulsion between masses. Then it emerged that de Sitter's calculation was erroneous and that the cosmological constant (the introduction of which Einstein called "my biggest blunder") was not necessary because the universe seems to be expanding. The "Einstein–de Sitter" solution of the equations is an amalgam of

Einstein's dust-filled universe and de Sitter's model without the cosmological constant.

25. See F. Shu et al., "Star Formation in Molecular Clouds," *Annual Review of Astronomy and Astrophysics* 25 (1987): 23–81.

26. In the real world, hot objects do not all behave in this regular way; most emit extra amounts of radiation at particular frequencies according to the atoms or molecules of which they are made. Mercury, sodium or neon street lamps are deliberately designed to be exceptions to the laws Planck's predecessors had established. They emit most of their radiation in a narrow band of frequencies. Planck's argument succeeded in 1900 because he defined an idealized "black body" as a source of radiation; that is now recognized to be the equivalent of empty space, or the vacuum, at the temperature concerned.

27. S. M. Pascarelle et al. *Nature* 383 (1996): 45–50.

28. Called "Lyman-α" radiation.

29. The Large and the Small Magellanic Clouds are the most familiar.

30. The value of the gravitational field at any point in space-time is not a single number, but a set of ten numbers, each of which can vary from place to place and from time to time; technically, it is known as a tensor field. (Each component links together two of the four dimensions of space-time, suggesting that there should be 16 in total, but the order in which particular dimensions are chosen is immaterial, so there are six pairs of equal components among the 16.) If the variations in space and time of these numbers are known ("solutions" of Einstein's equations), the curvature of space can be calculated from them. Maxwell's electromagnetic field is similarly represented by the variations in space and time of a set of four numbers (called the vector potential); tensors are a generalization of vectors.

31. See E. L. Wright et al., "Comments on the Statistical Analyses of Excess Variance in the COBE DMR Maps," *Astrophysical Journal* 420 (1994): 1.

32. According to Guth, the Universe doubled in size in an interval of about 0.1 billionth of a billionth of a billionth of a billionth of a second, and went through about 100 doublings.

33. In Guth's calculation, the inflationary expansion increases all distances by a factor of about 10^{50}, or 10 multiplied by itself 50 times. It is a huge number, 10 billion times greater than the ratio of the distance of the Virgo Cluster to the typical dimensions of an atomic nucleus.

34. Later versions of the theory make it possible to avoid the conclusion that the density should be critical, which fits more comfortably with observation.

35. In *The New Physics*, Editor Paul Davies (Cambridge University Press, 1989); see also Alan H. Guth in *The Inflationary Universe* (Cape, London, 1997).

36. See Guth, *The Inflationary Universe.*

37. Lee Smolin, *The Life of the Cosmos* (Oxford University Press, 1997).

38. The total energy of the cosmic rays reaching the upper atmosphere of the Earth is comparable with the energy of all the starlight reaching the Earth's surface.

Chapter 2. Simplicity Buried in Complexity

1. Hadrons are nuclear matter, leptons electronic matter, typified by the proton and electron respectively.

2. More correctly, the nucleus of a hydrogen atom, a proton, is 1,842 times as massive as an electron, but a nucleus of uranium is 4,765 times as massive as the 92 electrons in the periphery of the atom.

3. Neutrons are more massive than protons by roughly 1.35 parts in 1,000, which difference is the equivalent of roughly 2.3 electron masses; the result is that neutrons are unstable relative to protons, into which free (but not nuclear) neutrons decay radioactively with a probability of 50 percent every 12 minutes. By convention, the actual masses of nuclear particles are usually described in electron volts (the energy acquired by a particle carying an electronic charge in an electrical potential difference of one volt, written eV), in which notation the mass of the proton is 938 million eV (written MeV).

4. As science attaché at the British Embassy and, in 1942 as a member of the "Tizard" Mission. Sir Henry Tizard was the British government's Chief Scientist throughout the Second World War. His mission to Washington traded British information about the electronic device called the magnetron (used in radar) and the feasibility of separating the isotopes of uranium by a diffusion process for a U.S. undertaking to supply conventional military equipment.

5. The number of neutrons in a nucleus can vary to some degree. For example, atoms of heavy hydrogen (deuterium) differ from those of hydrogen itself in that their nuclei contain both a proton (as in hydrogen) and a neutron. (Versions of a nucleus with the same electric charge but different numbers of neutrons are called isotopes of that element.) The nuclei of radioactive elements such as uranium have either too many or too few neutrons to be stable, and so spontaneously decay into other nuclei.

6. The instability caused by an excess of neutrons leads to the emission of an electron (originally called a β = particle) from the nucleus, a deficiency of neutrons results in the emission of a helium nucleus (originally called an α = particle).

7. That is an overstatement. Calculating the properties of nuclei remains a black art, understanding the properties of nuclei with large amounts of intrinsic

spin and with nucleons (usually neutrons) lying a long way beyond their boundaries has recently become of great interest and may yet be of technological importance in the nuclear industry.

8. The radius of a proton, the nucleus of hydrogen, is roughly one part in 10,000 of the radius of the intact atom.

9. Pauli was awarded a Nobel Prize (in 1945) for his work on electron spin, but Goudsmit and Uhlenbeck were never so honored.

10. Another and more common way of putting this hangs on regarding the direction of the intrinsic spin as one of the quantities describing the motion, in which case the exclusion principle is that two electrons with the same dynamical properties cannot be in the same state.

11. At the age of 24.

12. Published in the *Dublin University Review* 1833 (pp. 795–826).

13. See Thomas S. Kuhn, *The Structure of Scientific Revolutions*, 3rd ed. (University of Chicago Press, 1996).

14. Those chiefly concerned with that development were Fermat, Lagrange, d'Alambert and Maupertius, all French.

15. The explanation is that there are stationary states of an electron so extensive that they encompass two or more atomic nuclei, binding them together; in the hydrogen molecule, for example, the two electrons (one from each atom) envelop both nuclei.

16. Dealing with real molecules, even those as simple as water (H_2O) or benzene (C_6H_6) was another matter. When the computational difficulties became plain, the sentence was dropped from the second edition of the book in 1931.

17. According to rules derived from Hamilton's mechanics.

18. So too are angular position and angular momentum and, interestingly, time and energy.

19. The arithmetical product of the two errors will certainly exceed Planck's constant.

20. The numerical result cannot be correct because the argument ignores the three-dimensional structure of real atoms.

21. Strictly speaking, what Dirac first predicted is that, for every state of an electron with positive energy, there would be a similar state with negative energy. Evidently, states of the second kind would usually be filled with electrons and would not be observable; only if one were removed from the negative-energy sea would the gap be apparent. Dirac's first hope was that such a gap would account for the proton, but J. Robert Oppenheimer quickly proved that to be impossible. Dirac reverted to the idea that the lack of an electron

from the sea of negative-energy electrons would have the properties of a positron.

22. The positron was discovered at Berkeley, California, by Carl Anderson in the summer of 1932, but his account was published only the following year, in *The Physical Review* 43 (1933): 491–494.

23. Charged particles will generate flashes of light if they are traveling not through empty space, but through a medium such as a gas or liquid with a velocity greater than the velocity of light in that medium.

24. The K-particles are 966 times as massive as an electron or roughly half the mass of a proton; their lifetime is roughly an eightieth of a microsecond.

25. The inference that the V and K particles have kinship with nuclear rather than electronic matter rests on the products of their decay, including positrons and mesons.

26. The other solutions, by Julian Schwinger at Columbia University (later at Harvard University) and Sanitiro Tomanaga at the University of Tokyo, were more abstract than Feynman's and less convenient for calculation.

27. The classical benchmark of success is the calculation of the so-called Lamb shift, which is the difference in the energy between the two states of a hydrogen atom in which the spins of the electron and the nuclear proton are respectively parallel and antiparallel.

28. Several physicists in Britain, Germany and Scandinavia were involved, but the general principle seems first to have been spelled out by Gustav Kirchhof of Heidelberg. Earth satellites have now made possible the direct investigation of solar material, which is continually flung off the surface of the Sun as what is called the solar wind; but there is no way of telling, as yet, how biased a sample this may be.

29. Specifically, a proton consists of two *up* quarks and one *down,* giving the correct charge of $2 \times \frac{2}{3} + 1 \times (-\frac{1}{3}) = +1$; the neutron, with one *up* and two *downs,* has the correct charge of zero. Yukawa's mesons, on the other hand, are composites of either an *up* and an anti-*down* (π^+) or a *down* with an anti-*up* (π^-).

30. The missing elements are gallium, germanium and scandium.

31. Predictions of the existence of novel quarks have been more or less tentative; for example, the prediction of *charm* (discovered in 1975) was less than confident, that of the *top* quark (for which there was only indirect evidence at the end of 1997) is almost certain to be borne out.

32. The proton is the least massive of the baryons; mesons (the π-meson, for example) may be less or more massive than the proton.

33. Plainly, this description does not exhaust the possibilities; for example, the π^+ meson, a combination of an *up* and an anti-*down* quark, has positive

electric charge, but may have color charges of *red* and anti-*red, green* and anti-*green* or *blue* and anti-*blue.* These three different possibilities are not represented in the real world as three recognizably different π^+ mesons; but are rather best regarded as three different stationary states of the particle which are mixed together in the real world to yield just one kind of particle.

34. Similar difficulties arise in telling the intrinsic spin of the composite particles. Thus it seems that more than half of the intrinsic spin of a nucleon, which is exactly half of the standard quantum of intrinsic spin is carried by gluons rather than quarks (see Ian Balitsky and Xiangdong Ji, *Physical Review Letters* 79 (1997): 1225–1228).

35. See P. W. Higgs in *Physics Letters* 12 (1964): 132.

36. Worse, there seems to be an empirical rule that the characteristics of each generation of leptons is preserved during their interconversion: for example, the μ^- particle could in principle decay into an electron and a photon, but does not. Instead, the process results in an electron, an anti-*electron*-neutrino and a regular μ-neutrino; the μ-neutrino is the survivor into the future of the μ-electron, while the creation of an electron and the corresponding antineutrino is an assurance that there has been no net creation of "electron-ness." This behavior is accommodated in the electroweak theory by the description of which particles interact with which, and how, but the physical meaning of the rule is, to say the least, a mystery (see Sidney Drell in *Physics Today* 50 (October 1997: 34–40).

37. An experiment at the German particle accelerator laboratory at Hamburg was reported to have shown anomalies in the collision of high-energy electrons and positrons that are best explained by supposing that electrons have an internal structure.

Chapter 3. Everything at Once

1. The most conspicuous (but not the only) professional offender is Professor Stephen Hawking, who now holds Isaac Newton's old professorial chair at the University of Cambridge, and who is confined to a wheelchair by motor-neuron or Lou Gehrig's disease. In his best-selling *A Brief History of Time* (Bantam, 1988), Hawking writes, "I believe there are grounds for cautious optimism that we may be near the end of the search for the ultimate laws of nature"; he concludes that then ". . . we would know the mind of God."

2. The magnitude of this "time dilation" depends on the ratio of the speed of the moving timepiece to the speed of light, and is usually very small; the speed of an Earth satellite in a low orbit is approximately 30,000 kilometers an hour, or 0.0028 percent of the speed of light; the time dilation amounts to 39

parts in a billion, which works out at 2.1 millionths of a second for each orbit, easily measured with atomic clocks.

3. See Albert Einstein, in *Albert Einstein: Philosopher-Scientist,* ed. P. A. Schilpp (Library of Living Philosophers, 1949).

4. Which may be zero.

5. See Chapter 2.

6. Photons have one unit of intrinsic spin, but because (by definition) they travel with the speed of light, their spin is most simply apparent in the rotation of the associated oscillating electrical and magnetic fields in either a clockwise or an anticlockwise direction, called circular polarization. These states are equivalent, in the sense of quantum mechanics, to (1) vibration of the electrical (or magnetic) field in a particular direction in a plane at right angles to the direction of travel and (2) vibration in the same plane but at right angles to (1). The permissible plane polarization of a photon may thus be described by an angle (ranging from 0° to 180°) in a plane perpendicular to the direction of travel.

7. The electrically neutral π-meson, called π^0, decays into two photons in just the way required.

8. Roger Penrose, in *The Emperor's New Mind: Concerning Computers, Minds and the Laws of Physics* (Oxford University Press, 1989), has devised a form of the EPR paradox in which two components of a quantum system are on different stars.

9. See, for example, J. S. Bell in *Speakable and Unspeakable in Quantum Mechanics* (Cambridge University Press, 1987).

10. See Penrose, *The Emperor's New Mind.*

11. One must suppose a live cat in a sealed cage that may be filled by the release of a toxic gas itself triggered by particles from the decay of a radioactive atom; Schrödinger invited his readers (in 1935) to ask whether the process is really analogous to the fixing of the polarization of a photon by a polarizing filter.

12. The restriction applies only to particles with half-integer intrinsic spin —those called fermions (see Chapter 2).

13. The Planck length is defined in terms of the gravitational constant, Planck's constant and the velocity of light as the square root of (Gh/c^3) and is roughly a hundredth of a billionth of a billionth of the dimensions of a proton; the corresponding energy is a billion billion billion electron volts, equivalent to the mass of a microgram.

14. The so-called Brans-Dicke theory is the best known of many elaborations of Einstein's theory of gravitation; its starting point is that as well as Einstein's ten-element gravitational field (technically called a tensor), there is

an extra unrelated component (called a scalar). Some of the consequences of the theory should be perceptible in the real world; for example, the supposedly universal constant describing the strength of gravitational forces should change with the passage of time. The theory was first published by C. Brans and Robert H. Dicke (of Princeton University) in *The Physical Review* 124 (1961): 925–935.

15. In *The New Physics,* ed. Paul Davies (Cambridge University Press, 1989).

16. The strings are supposed to be very small, comparable in dimensions with the Planck length, and are usually thought of as closed loops which are nevertheless capable of interacting with others, sometimes breaking apart and then being rejoined.

17. This was the year in which several competing versions of string theory were recognized to be equivalent.

18. As Archimedes is said to have done when he first understood that objects floating on water do so because their density is less than that of water.

19. The pressure at the center of a collapsing star must be the weight of material in the envelope as abated by the outward pressure of the outward radiation flux.

20. That may be an oversimplification; many white dwarfs are known to burst into life again if they happen to accrete from interstellar space material capable of causing thermonuclear reactions on their surface.

21. It should not have been a surprise; neutron stars were first suggested by Walter Baade in 1935, were more cogently predicted in 1938 by J. Robert Oppenheimer and, a little later, by Lev Landau in the Soviet Union.

22. The curvature of space-time is defined in terms of the trajectory that would be followed by a light beam; if trajectories form closed loops, radiation is permanently confined.

23. Now there is evidence that many quasars are radiation sources embedded within more normal galaxies, but their energy output substantially exceeds that of the remainder of the galaxy.

24. The first discovery of X-ray stars was made with captured German V-2 rockets launched from the White Sands Proving Ground in the United States and equipped with instruments devised by Herbert Friedmann, a scientist at the Naval Research Laboratory in Washington, D.C.

25. X rays are supposed to be generated as material pulled by gravitation away from the visible star reaches the surface of the companion at great speed, just as hospital X rays are generated by the impact of fast electrons on a metal target.

26. R. Narayan et al., *Astrophysical Journal* 492 (1998): 554–568.

27. The quantities concerned are not easily perceived, usually being dwarfed by the interactions between different chemical molecules.

28. This is a reference to Boltzmann's "H theorem"; given Boltzmann's definition of H, it is the negative of this quantity that equates with entropy.

29. Special relativity is more tricky in that a person's perception of phenomena, including the rate at which time passes, depends on the relative velocity, while Einstein's theory of gravitation is embodied in equations that do not refer directly to time as such, but to a "metric distance" of which time is a component; but in both cases, reversing the direction of time still yields a physically realistic motion.

30. Penrose, *The Emperor's New Mind*.

31. *The Anthropic Cosmological Principle* by John D. Barrow and Frank J. Tipler (Oxford University Press, 1986).

32. See Chapter 2; the evolution of sentient life on Earth has taken at least 4 billion years, but the manufacture of carbon and other elements necessary for our existence would have taken at least a further billion years, probably longer.

33. Anaxagoras (500–428 B.C.), who made the first estimate of Earth's circumference, held that the creation of an ordered universe could well have taken a long time.

34. Pierre Louis de Maupertius (1698–1759) was the first to show that Earth is flattened towards its poles, but is best known for having devised the "principle of least action" (see Chapter 2).

35. The critical density is $3H_0^2/8\pi G$, where H_0 is Hubble's constant.

36. In Barrow and Tipler, *Anthropic Cosmological Principle*, p. 21.

37. The project (under the supervision of the California Institute of Technology) is known as LIGO (for Laser Interferometry Gravitational Observatory); laser beams in two vacuum chambers at right angles to each other should be able to detect changes in the length of either light path caused by gravitational waves. If the first instrument, whose cost will exceed U.S. $150 million, functions well, it is hoped that companion observatories will be built elsewhere in the world so as to provide, by triangulation, information on the directional origin of the disturbances.

38. Even so, the research community is hard at work on the design of "the next" accelerator: it would consist of two linear accelerating machines arranged to fire beams of electrons directly at each other.

Chapter 4. The Likelihood of Life

1. The belief that the universe may have been set on its present course at the outset, and that continuing intervention by a deity is not necessary, has its

origins in the nineteenth century; it presupposes that God is immune from the laws of chaos, which make historical accidents unpredictable.

2. Interestingly, vitalism has recently been partially revived as the belief that living things (and some inanimate systems) are distinctive because of the complexity of their molecules and the complexity with which they coexist.

3. Haeckel was also the one who held that embryos, in the course of their development, "recapitulate" the course of the Darwinian evolution of the organisms concerned.

4. See A. I. Oparin, *The Origin of Life*, translated from the Russian text of 1924 (Macmillan, 1938).

5. Argument persists about the precise composition of the primeval atmosphere, in part because of uncertainties about the formation of Earth's largely iron core, but also on the grounds that radiation from the primeval sun would have reduced the proportions of ammonia (NH_3) and methane (CH_4) in the early atmosphere. Meanwhile, it has become apparent that the atmosphere of Jupiter contains much less water than had been expected, and than is supposed to have been present on Earth 4,000 million years ago.

6. See Stanley W. Miller, *Aspects of Chemical Evolution* (Proceedings of the 17th Solvay Conference on Chemistry, 1980), Ed. G. Nicolis (Wiley, 1984).

7. In this context only, a chemical reducing agent is one whose influence is to remove oxygen atoms, or to add hydrogen atoms, to other chemicals with which it interacts, usually with the formation of water molecules (or their equivalents, such as hydrogen sulfide) as by-products.

8. The result is 1.26 times 10 multiplied by itself 130 times. With a smaller protein made of 27 amino acids (much the size of the insulin molecule), the total weight of a collection of one of each of the possible molecules would weigh exactly one tonne.

9. Taking the average mass of an amino acid molecule as 150 times that of a hydrogen atom and the number of baryons (protons and neutrons) in the universe as 10^{79}.

10. In DNA, the four chemical units are called thymine, cytosine, adenosine and guanine, otherwise T, C, A and G, which belong in pairs (T and C, A and G) to the families of chemicals called *pyrimidine* and *purine* bases. Each base is chemically linked to a molecule of the sugar called deoxyribose which is, in turn, linked to a phosphate group familiar from inorganic chemistry. Nucleotides are strung together by chemical links between the phosphate end-group of one nucleotide and the sugar group of its successor, which endows a DNA molecule with directionality.

11. This assumption is not implausible; circumstances in which greater numbers of atoms or molecules induce a greater degree of order are familiar

(in, for example, the formation of crystals) and are called cooperative phenomena.

12. There is one important difference: in both RNA and DNA, the chemical units called nucleotides that constitute the polymer molecules include structures called purine and pyrimidine bases which are complementary to each other in the sense that unlike, rather than similar, units line up alongside each other.

13. See Stuart Kauffman, *At Home in the Universe* (Oxford University Press, 1995).

14. See D. S. McKay et al., *Science* 273 (1996): 924–930.

15. In 1986, the core of Halley's Comet was found to be made mostly of carbon.

16. See D. M. Mehringer et al., *Astrophysical Journal Letters* 480 (1997): L71–L74.

17. See P. Parsons, *Nature* 383 (1996): 221–222.

18. This heading is the title of a book by Erwin Schrödinger in 1947 which appears to have had a powerful influence on several of the founders of molecular biology.

19. The basis of the code is that each consecutive triplet of nucleotides along the length of a DNA (or RNA) molecule specifies a single amino acid unit in the corresponding protein; for example, the triplet AAA in DNA specifies the amino acid *lysine* (as does AAG). The exceptions to the universal code are the subcellular structures known as *mitochondria* (in animal and plant cells) and *chloroplasts* (in plant cells only), both of which are involved in energy production and conversion. There are good reasons for believing that these structures are present in the cells of higher organisms as a result of the incorporation into evolving lineages of single-celled organisms of energy-producing bacteria on which they had become dependent; if that is indeed what happened, the implication is that the first self-replicating organisms used a variety of genetic codes differing from each other in minor ways.

20. S. L'Haridon et al., *Nature* 377 (1995): 223–224.

21. See M. Magot et al., *FEMS Microbiology Letters* 155(2) (October 15, 1997): 185–191.

22. See Chapter 7 and the work of Doolittle referred to there.

23. RNA differs from DNA in that each nucleotide includes the sugar molecule ribose rather than deoxyribose and that the thymine of DNA is replaced by the pyrimidine base uridine.

24. The close similarity of the chemical structure of the RNA molecules of the ribosomes from organisms of the same class is persuasive evidence of the antiquity of ribosomes in the evolution of living things.

25. At the end of 1997, there were eight known planets in orbit around stars essentially similar to the Sun, all of them half as massive as Jupiter. But these are early days, when it is possible to detect only large planets.

26. Purine bases in nucleotides may, for example, simply be lost by the interaction of a water molecule with their standard chemical link to the deoxyribose part of the nucleotide; in a single day, it is believed that some 5,000 purine bases (A and G) are lost by this means in every cell of the human body.

27. Some viruses rely on a single strand of RNA rather than a double helix of DNA as their genetic material; repair mechanisms (if there are any) are therefore less effective for lack of a second strand to which to refer. This may be part of the explanation why HIV, the virus responsible for AIDS infection, changes its genetic character easily.

28. See A. Sancar, *Annual Reviews of Genetics* 29 (1995): 69–105; and R. Wood, *Annual Reviews of Biochemistry* 65 (1996): 135–167.

29. There is circumstantial evidence that certain kinds of cancers, of which familial bowel cancer is one, arise because of inherited defects of the DNA repair mechanism.

30. In the language of the physical sciences, cells and the organisms made of them are not *in equilibrium* with their surroundings, but are maintained in what is called a *steady state* by the continued inward and outward flow of energy and of simple chemicals. Technically, their entropy is not at maximum, as the second law of thermodynamics requires of isolated systems. Systems (such as cells) which are *open* to energy and materials from their environment are exempted from this strict requirement, but the degree of order they attain depends both on the throughput of energy and, more particularly, on the chemicals to which they have access. (The layered structure of Earth's atmosphere is its response to the inward flux of energy from the Sun.)

31. J. W. Schopf (ed.), *Earth's Earliest Biosphere* (Princeton University Press, 1983).

32. Thomas Gold, a retired professor at Cornell University and one of the originators of the steady-state theory of the universe in the 1940s, believes that geochemical life may be common. Based on the findings of a deep-drilling project in Sweden in the early 1980s, he claims the recovery of live bacteria from a depth of 2,000 meters in an old impact crater. Unfortunately, the risks of contamination by surface bacteria are high, but the investigation deserves to be repeated under more carefully controlled conditions.

33. A world in which all creatures consist of a single cell is not necessarily featureless; fungi (including all mushrooms), coral reefs and seaweeds are colonies of identical cells growing together.

34. The fossil record is not helpful on the age at which multicellular organisms first appeared, though they are predominant among the fossils of the

Cambrian Explosion 550 million years ago. Recently, fossils recovered from phosphate rocks in China dated at 600 million years old have revealed what appear to be the fossilized relics of early embryonic structures, consisting of small numbers of cells held together as in the early stages of the development of animals as different as frogs and mice. This evidence suggests that multicellular life predated the Cambrian Explosion by at least some tens of millions of years. There are also sketchy reports of fossil tubeworms from much earlier rocks. But single-celled organisms are likely to have had Earth to themselves for a good 2,000 million years, probably longer.

35. Gould's account of the Cambrian fossils of the Burgess Shale is in *Wonderful Life* (W. W. Norton, 1989).

36. In sexually reproducing organisms, this is now done by the union of male and female sex cells, but the viability of vegetative reproduction (in plants, for example) proves that it must be possible to recover genetically omnipotent cells from differentiated tissue.

37. The timing is more fully discussed in Chapter 7.

38. See *Symbiosis in Cell Evolution: Life and Its Environment on the Early Earth* by Lynn Margulis (W. H. Freeman, 1981).

39. The material used in human vision is called *rhodopsin.*

40. Where they are linked to the sugar units in each nucleotide, respectively ribose and deoxyribose.

41. The resulting material is called PNA (the P stands for peptide); matching strands of PNA are more strongly bound together than those of DNA itself.

42. What this means is that the significant amino and acid groups are as close together as they could be in the molecules.

43. Technically, it is the figure of a tetrahedron.

44. To check this, paint the four faces of two triangular pyramids with different colors. In each case, paint the base black, say, and then use red, yellow and blue to paint the three other faces consecutively, first by going from right to left and then in the opposite direction for the second pyramid. Simple inspection will show that the painted pyramids are intrinsically different from each other; no amount of rotation can make one like the other, which is its mirror image.

45. The capital letters stand for *laevo* and *dextra,* the Latin words for left and right, and derive from the mid-nineteenth century, when it was first recognized that asymmetric molecules can change the orientation of polarized light in one direction or the other, but now the usage has become conventional.

46. A. G. Cairns-Smith's theory of the origin of life in (and on) solid crystal particles such as those of clay has been popularized by Richard Dawkins in Chapter 3 of *The Blind Watchmaker* (Longmans, 1986).

47. See K. R. Popper, *Nature* 344 (1900): 387.

48. The bases adenosine (A) and guanine (G) are both used in modern cells for the transfer of energy for triggering chemical reactions of particular molecules, but chemicals closely related to the nucleotide bases are also involved in various chemical reactions in the role of coenzymes.

49. Beginning in 1957, Sidney W. Fox at the University of Florida began a series of experiments which showed that molecules resembling real-world proteins can be produced by, for example, heating mixtures of amino acids.

50. The synthesis has been given most attention in the past two years, but relatively little has been done to uncover the general principles on which other important chemicals such as fatty acids, sugars and even steroids are manufactured in cells.

51. Cech (pronounced as in "Czech") found that one end of a molecule of messenger RNA transcribed from a plant gene is capable of removing unwanted stretches (introns, see Chapter 5) from its own length.

52. The genetic information in the arrangement of the four DNA units in a single gene is first *transcribed* into the same arrangement of the corresponding RNA bases in the structure of an RNA molecule, called "messenger RNA" or (mRNA for short), which is then *translated* into the arrangement of amino acids in the encoded protein molecule.

53. The small molecules are called "transfer RNA" or "tRNA"; There is at least one kind of tRNA molecule for each of the 20 amino acids.

54. Interestingly, this series of experiments uses a form of clay much studied by Bernal, called montmorillonite, either as a catalyst or as a way of anchoring the molecules whose length is being extended—which may have the same effect.

55. See T. Tjivikua et al., *Journal of the American Chemical Society* 112 (1990): 1249; and G. von Kiedrowski, *Angewandte Chemie* (English Edition) 25 (1986): 932.

56. Known as *co-factors.*

57. RNA molecules differ from those of DNA primarily in the structure of the sugar molecules, ribose and deoxyribose incorporated in each nucleotide; deoxyribose has one oxygen atom fewer than ribose.

Chapter 5. Cooperation and Autonomy

1. *Genome* is both the name for the set of genes in a particular cell and for the set of genes that characterizes the species to which the cell belongs.

2. The nomenclature can be confusing: this maturation of a fertilized egg into an adult is variously called *ontogeny* and *development,* but this field of study is called *embryology.*

3. There are exceptions; some bacteria produce two differently shaped daughters at cell division. In eukaryotic cells, asymmetrical division is more common. In the course of development, for example, it is often the case that a cell will yield on division one like itself and one with more specialized characteristics. The production of germ cells is an example.

4. As will be seen, inherited or acquired defects of the signaling mechanism have important medical implications, notably in the causation and possibly the treatment of human cancer.

5. The human genome is supposed to contain the most genes, but the precise number will be known only when the Human Genome Project is finished. Current estimates range from 60,000 to 100,000. The uncertainty arises because genes are not uniformly spread through the genome.

6. The verbs to "replicate," "transcribe" and "translate" have become technical terms in the sense defined in the text. The RNA molecules intermediate between DNA and protein molecules are known as "messenger-RNA" (mRNA) molecules.

7. Called *retroviruses* for the obvious reason that the flow of genetic information runs "backward"; HIV, the infectious agent of AIDS, is such a virus.

8. Reverse transcriptase was discovered independently by David Baltimore and the late Howard Temin; the other enzymes referred to are respectively *restriction enzymes* and DNA *ligases.*

9. See R. K. Saiki et al., *Science* 239 (1988): 487–491.

10. See M. Krings et al., *Cell* 90 (1997): 19–30.

11. The technique is called *electrophonesis.* In effect it is an elaboration of chromatography devised in the 1950s for separating similar chemicals by exploiting the more rapid diffusion of those chemicals whose molecules are smallest. Electrophonesis relies on (1) the electrically charged character of proteins and DNA molecules, which makes their motion sensitive to electric voltages, and (2) specially constructed polymer materials whose molecular structure effectively distinguishes between biological molecules.

12. The "incubation period" preceding the appearance of the symptoms of infectious diseases is the time during which the bacteria (or viruses) multiply to become comparable in number with the numbers of cells of the immune system, which has evolved to combat infection.

13. Called the *endoplasmic reticulum.*

14. The two kinds of tubulin molecules are called α and β; they are assembled into hollow tubelike structures, called microtubules, which appear

to consist of 13 chains of alternating α and β units held together by sideways forces not yet well understood. A further complication is that the human genome contains half a dozen versions of the genes corresponding to the α and β units, presumably for use in different structures in the cell. The naming of the parts has not yet gone so far as to tell which versions are used where.

15. In engineering language, microtubules are stronger in compression than under tension—in other words, they are better able to push cell components apart than to pull them together.

16. If these cells had nuclei, they would be too large to traverse the capillaries of the mammalian bloodstream.

17. Plainly it is a major goal of modern research in surgery to find ways of making muscle regeneration more effective.

18. There are variations on the photosynthesis theme, but they are not so radical as to suggest that they evolved separately.

19. Taken together, these circumstances suggest that eyes evolved from a primitive light-sensing organ, and there are still extant organisms, including bacteria, which are called *phototactic* because of their tendency to move toward or away from light.

20. There has been much Russian work on blood substitutes.

21. In the same way, a single fertilized ovum contains the chemical specification of the intermediate stages as well as of the adult that eventually develops.

22. Technically, *Sacceromyces cervisiae.*

23. At different stages in this cell division cycle, chemically distinct cyclin molecules are involved. Moreover, three distinct forms of cyclin molecules are secreted in the first phase of the division of mammalian cells. The advantages of this apparent redundancy are not understood.

24. *Science, Nature, Cell, The Proceedings of the U.S. National Academy of Sciences* and the *EMBO Journal.*

25. Cyclins are protein molecules whose sequence of amino acids varies between one species and another, but not so much that the chemical similarity is unrecognizable; ordinarily, cells synthesize several cyclins whose amino acid structure differs significantly and which are involved in different stages of the cell-division cycle.

26. The proteins concerned are called *cyclin-dependent kinases,* or cdk for short.

27. This is not quite the whole story. The internal transformation of the complex entails the removal of a phosphate group from one place in the cdk molecule and the addition of a phosphate group to another place. Each

of these steps requires a specific enzyme, also encoded by the genome of the cell.

28. Many of the details of these processes are still to be worked out.

29. That is also a process whose details are still to be understood.

30. That process has another important function: checking that the newly constituted chromosomes are appropriate entails lining them up in pairs laid tightly side by side in a manner that allows corresponding parts to be interchanged. This *recombination,* in sexually reproducing organisms, is the origin of the reshuffling of maternal and paternal genes that complicates human inheritance and that is also the source of the genetic diversity on which Darwinian natural selection works.

31. It is striking that the effect of recombination in the formation of germ cells is to provide individuals with combinations of genes that are distinct from those of their relatives. The manner in which this mechanism evolved is not understood, nor is it known if recombination can occur everywhere in the genome.

32. The process, called *methylation,* consists of the addition of methyl groups to the nucleotide called cytosine.

33. The bacterial enzyme GroE1 is a ring-shaped structure (with a hole through the middle) which helps other protein molecules to assume the shape required of them; it is believed that the newly formed protein molecule is guided through the central hole of the donut-shaped molecule and, in the process, folded, but the effectiveness of this molecular machinery is not understood.

34. *Biochemical Oscillations and Cellular Rhythms* (Cambridge University Press, 1996).

35. Cyclic AMP has more recently been found to be one of several internal messengers in eukaryotic cells.

36. This is why the boiling point of water is much greater than that of comparable substances such as ammonia, and why the condensation of steam into water yields anomalously large amounts of heat.

37. The name "hydrogen bond" suggests a static structure, but that is inappropriate; whether the hydrogen is attached to one or other of the two atoms it holds together is uncertain in the sense of quantum mechanics.

38. The bacterial enzyme GroE1 referred to in note 33 is one such molecular device; many others, found in eukaryotic cells, are generically called *chaperone* molecules.

39. See W. Düchting, W. Ulmer & T. Ginsburg, *European Journal of Cancer* 32A (1996): 1283–1292.

Chapter 6. The Genome and Its Faults

1. E. Schrödinger, *What is Life?* (Cambridge University Press, 1944).

2. A is adenine, T thymine, G guanine and C is cytosine; A and G are generically known as *purines*, T and C are *pyrimidines*.

3. Watson and Crick suggested that there would be 11 nucleotide pairs to each turn of the helix, which is approximately correct.

4. That end of the molecule is called the "5'" or "five-primed" end in a reference to the position in the deoxyribose part of the nucleotide to which the phosphate is attached, and may be thought of as the "beginning" of the molecule in the sense that it is the end from which DNA is transcribed into RNA. The other end of the molecule is technically known as the "3'" end.

5. The sugar part of the nucleotides in RNA is also different, with an extra oxygen atom to make *ribose*. The distinctive RNA base U is the pyrimidine base *uracil.*

6. The original research paper (*Nature*, 19 April 1953) occupied less than a single page and included the sentence, "It has not escaped our attention that our proposed structure [of DNA] readily suggests a mechanism of inheritance."

7. Recognizing a gene in a stretch of DNA is easier in bacteria than in other organisms—genes in bacteria usually begin with the triplet of the three nucleotides AUG (corresponding to the amino acid methionine) and end with one of the three signals in the genetic code indicating an end to transcription —the triplets UAA, UGA and UAG. But it is essential that the *start* and *stop* signals should be accurately in register with each other, or separated by an integral number of nucleotide triplets, each corresponding to a particular amino acid. In eukaryotic genomes, the starting point is not so clearly defined, so that it is necessary to work backwards from potential *stop* signals, looking for what are called "open reading frames." Genes identified in ways like this will not be functional unless they are accompanied by appropriate control elements in the DNA, usually ahead of the beginning of the gene. A further complication is that genes may be found on one or the other of the two complementary strands of the double helix.

8. In plants, the cells of a growing tip can be converted into germ cells by plant hormones and environmental clues.

9. Plants produce pollen and ovules instead.

10. As noted in the text, female mammals have two similar sex chromosomes (called X) while males have a pair of visibly different sex chromosomes (X and Y). The situation is reversed in, for example, birds, where the males have similar chromosomes. In some animals, there is a single sex chromosome that determines male or female character as the case may be (and which are

said to be X, O in type). In mammals, the female condition is the default condition because only the Y chromosome carries the *sry* gene that, during development, gives the reproductive organs their male character. For reasons not fully understood, in female mammals one of the two X-chromosomes is permanently inactivated in the course of development; unless that process is completed very early in the cell division of the embryo, the result may be an adult who is a genetic mosaic, with some tissues expressing maternal genes and others, those from the father. Further complications can arise when individuals inherit three sex chromosomes, say two X's and a Y or two Y's and an X, as well as with the reproduction of the species called "polyploid" because their genomes contain several complete sets of chromosomes. It seems to be agreed that the sex chromosomes play an important part in the emergence of new species, presumably because they influence the development of reproductive mechanisms and thus of the mutual sterility of animals and plants from different species, but the manner in which this comes about is far from understood.

11. This was first shown in mice.

12. "Vertical" transmission is the intergenerational transfer of genetic information at cell division for single-celled organisms, and the formation of embryos by fertilization for other creatures.

13. For a popular reference, see *The Hot Zone* by Richard Preston (Random House, 1994).

14. There is no doubt that eggs rather than sperm are the chief source of mitochondria for newly fertilized embryos, which means that the mitochondria in adults are of maternal origin. Mitochondrial DNA is also a convenient means of investigating a person's relationships with others, both in forensic science and in, for example, the investigation of the fate of Tsar Alexander II and his family. The maternal origin of mitochondria has led to the notion that all human beings now alive have mitochondria derived from the relatively small number of women in the founding population of the human stock, whence the notion that we are all descended from some prehistoric "Eve." Yet the exclusive dependence of living animals on the inheritance of maternal mitochondria has not been rigorously tested, while even the most refined electron microscopes cannot establish the *absence* of mitochondria in sperms. On the other hand, there are now persuasive rumors that unpublished genetic research has established that mitochondria of paternal origin do occur in modern adults.

15. Corresponding either to the amino acid *methionine* or to that called *valine*.

16. Setting aside the question of a Creator, of course.

17. By convention, the usually symbolic names of genes are printed in italic type so that the same names can be used in Roman type to refer to the proteins the genes produce.

18. One of the longest genes so far identified is that for the protein called dystrophin, mutations of which are responsible for inherited muscular dystrophy; the whole gene (exons and introns) occupies a stretch of DNA more than 2 million nucleotide pairs long.

19. The BRCA1 gene was linked with inheritable breast and ovarian cancer in 1995 on the strength of the association of aberrant alleles with inheritable disease; the function of the gene in people who do not inherit susceptibility to disease appears linked with surveillance of the integrity of DNA replication in the cell cycle.

20. The beginnings and ends of introns are marked by characteristic pairs of nucleotides.

21. Each nucleosome is made from two each of four different histone molecules.

22. It is far from clear how the necklaces of nucleosomes are packaged into chromosomes. The diameter of the standard nucleosome is about 11 millionths of a millimeter—five times greater than the diameter of the DNA helix— but the chromosomes visible at the penultimate stage of cell division, called metaphase, are at least 50 times thicker. A better understanding of how this material is arranged is essential to an understanding of what happens when genes are activated for transcription. A further question is not yet answered: why are some parts of the nuclear chromatin active or inactive? Different parts of the same chromosome may be in one condition or the other at the same time. There is also the puzzle that in female mammals which have inherited an X chromosome from each parent, only one of which is needed for the ordinary life of the cell, one of the X chromosomes is inactivated early in the development of the embryo, but in a random fashion, so that some cells of the embryo (and the adult to which it leads) have an active paternal X chromosome and others have an active maternal version. In circumstances in which part of an X chromosome may be moved to the end of another, neighboring genes in the target chromosome may also be inactivated, affecting intergenerational inheritance.

23. The idea that telomeres shorten as organisms age first emerged in 1994, on the basis of the molecular analysis of rat cells grown artificially. See S. Henderson et al., *Journal of Cell Biology* 134 (1996): 1–12.

24. The issue will not be settled until some years after the Human Genome Project has been completed.

25. By F. H. C. Crick.

26. The influence of transposons is especially important in the crop plant maize.

27. See *Wonderful Life* by Stephen Jay Gould (W. W. Norton, 1989).

28. Garrod's work was founded on his study of patients with the inherited disease alcaptonurea, recognized in young children by the way in which the urine turns black after exposure to the air; if untreated (by special diet), children who inherit this condition develop serious mental deficiency.

29. There are 64 distinct triplet sequences (4 × 4 × 4), but only 20 amino acids, which means that several triplets may specify the same amino acid. For example the simplest of all the amino acids, glycine, is specified by all four of the triplets that begin GG, so that the presence of any one of GGG, GGC, GGA or GGT in the genetic DNA will yield exactly the same protein.

30. In principle, these variations between families could be used for forensic purposes, but "DNA fingerprinting," now widely used as evidence in criminal trials, instead relies on parts of the human genome where variations of nucleotide sequence are expected to be strictly irrelevant to its overall function.

31. The implied exceptions are when particular genes are entirely absent from the genome.

32. The technique requires the laboratory synthesis of short stretches of DNA, called "probes," spanning the sites of mutations known to be linked with disease and corresponding both to the mutated and unmutated alleles; diagnosis then entails telling which probe sticks to the DNA and which does not.

33. BRCA stands for "BReast CAncer"; two separate genes have now been found in the human genome by searching for genetic mutations linked with breast or ovarian cancer that are inherited. Each seems to function as a "tumor suppressor gene," perhaps by triggering the death of a cell (or apoptosis) if an excess of genetic mutation has accumulated in it. Mutations of the BRCA genes then make them ineffectual in this role. Disconcertingly, both genes are very large, and several distinct mutations are linked with hereditary breast and ovarian cancer, which complicates genetic diagnosis. See R. Wooster et al., *Nature* 278 (1995): 789–792.

34. The most common is "fragile-X syndrome," characterized by severe congenital mental defect and early death.

35. See Jean Weissenbach, *Nature* 380 (1996): 152–154; and "Landing on the Genome" *Science* 274 (1996): 479.

36. The number is only approximate: for one thing, the X and Y chromosomes differ in their DNA content, for another, there are variations between one person's genome and another's due to such idiosyncrasies as the number of repetitive triplets of nucleotides in the *huntingtin* gene.

37. Nobody knows what the true cost of the project will be, given the variety of funding agencies in a score of countries contributing to various aspects of the project, often with overlapping efforts.

38. Mycoplasma are bacteria lacking essential genes that must live parasitically on other cells.

39. In fact there are two of them, close together on chromosome 11.

40. See J. A. Rafael et al., *Nature Genetics* 19 (1998): 79–82.

41. If the nucleotides of human DNA were randomly distributed (which they are not) an arbitrary 12-nucleotide-long sequence would be expected to occur by chance 179 times.

42. What this means is that the cells of the developing embryo oscillate between the S phase of the cell-cycle in which the chromosomes are duplicated and the M phase in which the cell divides into two, omitting the G_1 phase, which enters into its own only when the initial endowment of nutrients has been consumed.

43. The classical organisms of embryology are the chicken, the amphibian *Xenopus,* the sea urchin and the fruit fly *Drosophila* (often the species *Drosophila malanogaster;* more recently, there have been added to this menagerie the nematode worm *Caenorhabditis elegans,* various mammals (particularly the mouse) and the zebra fish.

44. See W. McGinnis and R. Krumlauf "Homeobox Genes and Axial Patterning" *Cell* 68 (1992): 283–302.

45. See E. Hafen et al., "Regulation of Antennapedia Transcript Distribution by the Bithorax Complex in Drosophila," *Nature* 307 (5948), (January 19, 1984): 287–289.

46. See R. D. Riddle et al., "Sonic Hedgehog mediates the polarizing activity of the ZPA" *Cell* 75 (1993): 1401–1416.

47. The phenomenon is called "apoptosis."

48. The truncated form of chromosome 22 is called the Philadelphia chromosome; the leukemia is chronic myelogenous leukemia.

49. What happens is that Rb protein binds to some of the protein enzymes involved in DNA replication, meaning that cells do not embark on the replication of their chromosomes; but if the Rb is phosphorylated, which can be brought about by the action of growth factors on the cell, the inhibition is removed.

50. The number simply refers to the size of the protein product of the gene, which has a *molecular weight* of 53,000, or is that many times heavier than a single atom of hydrogen.

51. AIDS is indeed caused by a retrovirus, called HIV, but does not cause cancer directly; the association of AIDS with the distressing and frequently fatal skin cancer called Kaposi's syndrome has recently been shown to be due to a previously unrecognized form of herpes virus that is normally kept in check by the body's immune system (which is undermined in AIDS).

52. See S. Pulciani et al., *Nature* 300 (1982): 539–542.

53. Alzheimer's disease, for example, affects a large proportion of people in their seventies, but (like cancer) has familial tendencies. An onset at much earlier ages is about as well understood in 1997 as cancer was in the late 1970s. Genes have been identified as responsible for the familial (early onset) disease, while others are known to control the formation of the damaging protein deposits in nerve cells. Effective treatment should not be delayed more than a decade or two: cure is much further off.

54. IVF was formally outlawed by the Vatican's encyclical on embryology in 1992.

55. Investigations would entail the removal of one cell and its growth in an artificial medium before testing for the presence of particular genetic alleles with appropriate probes. Under British law, these procedures are not explicitly prohibited, but would require a license (none of which has yet been granted).

56. To the extent that the assembly of the brain from its component nerve cells (see Chapter 8) entails an unknown random element, adults produced in this way would be expected to be less alike in intellectual function than are identical twins.

57. The technique involves the removal of the nucleus from a single-celled embryo and its replacement with the contents of the nucleus of the donor sheep, followed by normal gestation in the uterus of an unrelated sheep. At the end of 1997, it was not known whether the injected nucleus was that of a stem cell or of a fully differentiated cell.

58. This effect is called "heterozygous advantage."

Chapter 7. Nature's Family Tree

1. A study of two fossilized skulls (from a Latvian quarry) of an animal called *Panderichthys* from early in the Late Devonian period has now revealed an intermediate snapshot of the anatomical changes accompanying this event, among which a reorganization of the braincase of the animal is conspicuous.

2. Richard Dawkins, the Oxford evolutionist, sets out to prove that there has been time for even complicated structures such as eyes to evolve in his book *Climbing Mount Improbable* (W. W. Norton, 1996).

3. The statement that Darwinism is not a predictive theory is not the same as saying that it is a "nondeterministic theory," like quantum mechanics, but amounts simply to the statement that Darwinism is incomplete: it says nothing about the ways in which the environment changes. But there is another important source of unpredictability in Darwinism; for sufficiently small populations of evolving organisms, the raw material with which natural selection has to work are the genetic variations arising in the first few generations, which

must be a small fraction of all the variations of which the genetic potential of the organism is capable. The end result of evolution may therefore be a population biased by the chance variations arising in the first few generations; for obvious reasons, that is called the "founder effect."

4. Even as things are, it seems reasonable to guess that species of fish hunted commercially as food for people would quickly become more numerous, as would forest trees, and that feral versions of some but not all domesticated animals (dogs, cats and horses, for example) would have a good start in post-human evolution; whether the now much-reduced populations of the great apes are large enough to resume the course of evolution from the point at which the human lineage separated 4.5 million years ago is another matter.

5. See *Life's Grandeur* by Stephen Jay Gould (Cape, 1996).

6. The formal study of the mechanism of inheritance was christened genetics only in 1906, by William Bateson at Cambridge, England.

7. Muller spent only the years 1912–15 at Columbia; he moved to the Rice Institute in Texas in 1920 and, after a spell in Germany and Russia beginning in 1935, moved to the University of Indiana in 1945.

8. Alleles whose effect takes precedence are called dominant, while the others are recessive. Because there are two genes of each kind in somatic cells, recessive alleles are usually those whose protein products have lost a function normally required of them.

9. Morgan moved to the California Institute of Technology in 1928; by then the essence of their work was complete.

10. False conflicts of this kind arise frequently in science. In quantum mechanics at about the same time, there had emerged two novel but seemingly different ways of formulating a new description of matter, due to Heisenberg and Schrödinger respectively (see Chapter 2). In any event they were shown to be equivalent; neither of the innovators was wrong. Just like space and time, the description of nature is not absolute.

11. Cited in *Time Frames* (Heineman, London, 1986) by Niles Eldredge; originally published by E. S. Vrba in *The South African Journal of Science* 76 (1980): 61–84.

12. John Maynard Smith in *The Theory of Evolution* (Cambridge University Press, Canto Edition, 1993).

13. See "Wing Upstroke and the Evolution of Flapping Flight" by S. O. Poore, et al., *Nature* 387 (August 1997): 799–802.

14. See *The Call of Distant Mammoths; Why the Ice Age Mammals Disappeared* by Peter D. Ward (Copernicus, 1997).

15. The fossil record of the Great Apes over the past 4.5 million years is almost nonexistent, but the similarity of the oldest hominid fossil *(A. afarensis)*

to modern chimpanzees is indirect evidence that there has been little change in the most probable origin of the hominid lineage.

16. There are exceptions; the hybrid of a horse and a donkey is a mule, for example.

17. See D. A. Wheeler et al. in *Science* 261 (1991): 1082–1085.

18. See H. A. Orr, *Evolution* 45 (1991): 764–769.

19. See Jerry A. Coyne, *Nature* 355 (1992): 511–515.

20. H. A. Orr has also played a prominent role in the investigation of fruit-fly sterility, see *Genetics* 115 (1987): 555–583.

21. "Enhanced" does not mean "better."

22. The word "sum" in this sentence should not be understood arithmetically. Reduced fitness of an organism is usually measured by the percentage of the offspring surviving, or being handicapped in some way (by infertility, for example), relative to organisms with unmutated genes. If it is assumed that several genetic changes in different genes act independently of each other, the reduced fitness arising from their combined effect will be the appropriate percentages for the separate mutations multiplied and not added together.

23. S. F. Elena and R. E. Lenski, in *Nature* 390 (1997): 395–398; describe how the effects of up to three mutations at random places in the *E. coli* genome on the fitness of the bacteria were sometimes greater than expected, sometimes less and on balance made no difference.

24. Bees and wasps belong to the genus *Hymenoptera;* termites, whose evolutionary history is different, show behavior similar to that described in the text.

25. W. D. Hamilton's account of his theory of kin selection was published in 1964, long before molecular biology offered the promise of telling the detailed structure of genes, let alone that of the chromosomes in which they are embedded. Under the title "The Genetical Evolution of Social Behaviour," his two formal papers on the topic occupied the first 52 pages of volume 12 of the *Journal of Theoretical Biology.* For many years their importance was not appreciated.

26. Richard Dawkins in *The Selfish Gene* (Oxford University Press, 1976 and 1989).

27. In the preface to the second edition in 1989.

28. See *Against Biological Determinism: The Dialectics of Biology Group* by Steven Rose (Allison & Busby, 1982).

29. See Edward O. Wilson in *Sociobiology* (Harvard University Press, 1975).

30. See *On Human Nature* by Edward O. Wilson (Harvard University Press, 1978).

31. See *Lifelines: Biology Beyond Determinism* by Steven Rose (Oxford University Press, 1998).

32. Early versions of this theory appeared in 1962, but the most comprehensive account of it is in V. C. Wynne-Edwards *Evolution Through Group Selection* (Blackwell Scientific Publications, 1986).

33. Because of the triplet structure of the genetic code and because much of its redundancy rests in the third of the three nucleotides in each codon, mutations of the third nucleotide are more common than elsewhere; that is not so much a complication as a kind of internal standard by which the mutability of a stretch of DNAs can be estimated.

34. See *Science* 271 (1996): 470–478.

35. Three different ways of making the comparisons yield estimates of 2.142, 2.017 and 1.873 billion years ago.

36. See *Science* 274 (1996): 1750–1753 for criticism and rebuttal.

37. Some may see this as merely a truism, but that the transfer is episodic is more surprisingly consequential than it may at first appear.

38. "Hominid" refers to any human precursor not belonging to *Homo sapiens* itself.

39. The others include the gorilla, baboon and the orangutan.

40. "Neandertal DNA Sequences and the Origin of Modern Humans," by M. Krings et al., *Cell* 90 (1997): 19–30.

41. See Timothy D. Howard et al. and Vincent El Ghouzzi et al., *Nature Genetics* 15 (1997): 36–41 & 42–46 respectively. The function of the twist protein, which is constructed so that one end can bind to DNA molecules, appears to be to regulate in embryonic development a number of genes whose products are known as *fibroblast growth-factor receptors*.

42. The prime mover of this enterprise is Professor G. Cavalli-Sforza from Stanford University, California.

Chapter 8. Thinking Machines

1. Cajal and Golgi were jointly awarded one of the first Nobel Prizes in 1906; that the junctions between neurons do not entail communication between the cytoplasm of cells is entirely consistent with the modern view of neuronal structure (see below).

2. The sense of touch is the most delicate of biochemical mechanisms; it depends on the presence in touch-sensitive neurons of molecules of the protein actin, best known as a component of muscle fibers. In touch cells (called *mechanoreceptors*) actin is coupled with molecules of ATP, the universal cur-

rency of energy, which are released when the actin molecules are bent, activating the neurons in which they are placed.

3. The cortex appears as a thin sheet of "gray matter" less than a centimeter thick and containing 10 million neurons in every square centimeter. In primates, there are many more than six or seven patches of *visual cortex* on each of the two hemispheres of the cortex.

4. The Latin verb *transducere* means "to lead across."

5. Long-term memory is widely believed to be represented by the strength of the interconnections between neurons.

6. Animals that do not move, marine sponges for example, have not evolved nervous sytems.

7. The target organs are recognizable only because their cells carry receptors specific for the hormones concerned.

8. This is the step called *gastrulation.*

9. The cells of the neural crest also populate other parts of the body to form other organs, such as muscles.

10. Uterine deprivation is known to damage the capacity of a growing brain, for example.

11. In the human embryo, neurons make their first appearance about 20 days after fertilization.

12. The sense of pain in the periphery is gathered by neurons originally in the neural crest which have migrated to the tissues of the hands.

13. See P. J. Garron *et al.* in *Science* 279 (1998): 220–222; this finding does not imply that chimpanzees are capable of speech, for the asymmetrical structure may be used for some other purpose in modern chimpanzees.

14. See T. J. Crow, *Schizophrenia Research* 28 (1997): 127–141.

15. Multiple sclerosis is a degenerative disease whose underlying cause is the failure of the myelin-wrapping process.

16. Exceptions to this rule have recently come to light; recipient dendritic connections can influence the upstream neurons.

17. Normally, cells have positive charges on the outside of the membrane and negative electric charges on the inside.

18. The decrease of the voltage (known as *depolarization*) is brought about by the ingress of sodium ions through molecular channels in the membrane of the cell; the patch of depolarization on the cell surface moves along the axon as a *soliton* or *solitary wave,* the archetype of which is the movement of a hump of water into a tidal estuary.

19. See E. Neher and B. Sakmann, *Nature* 260 (1976): 799–802.

20. Experiments with kittens have shown that they have to "learn to see" from their visual experience in the first few days of life.

21. Repairing damaged connections in the spinal cord is similarly a matter of great medical interest. The feasibility of inducing even damaged axons to regrow seems not to be beyond the bounds of possibility.

22. The phenomenon, called *long-term potentiation,* has been mostly studied in rats.

23. There is a difficulty about the "exclusive *or*" operation (called XOR in computer language), which generates a signal when either one or the other, but not both, of two inputs to a cell is positive. A combination of two cells can in principle perform that function.

24. See Steven Pinker in *How the Mind Works* (W. W. Norton, 1997).

25. The first scanning device, developed at the U.S. National Institutes of Health, used a chemical analog of glucose carrying a radioactive fluorine atom; because the material is not metabolized in cells, it accumulates there, serving as a marker for cells that are or have recently been active.

26. See *How the Mind Works* by Steven Pinker (W. W. Norton, 1997).

27. The mechanism of the bending of a leech in response to an external stimulus is probably a good model for the behavior of the autonomic nervous system in more complex animals—the "instinctive" withdrawal of a human hand from a hot object touched by accident, for example. The successful dissection of the leech's bending response has revealed a constant and somewhat stylized arrangement of neurons that bears out engineers' claims that some functions of the brain are "hard-wired," no doubt by the control exercised by the genes regulating embryonic development. But it is also reasonable to suspect that many of the functions of the brain that are not automatic will turn out to be determined by groups of neurons arranged on similar principles. See J. E. Lewis and W. B. Kristan, *Nature* 391 (1998): 18–19.

28. See K. C. Schmidt, G. Lucignani, and L. Sokoloff, *Journal of Nuclear Medicine* 37 (1996): 394–399.

29. See Karl H. S. Kim et al. *Nature* 388 (1997): 171–174.

30. A flat array of elements failed to carry out the "exclusive OR" operation referred to earlier in this chapter.

31. This example is meant for illustrative purposes only; the elements of real neural networks usually work on more complicated rules. One obvious difficulty with the arrangement described is that only two of the arrangements of digits are truly independent from each other.

32. This is where the illustration used is oversimple: the response of the

network to an arbitrary input would be indeterminate because only two of the four sets of symbols are truly independent of each other.

33. The use of the word "device" is not necessarily to be taken literally; it is standard practice to simulate neural networks on computers before building them from silicon or other components.

34. See T. J. Sejnowski and C. R. Rosenberg, *Complex Systems* 1 (1987): 145–168.

35. It is extremely improbable that the development of either human faculty was an all-or-nothing event comparable with the effect of a single genetic mutation in a single gene. On the other hand, there are many in linguistics who argue that because speech is intelligible only if both syntax and language are fully developed, language arrived in one fell swoop (See Bickerton . . . ref), perhaps by the subordination of some set of neural circuitry to this novel use. But that seems improbable also.

36. In *The Computer and the Mind* (Harvard University Press, 1988).

Chapter 9. The Numbers Game

1. One of the crucial developments was the introduction into mathematics of the concept of "complex numbers," essentially an extension of familiar arithmetic to include numbers, previously undefined, which are the square roots of *negative* numbers. Those responsible for the later development of the calculus (also called analysis) were Augustin Cauchy (1789–1857) in France and Carl Friedrich Gauss (1777–1855), one of Hilbert's predecessors at Göttingen.

2. Early in this century, for example, group theory was used to classify (and to limit) the kinds of symmetry displayed by crystals of solid materials as revealed by X-ray analysis.

3. Electronic computers had not, of course, been conceived of in 1900.

4. Readers should not spend time checking: the number is for illustration only, and is not a prime number but divisible by 3 (twice).

5. In *Nature's Imagination*, p. 23, ed. John Cornwell (Oxford University Press, 1995).

6. Turing afterward became the co-designer of the first electronic computer to be built in Britain, at the University of Manchester; tragically, he committed suicide in 1964.

7. Its capacity to absorb the instructions would also become exhausted.

8. In the computers now on our desks, data are coded in the binary system, using only the digits 0 and 1; the different states of the computer are defined by the numbers stored in parts of the immediate memory called *registers* and *stacks*. The details are not relevant to Turing's argument, which is abstract.

9. Roger Penrose, in *The Emperor's New Mind* (Oxford University Press, 1989) gives an illustration of a task in whose performance a Turing machine would never halt—that of trying to prove by computation that Fermat's last theorem is correct. The theorem asserts that there is no set of whole numbers a, b, c and n (where n is greater than 2) such that $a^n + b^n = c^n$. If the theorem is true (which has now been proved), a computer programmed to try all possible combinations of four integers would plainly never come to the end of its task. Penrose's book includes perhaps the most detailed account of Turing machines ever written for a general audience.

10. There are several ways of doing this, but one of the simplest is to arrange the rational numbers according to their denominators such as $1/2$; $1/3$; $2/3$; $1/4$; $3/4$; $1/5$, $2/5$, $3/5$, $4/5$; $1/6$, . . . etc. (Duplicates have been omitted from that list.) Because each number in the list can be considered to correspond to one of the whole numbers, the total number is called a *countable infinity*, usually written as $\aleph_0$.

11. A little thought will show that there can be no explicit response to the obvious demand, "Show me a noncomputable number"; there is, however, a now-famous demonstration by construction for doing this due to the French mathematician Cartan at the end of the nineteenth century. First, put all the computable numbers in some order, then form a new number by taking as its first digit (after the decimal point) the first digit *plus 1* of the first number in the list; as the second digit, the second digit of the second number, again *plus 1*; as the third digit, the third digit of the third number *plus 1*; and so on indefinitely. (A rule such as $9 + 1 = 0$ may be a convenience.) The result is a number that differs from all the numbers in the original list in at least one of its decimal places; by definition, it cannot therefore be a computable number. Other noncomputable numbers can be constructed simply by changing the successive digits in the original list in arbitrary ways.

12. A computer program intended to verify Fermat's theorem (see note 11, this chapter) by testing all possible combinations of integers would plainly never be completed, and so would never halt. The same is notoriously true of attempts to divide any number by zero: machines usually do division by repeated subtraction, meaning that the dividend never changes.

13. The circumstances are a little more complicated; Euclid's axiom is essential as stated, but he also postulated that if two lines intersect a third in such a way that the sum of the internal angles on the same side of the third line is less than two right angles, the two lines will eventually intersect. The status of the second postulate was controversial even in antiquity, but stimulated energetic activity among European mathematicians from early in the seventeenth century. By the mid-nineteenth century the feasibility of non-Euclidian geometry had been established by Carl Friedrich Gauss and, independently, by Janos Bolyai, a Hungarian. Unknown to them, the Russian Nicolai Lobachevskii had

published a pamphlet making the same case in a language few people read. It is curious that the techniques of so-called spherical geometry, developed primarily for navigation, were not recognized to be a particular kind of non-Euclidian geometry until late in the nineteenth century.

14. In Euclid's geometry, the three angles of a triangle sum to exactly 180°. It is easily appreciated that the rule is not followed on a curved surface, such as that of Earth.

15. The *integral sign* was invented by Leibnitz early in the eighteenth century.

16. In *Gödel, Escher and Bach* (Basic Books, 1979); this is not so much a book as a workbook for those with a miscellaneous curiosity in the author's own interests.

17. The equations concerned were formulated (but not solved) by D. J. Kerteweg and G. de Vries in 1895.

18. Quoted in *Solitons and Particles,* eds. C. Rebbi and G. Soliani (World Scientific Press, Singapore, 1984).

19. In Britain, the best known example is the Severn "bore," which appears on several days each year.

20. This is the transition from heat conduction to turbulent convection (or the bodily movement of hot liquid).

21. In 1905, Einstein had tackled the problem of the random so-called Brownian motion of tiny particles (such as pollen grains) suspended in a liquid, which are repeatedly hit by invisible molecules of the liquid, so that they move in a random fashion. He concluded that the average distance moved by a pollen grain would be proportional to the square root of the total time—but some grains would move much further than that and even more would move a lesser distance. That, too, is chaotic motion.

Chapter 10. Avoidance of Calamity

1. A phrase used by Adlai Stevenson in 1961 when serving as U.S. Ambassador to the United Nations.

2. Remaining stocks of the virus, previously held at laboratories designated by WHO, were destroyed in 1996; if necessary, the virus can be reconstituted from its nucleotide sequence.

3. One such was Dr. Ephraim Anderson from the British National Institute for Medical Research.

4. The face that Kaposi's sarcoma was first recognized by a Viennese physician of that name in 1907 does not mean that AIDS as an infection goes back that far; sporadic cases still occur in people without AIDS.

5. The complex influenza virus carries in its outer coat an enzyme called neuraminidase that is essential for the infection of susceptible cells; a drug designed to interfere with the active site of the enzyme has been designed (in an Australian government laboratory in Melbourne) on the basis of the structure of the molecule concerned, and is now in clinical trials in several countries.

6. The papilloma virus, presumably transmissible at sexual intercourse, is known to be a cause of human cervical cancer.

7. One tonne equals one metric ton.

8. This estimate derives from the rate at which radioactive carbon, originally formed during nuclear weapons tests, is gradually being transferred to the oceans; a similar transfer between the atmosphere and the biosphere continues.

9. The CFCs originally used as refrigerants are not now thought to be greenhouse gases because their effect in the lower atmosphere is canceled by their effects on the ozone layer in the stratosphere; but the substitutes for them are believed to have a warming effect.

10. All temperatures are given in degrees Centigrade or Celsius. Most of the data quoted here are taken from *Climate Change 1995* (Cambridge University Press, 1996), the second assessment of Working Group I of the Intergovernmental Panel on Climate Change (IPCC), whose work is discussed in more detail below.

11. Past temperatures can be derived from measurements of the ratio of the concentrations of different oxygen isotopes in the fossils of marine plankton.

12. See J. H. Ausubel, *Nature* 350 (1991): 649–652.

13. Traditionally, sea level has been estimated from the readings of tide gauges, sited principally at commercial harbors around the world; the U.S. Earth satellite *Poseidon* has broken new ground since 1993 by demonstrating that downward-pointing radar can measure the height of the open sea to within a few millimeters.

14. The measurement of global sea level is far from simple; long-term changes are tiny compared with the changes of sea height caused by tides, while the estimates of sea level derived from any one measuring instrument (usually tidal gauges in commercial harbors) may be affected by local upward and downward movement of the land on which they are sited.

15. If all the ice were to disappear, Antarctica would be revealed as an archipelago of half a dozen large mountainous islands.

16. The process would begin suddenly as the buoyant separating ice sheet drifted free of the underlying Antarctic rock, for ice is not as dense as water.

17. Even the estimation of the average temperature of the surface of the Earth has until recently been biased by the presence of weather stations only in inhabited regions of the world, so that comparisons with the model predictions have been uncertain; direct measurements by Earth satellites should resolve the difficulty in the future.

18. These are droplets of sulfuric acid formed by the oxidation of the atmosphere of sulfur dioxide from the burning of fuel, almost always associated with small particles of dust.

19. Formally, the U.S. Senate ratified the treaty in 1992, but the U.S. government resists restrictions on the use of fossil fuel; China has also ratified the treaty, but as a "less developed country," meaning that it is temporarily free from restrictions.

20. Quoted from Sir John Houghton, chairman of Working Group I, in *Nature* 383 (1996): 572.

21. The ICRP is now constituted as a charity registered under British law and has an annual budget of $300,000 contributed mostly by charitable foundations.

22. In 1996, it came to light that the finality of one of the conclusions in the 1996 assessment was (correctly) blunted at the request of the U.S. State Department after the formal negotiation of the text had been completed.

23. In its second assessment in 1996, IPCC acknowledged that the grid used in the computer models is too coarse for accurate modeling of the el Niño phenomenon.

24. In its 1990 assessment, IPCC predicted (in an "extensive summary") that, without restrictions on greenhouse gases, there will be "a rate of increase of global mean temperature during the next century of 0.3 degrees per decade (with an uncertainty range of 0.2 degrees to 0.5 degrees per decade)"; the 1996 assessment says that the newer estimates are "approximately one third lower" than the earlier estimates.

25. Duncan Steel (see following note) points out that the U.S. astronomer Jack Hartung has linked the formation of the lunar crater Giordano Bruno with a report in 1178 that five people at Canterbury, England, observed a flash of light on the surface of the Moon in that year.

26. Steel's book, *Rogue Asteroids and Doomsday Comets* (Wiley, 1995), is a readable and comprehensive account of the impact problem whose purpose is much more serious than its title suggests.

27. The damage done by the impact of a meteoritic object is entirely determined by its kinetic energy, which is proportional to its mass and which

increases as the square of its velocity relative to Earth. Unlike nuclear weapons, used as a convenient yardstick of energy in this discussion, meteoritic impacts do not produce ionizing radiation or radioactivity.

28. The fate of an object in the atmosphere depends on its speed, the inclination of its track to the surface and its chemical composition; metallic meteorites are more likely to reach Earth's surface than those made mostly from stony material.

29. This idea is due to W. Napier and V. Clube, at the Royal Edinburgh Observatory in the 1970s, who suggest that the interval between successive passages through the plane of the galaxy is about 30 million years; the idea has been echoed by J. Raup and J. Sepkowski at the University of Chicago, who cite evidence that the extinction of species of marine plankton recur at intervals of 26 million years. The reality of this periodic behavior is not widely accepted. A prediction by the British astronomer-journalist R. A. J. Matthews, that the nearest approach of the star Alpha Centauri to the Sun 30,000 years from now will disturb the outer clouds of comets, should however be a test of whether the cometary cloud indeed exists.

30. Duncan Steel (see note 26), who participated in a meeting in 1992 at the Los Alamos National Laboratory to consider how nuclear explosions could be used to avoid impacts, argues persuasively that the safest method is to use a "neutron bomb" that would heat a predefined area of the surface of the asteroid, causing it to evaporate and to convert part of the explosion's energy into the changed momentum of the asteroid as a whole.

31. In the 85 million years of mammalian evolution, *Alu* sequences have accumulated at an average rate of one every 170 years or, say, 10 generations, but there is no way of telling whether the process continues; it may, for all we know, have all happened at an early stage.

32. The obvious candidates for condition X, such as early-onset dementia and infertility, are unlikely to be those that eventually create a sense of crisis.

Index

link between full-sized world and
atomic world, 96–97
mathematics and, 319
and observer influences on event
observed, 96
properties of matter and, 16
relativity and, 59
special theory of relativity and, 73
uncertainty principle and, 16
Quark confinement, 80
Quarks, 78–82, 84, 88, 89, 99, 102–03,
121, 386n29, 386n31, 386–87n33
Quasars, 41–43, 58, 110, 112, 382nn17–
19, 389n23

Radiation
big bang theory and, 30–32, 99
Chadwick's research on, 64
cosmic microwave background
radiation, 31–32, 35, 43, 52, 381n8,
382n20, 384n38
discovery of, and early studies on, 8,
12–13, 15, 63, 379–80n7, 383n26
in Earth's atmosphere, 151–52, 343–44
from galactic clumps, 50, 383n28
galactic measurement and, 34
Lyman-α radiation, 383n28
measurement of extragalactic distances
and, 34
properties of, 29–30
from quasars, 42, 110, 389n23
Ras-A gene, 226
Rational numbers, 316, 411n10
Raup, J., 415n29
Rb gene, 223–24
Real numbers, 315–17
Realism, as guiding principle of science,
371–74
Receptors, 181, 185
Recombination, 178, 398nn30–31
Red blood cells, 171–72, 397n16
Redshift, 34, 36, 40, 381n11
Reductionism versus holism, 372
Relativity
Einstein's general (gravitational) theory
of, 11, 17–18, 55, 58–59, 73, 85, 93,
97, 98, 100–01, 109, 110, 112, 309,
321–23, 368, 373, 374, 379n6,
390n29
equivalence principle and, 3
inflationary universe and, 98
intrinsic spin and, 73
Newtonian mechanics and, 11, 17

quantum electrodynamics and, 77–78
quantum mechanics and, 73
space-time and, 93
special relativity theory, 11–12, 17, 73,
77, 85, 93, 325, 390n29
Replication, fidelity of, 141, 143–44
Reproduction. See Sexual reproduction
Restriction enzymes, 396n8
Retinoblastoma, 223–24, 403n49
Retroviruses, 167, 225–26, 338, 396n7
Revelle, Roger, 340
Reverse transcriptase, 167, 396n8
Rhodopsin, 394n39
Rhythmic processes, 186–87
Ribosomes, 146, 161, 162, 170–71, 180,
185, 392n24
RNA
Alu sequence and, 206
Cech's studies of, 160–61, 395n51
cell division and, 175
chemical units of, 160
chromosomes and, 162
compared with DNA, 146, 198,
392n23, 395n57
concatenation of, into longer
molecules, 161
editing of, 179–80
in eggs, 220
genetic information storage of, 161,
162
genetic time and, 260
HIV and, 207, 338, 393n27
identification of human genes through,
214
mRNA (messenger RNA), 395nn51–
52, 396n6
nucleotides in, 161, 198, 392n12,
392n19, 392n23, 395n57, 399n5
origin of life and, 203
phosphate groups and, 156–57
relationship with DNA, 138, 161,
395n52
retroviruses and, 167, 225–26
reverse transcriptase and, 167, 396n8
ribosomes and, 146, 161, 162, 392n24
role of, 195–96
somatic cells and, 198
translation of, 167
tRNA (transfer RNA), 161, 180,
395n53
in viruses, 393n27
Röntgen, W. K., 8
Rosen, Nathan, 94